John Rawles

Atrial Fibrillation

With 109 Figures

Springer-Verlag
London Berlin Heidelberg New York
Paris Tokyo Hong Kong
Barcelona Budapest

John Rawles, BSc, MB, BS, FRCP, FRCP (Edin)
Department of Medicine and Therapeutics, University of Aberdeen,
Foresterhill, Aberdeen AB9 2ZD, UK

ISBN-13:978-1-4471-1900-5   e-ISBN-13:978-1-4471-1898-5
DOI: 10.1007/978-1-4471-1898-5

Cover illustration: Fig. 2.1. The electrocardiogram in atrial fibrillation

British Library Cataloguing in Publication Data
Rawles, John Michael *1938–*
Atrial fibrillation.
1. Arrhythmia
I. Title
616.128
ISBN-13:978-1-4471-1900-5

Library of Congress Cataloging-in-Publication Data
Rawles, John, 1938–
Atrial fibrillation / John Rawles.
  p. cm.
Includes index.
ISBN-13:978-1-4471-1900-5
1. Atrial fibrillation.   I. Title.
RC685.A72R39 1991   91-14568
616.1'28–dc20   CIP

© Springer-Verlag London Limited 1992
Softcover reprint of the hardcover 1st edition 1992

Typeset by Wilmaset, Birkenhead, Wirral
2128/3830-543210 Printed on acid-free paper

Preface

Atrial fibrillation poses problems in many areas – electrophysiology, haemodynamics, autonomic function, control theory, mathematics, computer modelling, and not least, clinical management. Advances have been made in all these domains in recent years, but such is the specialisation even within a subject as circumscribed as cardiology that experts in one area may be unaware of developments elsewhere within the subject. This book is one man's attempt to understand this new knowledge, and relate advances to each other and to what was known before. I run the serious risk of misunderstanding and misrepresenting the subject matter in all these specialised fields, since I am an expert in none of them. Nevertheless, I have enjoyed the exploration of new territories and I would like to convey to readers some of the pleasures of insight and discovery that I have experienced.

The idea of writing the book arose after I had experienced great difficulty in getting original work on the haemodynamics of atrial fibrillation published. It was then that I realised how fragmented the subject of cardiology has become; for example, it is possible to attend a meeting on arrhythmias and never once hear mention of cardiac output. I also realised the limitations of the scientific paper as a vehicle for the development and presentation of ideas, rather than the recording of observations.

This book is a blend of review, original observation, and conjecture, in which I have indulged in developing the themes, in particular, of the atrioventricular node as a biological oscillator, cardiac output as a linear measure, and "control" of ventricular rate.

I have been furnished with raw material by a succession of very able research students, fellows and registrars – Donald Mowat, Neva Haites, Ramdas Pai, Susan Dewar, Shirley Copland, Stephen Cross, Susan Reid, Malcolm Daniel, Martin Cowie and Amalia Mayo. Throughout, Joy McKnight has provided unfailing technical assistance of the highest order; she has been supported by a research grant from Grampian Health Board. To all of them I give thanks for their careful, thorough work, and for their good humoured tolerance and constructive comment when I have used them as sounding boards.

At a critical time financially, Dr. Andy Millar and Dr. Denis Lockhart of the Merrell Dow Research Institute gave me generous financial support for the development of non-invasive methods of drug evaluation, and some of the results of the work they sponsored are published for the first time in this book. I shall ever be grateful to them for their vote of confidence, and hope they feel it was not misplaced.

I am grateful too, for helpful comments on the manuscript from Drs. Richard Vincent, Ben Benjamin, and Stephen Cross. Any deficiencies in the text remain, of course, my sole responsibility.

The better figures in the book are the skilful work of Nigel Lukin and the Department of Medical Illustration, University of Aberdeen; many of them have appeared in the *British Heart Journal, Clinical Science, International Journal of Biomedical Computing* and the *Scottish Medical Journal*. I am grateful to the Editors of these journals for permission to reproduce figures and rework previously published material.

Throughout the endeavour, Katharine Mair has provided unending support and encouragement for which thanks are quite inadequate.

John Rawles   March 1991

Contents

Chapter 1

The Cardiac Conducting System and Its Autonomic Control

Anatomy of the Conducting System

The specialised cardiac conducting system is responsible for the initiation and propagation of the impulse that results in the heart beat. It includes the sinoatrial node, the internodal tracts, the atrioventricular node, the bundle of His, the right and left bundle branches, and the Purkinje network.

The sinoatrial node is situated in the anterolateral wall of the right atrium at the junction of the superior vena cava and the right atrium. The junction is marked by a groove externally, the *sulcus terminalis*, and a ridge on the endocardial surface, the *crista terminalis*. The node is oval in shape, measures about 15 mm × 5 mm × 1.5 mm, and is located just under the epicardium. The sinoatrial node is composed of small, fusiform, poorly striated cells with large nuclei. These cells, embedded in interstitial tissue rich in nerve fibres, are disposed longitudinally in the sulcus terminalis and circularly around the artery that runs through the middle of the node in its long axis. The sinus node artery arises from the right coronary artery in 55% of cases, and from the left circumflex in 45% (James 1968).

Three tracts that preferentially conduct impulses from the sinoatrial to the atrioventricular node are recognised (James 1963), though their anatomical differentiation from atrial myocardium is not very distinct. Various orifices occupy the right atrium between the nodes, and internodal tracts are found in each of the main routes around these natural barriers (Fig. 1.1). The anterior and middle internodal tracts run on either side of the opening of the superior vena cava and then across the interatrial septum to converge on the atrioventricular node from above. The posterior internodal tract runs inferiorly from the sinoatrial node, below the valve of the inferior vena cava, between the orifices of the vein and the coronary sinus, to enter the atrioventricular node from below. Bachmann's bundle is a subdivision of the anterior internodal tract that leaves the sinoatrial node to ramify over the left atrium. The internodal tracts consist of working atrial muscle cells, smaller node-like cells, and dispersed Purkinje-like cells. However, in the atria there is no true Purkinje network such as is found in the ventricles.

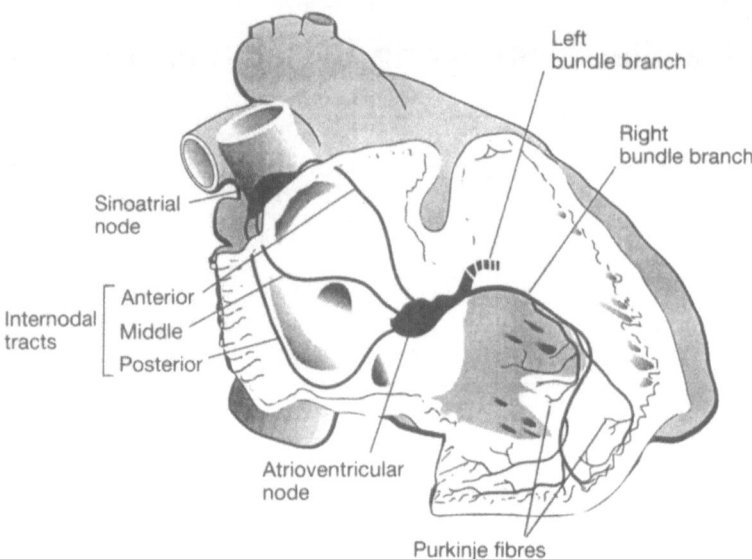

Fig. 1.1. The specialised conducting system of the heart.

The atrioventricular node is found just beneath the endocardium of the right atrium, in the muscular atrioventricular septum between the orifice of the coronary sinus and the origin of the medial leaflet of the tricuspid valve. It is oval in shape and about 6 mm × 3 mm in size. The node consists of a network of small, lightly striated cells with pale cytoplasm, in an interstitium of collagen and elastic tissue that is rich in nerve fibres, though they are not so plentiful as in the sinoatrial node. The artery to the atrioventricular node arises from the right coronary artery in 90% of cases, and from the left circumflex in 10% (James 1968). The anterior and middle internodal tracts fuse to approach the atrioventricular node from above, and the lower internodal tract approaches the node from below.

At the distal end of the atrioventricular node the cells converge into fascicles to form the bundle of His, which passes along the edge of the membranous septum and then into the apex of the muscular interventricular septum before dividing into right and left bundle branches. Peripherally both bundle branches subdivide into the network of Purkinje fibres which connect with the fibres of ventricular muscle.

In the normal heart, the atrioventricular node in continuity with the bundle of His is the only structure to conduct impulses from the atrial to the ventricular myocardium. Elsewhere the atrioventricular junction provides anatomical continuity but electrically acts as an insulator (Anderson and Ho 1990). In the Wolff–Parkinson–White syndrome, accessory atrioventricular connections in the form of strands of myocardial tissue may be present as congenital malformations and bridge the atrioventricular insulation at almost any point around its circumference (Wallace et al. 1976); some patients have multiple accessory atrioventricular connections (Gallagher et al. 1976).

Basic Cardiac Electrophysiology

A myocardial cell at rest has a potential difference across its cell membrane of about 85 mV, the interior of the cell being negative with respect to the extracellular space. This resting potential is the result of a concentration gradient for sodium (Na^+) and potassium (K^+) ions across the membrane, Na^+ being 10 times more concentrated outside, and K^+ 30 times more concentrated inside the myocardial cell. The maintenance of these concentration gradients, and of the resulting resting potential, requires energy. This is obtained from the hydrolysis of adenosine triphosphate (ATP), catalysed by ATPase. The ATPase liberates the energy to drive the Na^+/K^+ pump, that pumps Na^+ out of the cell and K^+ into the cell in the ratio of 3 : 2; the pump is inhibited by cardiac glycosides.

Excitation of the cell starts with the rapid abolition and then reversal of the transmembrane potential, all in the course of a few milliseconds. This is brought about by a sudden increase in cell permeability with a fast influx of Na^+ down its concentration gradient. There is also some movement of K^+ out of the cell. At a critical level of membrane potential a slower inward flow of calcium ions (Ca^{2+}) is activated. This Ca^{2+} current outlasts the fast Na^+ current and continues into the next phase of the action potential. Having reached its maximum positive potential, the transmembrane potential is restored to its negative resting value in three phases: rapidly (phase 1), followed by a plateau (phase 2), and then more rapidly again (phase 3) (Fig. 1.2). Propagation of a wave of excitation through the ventricular myocardium occurs as depolarisation spreads through the myocardium from one cell to the next. The action potential results in a mechanical response – that of shortening the fundamental unit of muscle structure, the sarcomere. The wave of depolarisation that spreads through the myocardium results in a coordinated, effective contraction of the ventricle.

A pacemaker cell, such as in the sinoatrial node, differs in a very important respect from a myocardial cell in that it shows spontaneous depolarisation during diastole. Instead of a steady resting transmembrane potential there is a slow (phase 4) depolarisation due to the passage of K^+ out of the cell and Ca^{2+} into the cell. There is also no obvious plateau phase of the action potential (Fig. 1.3). The

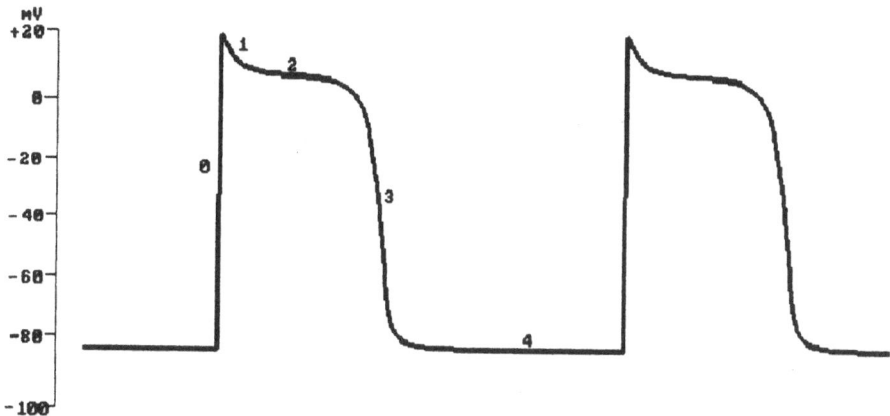

Fig. 1.2. The transmembrane potential during two cycles of a myocardial cell. The numbers refer to the phases of the action potential.

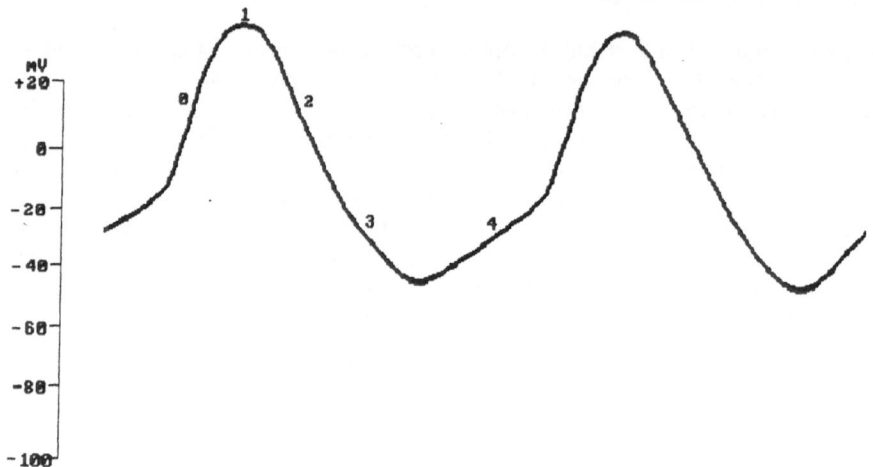

Fig. 1.3 The transmembrane potential during two cycles of a pacemaker cell.

rate of diastolic depolarisation depends on the location of the cell, being highest in cells of the sinoatrial node and lowest in the spontaneously excitable cells of the Purkinje network. At the end of phase 4, when the membrane potential reaches a critical value, rapid (phase 0) depolarisation occurs and an action potential is initiated, though the speed of depolarisation is not as fast as in myocardial cells.

During the earlier part of the action potential the cell is unresponsive to further stimuli, however strong. This is known as the absolute refractory period. During phase 3 of the action potential the cell is responsive once more, but only to stimuli that are greater than normal. This is the relative refractory period. During most of the cycle the cell's behaviour may be modified by sub-threshold stimuli which may alter the timing of the next spontaneous depolarisation.

The Initiation and Propagation of the Cardiac Impulse

In a healthy subject at rest the heart rate is 60–100 beats per minute, the cardiac impulse originating in the sinoatrial node. This consists of thousands of discrete cells arranged in clusters that have been likened to bunches of grapes. The nodal cells, which exhibit phase 4 depolarisation, are connected to each other through gap junctions – small openings in the cell membrane connected to extracellular bridges with a low electrical resistance (Bleeker et al. 1980). The rate of discharge of the sinoatrial node is intermediate between that of the cells with the lowest and highest rates of discharge. The passage of a wave of depolarisation through the node gives the appearance of conduction of the impulse. However, the impulse is not conducted, but the cells of the node discharge spontaneously, the timing of each cell's discharge being modulated by electrotonic interaction with neighbouring cells connecting with each other through gap junctions. This process of mutual entrainment results in all the cells adopting the same frequency of discharge, though there are small phase differences between different parts of the node (Jalife et al. 1988).

Resetting the Sinoatrial Node Pacemaker

An extraneous depolarising stimulus applied to the sinus node experimentally, or occurring naturally as an ectopic beat, may alter the timing of the node's discharge. The response obtained depends on the strength of the stimulus and its timing in the node's spontaneous cycle of discharge and recovery. A very weak stimulus will not alter the timing of the next beat at all, as shown schematically in Fig. 1.4, (dashed line). By convention, resetting curves show the relationship between the phase of the cardiac cycle when the stimulus was delivered (old phase) and the phase of the stimulus in relation to subsequent beats (new phase), phase being expressed in degrees; the basic cycle length is considered unchanged after the stimulus (Winfree 1980). A weak depolarising stimulus may alter the timing of the subsequent beat, retarding it if occurring in the first half of the cycle, and advancing it if occurring in the second half. This is shown as the sinuous phase response curve passing below and above the line of identity in Fig. 1.5.

With a strong stimulus we might expect the phase response curve to have a very different appearance, (Fig. 1.4, continuous line). An idealised stimulus delivered at the earliest phase, say at the time of rapid depolarisation, would occur in the refractory period and would not alter the timing of the next beat. The phase of the stimulus in relation to the next beat is therefore unchanged at 0°. Once the very strong stimulus is delivered away from the refractory period it causes an almost immediate depolarisation so that its phase in relation to the beat it elicits is just less than 360°. The phase resetting graph for an idealised strong stimulus is therefore a horizontal line in two parts, the first with a new phase of 0°, and the second with a new phase of 360°.

In fact, a schematic phase response curve obtained with a strong stimulus looks

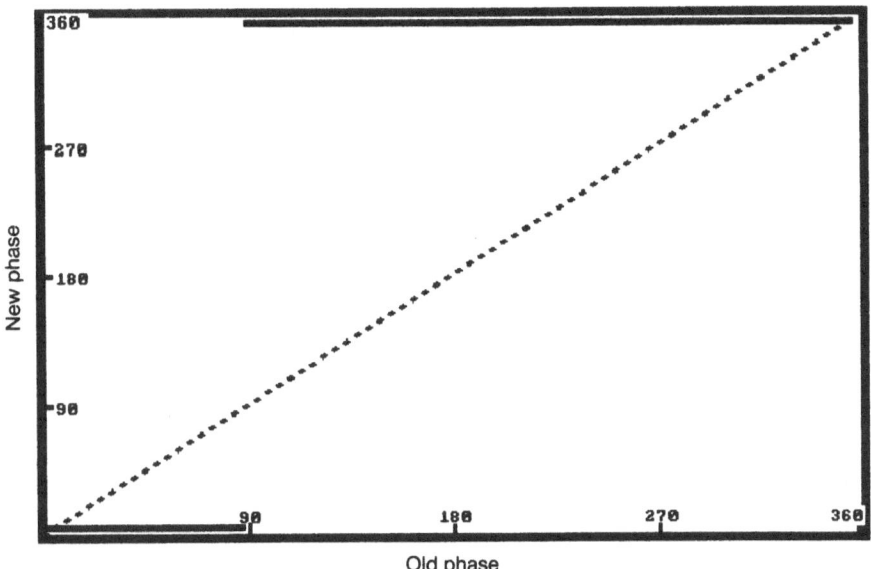

Fig. 1.4. Schematic phase resetting curves for a pacemaker exposed to a stimulus of zero strength (*dashed line*), and an idealised strong stimulus (*continuous line in two parts*).

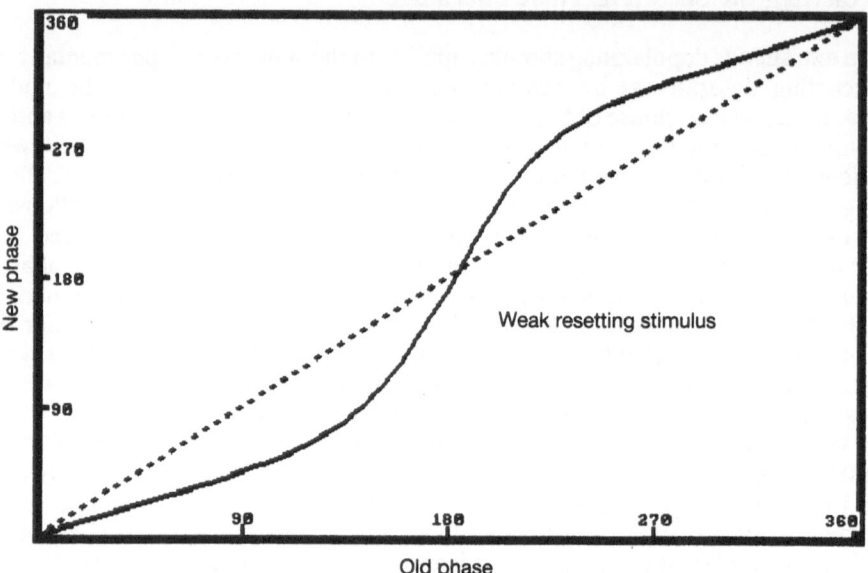

Fig. 1.5. Phase resetting curve for a pacemaker exposed to a weak stimulus.

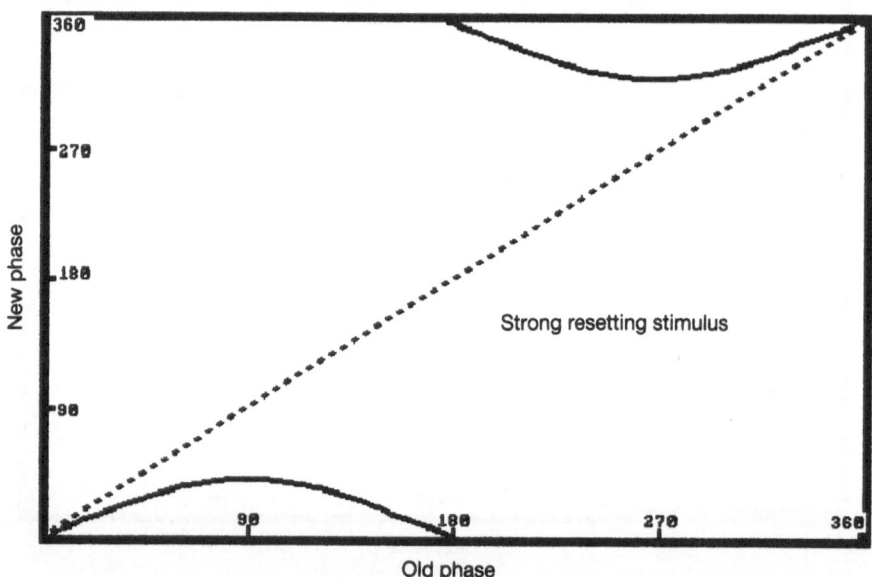

Fig. 1.6. Phase resetting curves for a pacemaker exposed to a strong stimulus.

like the continuous line in two parts shown in Fig. 1.6. Depolarising stimuli delivered in the first half of the spontaneous cycle tend to delay the occurrence of the next beat, but those delivered in the second half of the cycle bring the next beat forward. These two types of resetting curves, in response to weak and strong stimuli respectively, and especially the transition between them, are of great theoretical interest in relation to fibrillation (Winfree 1987). The topic will be discussed further in Chapter 3.

The Spread of the Impulse Through the Atria

The impulse originating in the sinoatrial node spreads through the right atrium via the internodal pathways, and to the left atrium by Bachmann's bundle. Conduction takes place rapidly along the internodal pathways, then more slowly away from the pathways into the atrial myocardium.

The Functioning of the Atrioventricular Node

Functionally, the atrioventricular node consists of three regions: AN, N and NH (Paes de Carvalho and de Almeida 1960; Meijler and Janse 1988). Although these regions correspond loosely to the transitional, central and junctional portions of the node, there is no close correspondence between these functional zones and any histologically distinct structures. Neither is there any histological evidence of dual pathways within the node, in spite of there being much electrophysiological evidence for them (Moe et al. 1956; Olsson et al. 1986).

The passage of the impulse from sinoatrial node to ventricular muscle is marked on the surface electrocardiogram by the P-R interval, which has a typical value of 160–200 milliseconds. With intracardiac recording of the electrogram from the His bundle the atrioventricular delay may be subdivided into A-H and H-V intervals. The A-H interval is normally 50–120 milliseconds, compared with 30–55 for the H-V interval. The greater portion of the atrioventricular delay therefore takes place in the atrioventricular node. Within the node, 25% of the delay occurs in the AN zone, but the main delay is in the N zone, which is responsible for the increment in conduction time in a Wenckebach cycle and is also the site where atrioventricular block is most likely to occur (Janse et al. 1976; Meijler and Janse 1988; Simson 1988).

Although speed of conduction is related to fibre diameter, the delay within the node is only partially explained by the small size of the fibres entering the node. The intercellular resistance is higher in the atrioventricular node than in other parts of the specialised conducting system, and this may also result in slow conduction. Other contributory factors may be the existence of dual pathways and of cul-de-sacs that act to diminish stimulus strength. The speed of conduction within the AN region of the node is also greatly influenced by the source of the stimulus, whether from the conjoined anterior and middle internodal tracts or the posterior tract (Janse 1969).

Propagation of the impulse into the node also depends on the strength of the excitation wave. After excision of the sinoatrial node in the dog, new pacemaker activity emerges low in the right atrium close to the atrioventricular node. However, control of cardiac rhythm is only obtained if the impulse from the new pacemaker is summated with a response from the right atrium. If most of the right atrium except the pacemaker is excluded by an incision, junctional rhythm emerges (Sealy and Seaber 1979).

The complex anatomical and functional arrangements within the atrioventricular node provide the basis for understanding two important electrophysiological phenomena. Decremental conduction is said to occur when an impulse penetrates into the node with a progressively diminishing action potential until the impulse fails to propagate. Concealed conduction is inferred when an impulse is not itself propagated but subsequent electrophysiological behaviour is modified in a way that can be accounted for by the non-propagated impulse (Fisch 1988). Thus if the P-R interval is prolonged after an interpolated ventricular ectopic beat, concealed retrograde conduction of the ectopic impulse into the atrioventricular node may be inferred.

The conduction time of the atrioventricular node is not fixed but varies with autonomic influence and with heart rate, variation with heart rate occurring even after autonomic blockade. There is a negative beat-to-beat relation between P-R and P-P intervals on the surface electrocardiogram, or between A-A and A-V intervals in intracardiac recordings. With a changing heart rate, A-V intervals are longer if the heart rate is decreasing rather than increasing, and longer if heart rate is increasing slowly rather than quickly. If the heart rate is increased stepwise there is an immediate lengthening of the A-V interval, and a further lengthening during the first few beats after the increase (Loeb and de Tarnowsky 1988). On the other hand, if the heart rate is decreased there is immediate downward adjustment of the A-V interval which is complete within the duration of one cycle (Harms et al. 1980). This short time constant for the change of A-V delay with change of heart rate is relevant to the irregularity of the pulse in atrial fibrillation (Meijler et al. 1982).

The effective refractory period of the atrioventricular node is also variable, being negatively related to cycle length (Josephson and Seides 1979, p.44). In this respect the atrioventricular node behaves in the opposite fashion to atrial, His–Purkinje, and ventricular tissue, where refractory periods tend to decrease with decreasing cycle lengths (Mendez et al. 1956), the phenomenon being most marked in the His–Purkinje system.

Most of the cells of the atrioventricular node do not show phase 4 depolarisation, but in the N region pacemaker cells have been identified histologically and by their electrophysiological properties (James et al. 1979). It is these cells that act as the pacemaker in junctional rhythm.

Does the Atrioventricular Node Conduct?

We speak of the atrioventricular node as "conducting" the impulse from atria to ventricles. However, as a conductor the atrioventricular node has some unusual properties. In sinus rhythm the A-H interval is about 100 milliseconds, giving a conduction velocity along the 5 mm axis of the node of 5 cm per second. This contrasts with a conduction velocity of 1 metre per second in the atria, and 2.5 metres per second in the Purkinje system supplying the ventricles (Durrer et al. 1978, p. 57). As a conductor, then, the atrioventricular node is extraordinarily slow.

A second peculiar property of the node is its ability to "filter" incoming stimuli, so that fewer stimuli emerge from the node than enter it. This is particularly striking in atrial fibrillation, where the ratio is of the order of 4 : 1 (Kirsh et al. 1988), but it is also seen in second degree heart block.

A third feature of interest is variability of the P-R interval, particularly in relation to heart rate; this may serve to optimise left ventricular performance at different rates (Meijler 1986).

These distinctive properties of the atrioventricular node are difficult to understand in terms of conduction of impulses through the node, even when invoking decremental or concealed conduction. In fact, the idea that the atrioventricular node is a through-conductor, connecting atria to ventricles, has grown up over many years but has little experimental support. The atrioventricular node cannot be mapped with the same precision as other parts of the specialised conducting system, and sequential conduction of an impulse through the node has not been demonstrated. The historical development of the idea has been reviewed by Meijler and Fisch (1989), who challenge the notion of one-to-one conduction. Alternative models that explain the atrioventricular node's behaviour have been developed by Van der Pol and Van der Mark (1928), Grant (1956), Robergé et al. (1971), Guevara and Glass (1982), and Cohen et al. (1983), amongst others. These models emphasise the pacemaker role of the atrioventricular node, and show how the sinoatrial and atrioventricular nodes may be considered as two coupled oscillators, the sinoatrial oscillator normally entraining the atrioventricular oscillator to run at the same frequency, though with a phase lag. Much of the atrioventricular delay occurs while the node is reset following arrival of a stimulus (Robergé et al. 1968). The variability of the A-V interval reflects the nodal pacemaker's response to stimuli of differing strengths at different phases of its intrinsic cycle. The "filtering" property of the node is readily understood once it is appreciated that the impulse emerging from the atrioventricular node is not the same one that entered it. The word "conduction" is no longer an accurate or a sufficient description of the way in which the node performs. However it remains a convenient shorthand description of the function of the atrioventricular node in passing the cardiac impulse from atria to ventricles.

Guevara and Glass's elegantly simple model of the atrioventricular node is developed in Chapter 3.

Autonomic Innervation of the Heart

The heart is richly innervated with sympathetic and parasympathetic nerves. The preganglionic sympathetic fibres originate in the first four thoracic segments of the spinal cord, relaying with the postganglionic fibres in the cervical and thoracic sympathetic ganglia, then forming the sympathetic cardiac nerves which join the cardiac plexus. The preganglionic parasympathetic nerves originate in the dorsal vagal nucleus, run in the vagal trunk, and leave it as the cervical and thoracic cardiac nerves, to relay in the ganglia dispersed in the cardiac plexus or in intrinsic cardiac ganglia. The postganglionic parasympathetic nerve fibres are therefore shorter and less well defined than postganglionic sympathetic nerves. The cardiac plexus lies between the concavity of the aorta and the tracheal bifurcation, receiving contributions from all the vagal and sympathetic cardiac nerves, and containing parasympathetic ganglia. Both sympathetic and parasympathetic cardiac nerves carry afferent as well as efferent fibres (Fig. 1.7).

At a microscopic level both sinoatrial and atrioventricular nodes contain many nerve fibres, and nerve cells are found nearby, their presence constituting the histological basis for the intimate interrelation between autonomic function and the working of the specialised conducting system of the heart.

There is much experimental evidence for parallel but functionally distinct vagal and sympathetic supplies to the sinoatrial and atrioventricular nodes. Further, there is asymmetry of action, sympathetic and vagal cardiac nerves from the right-hand side having a predominant action on the sinoatrial node and those on the left acting more strongly on the atrioventricular node. In the dog the vagal supply to the atrioventricular node can be isolated from that to the sinoatrial node as the former runs through a fat pad located at the junction of the inferior vena cava and left atrium, while the latter is found in a fat pad on the right pulmonary vein. This arrangement allows selective denervation to be carried out experimentally (Randall et al. 1983). There is also asymmetry between the nodes, the sinoatrial node being more responsive to vagal than sympathetic stimulation and the reverse being true for the atrioventricular node.

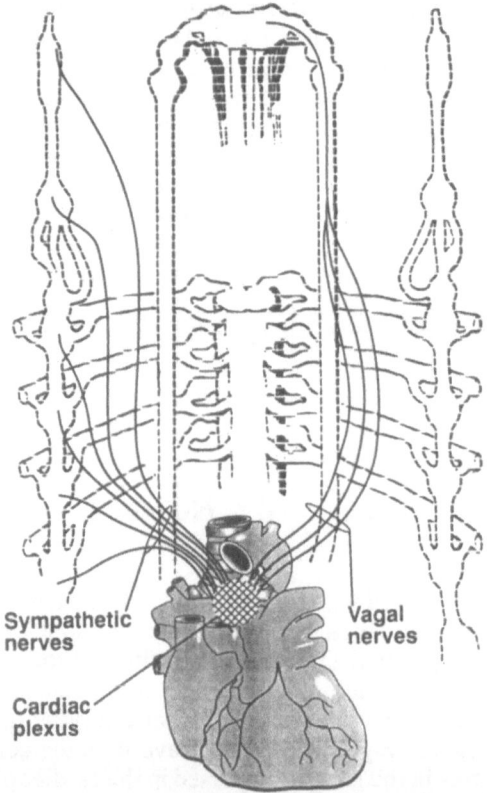

Fig. 1.7. The autonomic supply to the heart.

Autonomic Control of the Sinoatrial Node

Following autonomic blockade with atropine and propranolol the heart beats at a rate that is determined by the intrinsic discharge rate of the sinoatrial node. This declines with age according to the formula IHR $= 117 - (0.53 \times$ age) (Jose 1966). The intrinsic heart rate, while less than 100 beats per minute, is generally slightly higher than the resting, unblocked rate, vagal influence predominating over the sympathetic in resting subjects. However, in cross-sectional population studies heart rate is not related to age, presumably because the decline in intrinsic heart rate is matched by a decline in autonomic function (Smith 1982).

The effect of a short burst of vagal activity, such as might follow stimulation of the arterial baroceptors during systole, depends on the duration of the burst, its amplitude, and its timing during the sinoatrial node's cycle.

If the train of vagal stimuli is short in relation to the cycle length a triphasic response is obtained, as depicted in Fig. 1.8 (Brown and Eccles 1934; Jalife and Moe 1979). After a latent period, required for the release, diffusion and receptor binding of acetylcholine, hyperpolarisation of pacemaker cell membranes occurs. The degree of hyperpolarisation is dependent on the strength of the vagal stimulus. For example, hyperpolarisation is twice as great if both vagi are stimulated rather than one – evidence also of the independence of the vagal nerves. There is a further short latency after which there is prolongation of the first, and sometimes the second, pacemaker cycle, in proportion to the degree of hyperpolarisation. There follows a postinhibitory rebound phase in which there is an acceleration of the pacemaker rate compared with its slowest during the first inhibitory period. During this rebound phase the instantaneous heart rate may actually exceed the rate present before the vagal stimulus was delivered. There is then a more prolonged period lasting several seconds during which the pace-

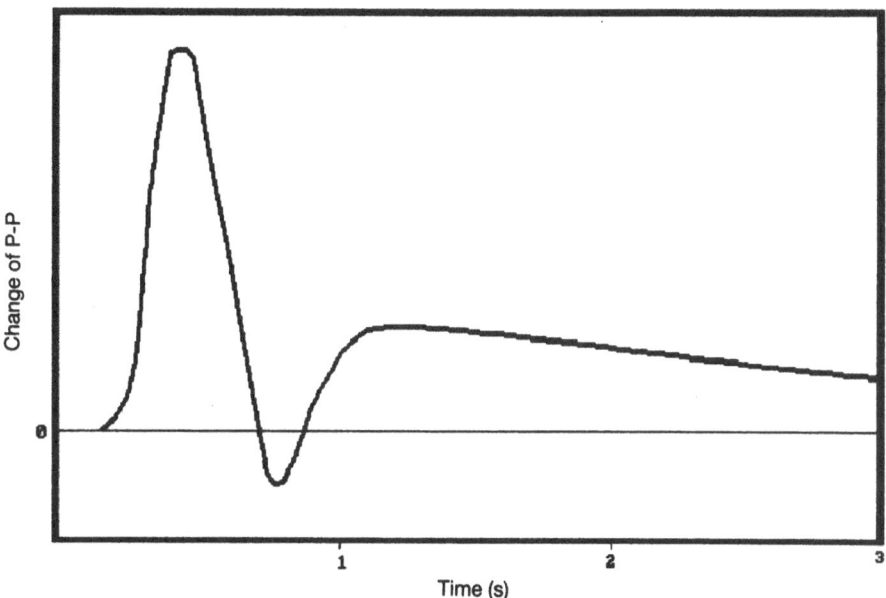

Fig. 1.8. The response of the sinoatrial node to a short burst of vagal stimulation.

maker rate is depressed but accelerating back to its initial rate. Whereas the primary inhibitory phase is due to hyperpolarisation, the secondary phase is due to a reduced rate of phase 4 depolarisation and probably outlasts the presence of acetylcholine (Jalife and Moe 1979). It should be noted that a short burst of vagal stimulation acts as a weak resetting stimulus to the sinoatrial node.

The triphasic pattern of response of the sinoatrial pacemaker to efferent vagal stimulation may also be evoked by stimulation of afferent vagal fibres in the carotid sinus (Levy and Zieske 1972), or by intermittent neck suction in human subjects (Eckberg 1976).

The presence of a rebound acceleration phase in the pacemaker response paves the way for a paradoxical increase in heart rate with an increasing frequency of vagal stimulation. In an experiment on a dog, Levy et al. (1972) increased the frequency of short bursts of vagal stimulation from 50 to 120 bursts per minute (Fig. 1.9). Over the range of 50–75 bursts per minute the sinus rate increased in step with the vagal stimulation rate. At 75 bursts per minute the sinus rate halved, but as the vagal stimulation rate increased further to 120, the sinus rate also increased up to 60, in a ratio of 1 beat to 2 bursts per minute. The ratio then increased to 1 : 3, the sinus rate falling to 40. Thus, overall, an increased rate of vagal stimulation reduced the rate of sinus discharge from 50 to 40, but over some ranges the vagal stimulation entrained the sinus pacemaker to discharge in step with vagal stimulation.

One possible consequence of this triphasic response is as follows. In the intact subject, short vagal bursts may be temporally related to the previous beat because of stimulation of the baroceptors in systole. If the next sinus beat falls due at the time of the down stroke of the primary inhibition curve, its timing will be locked to that of the previous beat. Any tendency of the beat to delay will result in less inhibition and less delay; any tendency to advance will result in more inhibition and more delay. This may well be the explanation for the characteristically jerky changes of heart rate that are seen at times of high vagal tone, such as during sleep (Ewing et al. 1984).

The effect of sympathetic stimulation on the sinoatrial node is to increase its discharge rate – the opposite of the effect of vagal stimulation. The nature of the

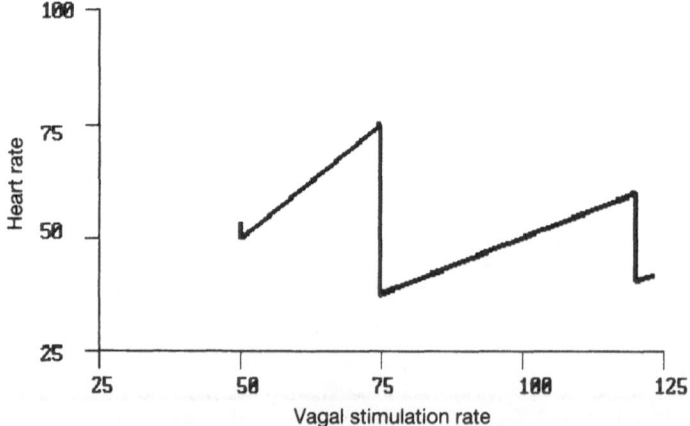

Fig. 1.9. The response of the heart rate in a dog experiment to short bursts of vagal stimulation over a frequency range of 50–120 bursts per minute. (After Levy et al. 1972.)

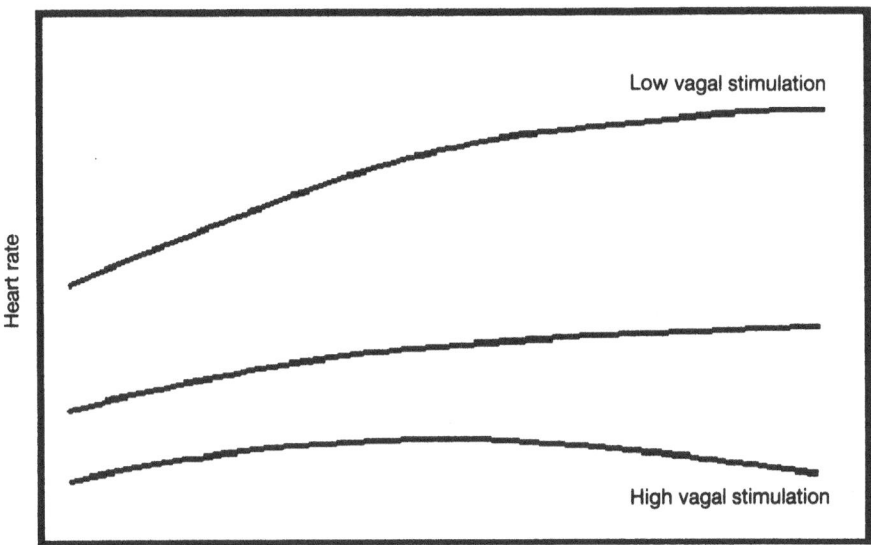

Fig. 1.10. The heart rate response to sympathetic stimulation at various levels of vagal stimulation, showing accentuated antagonism.

vagal–sympathetic interaction in the sinoatrial node is seen in the response curves showing the sinus rate at different sympathetic and vagal stimulation rates. Parallelism of sympathetic response curves at different vagal stimulation strengths would imply that the effects of sympathetic and vagal stimulation are algebraically additive. In fact, as shown schematically in Fig. 1.10, the curves are parallel over much of their course but diverge at high sympathetic stimulation rates, vagal stimulation having an augmented inhibitory effect at times of high sympathetic activity – so-called accentuated antagonism (Levy and Zieske 1969).

These peripheral vagal–sympathetic interactions may result from inhibition by acetylcholine of noradrenaline release from neighbouring nerve terminals, and inhibition of acetylcholine release by neuropeptide Y released along with noradrenaline from sympathetic nerve terminals. At a cellular level noradrenaline increases, and acetylcholine decreases, the concentration of the cellular messenger, cAMP.

Autonomic Control of the Atrioventricular Node

The atrioventricular node has several functions: to convey impulses from atria to ventricles, to protect the ventricles from excessive atrial rates, to adjust the relative timing of atrial and ventricular contraction for optimal ventricular performance, and to act as a subsidiary pacemaker in the event of non-arrival of the impulse from the sinoatrial node. Much of the published work on atrioventricular function is concerned with the first of these, the so-called conducting properties of the atrioventricular node.

Atrioventricular delay is less at slow heart rates, there being a strong negative correlation between P-P and P-R intervals. Vagal stimulation has a direct and an indirect action on atrioventricular conduction, the two actions opposing each other. Direct vagal action results in slower atrioventricular conduction. It also leads to a slower sinus rate and so, indirectly, to faster conduction; overall, conduction may be little changed. Study of the direct vagal effect on atrioventricular conduction therefore requires the atrial rate to be held constant by atrial pacing.

The time-course of the effect of a short vagal burst on atrioventricular conduction is very different from the triphasic response of the sinoatrial node. A short burst of postganglionic vagal stimulation results in a rapid lengthening of A-V conduction, which then returns to normal over the next 5–10 beats (Fig. 1.11) – a monophasic response (Mazgalev et al. 1988). The effect, which is due to hyperpolarisation of cell membranes in the N region of the atrioventricular node, is most marked if the stimulus is delivered in midcycle. The net effect of a vagal discharge affecting both sinoatrial and atrioventricular nodes is, therefore, a complex function of sinus rate and vagal action, sinus rate being affected by vagal action and itself affecting conduction directly, as well as determining the phase and hence effectiveness of vagal action on the atrioventricular node.

Following a short hyperpolarising vagal burst delivered early in a long cycle, a rebound phenomenon may be seen in which there is an increased rate of phase 4 depolarisation of atrioventricular nodal cells. This may result in the appearance of junctional escape beats, and an acceleration of ventricular rate (Mazgalev et al. 1988).

In the N region of the atrioventricular node are found groups of pacemaker cells which appear to be distinct from those cells in the same region that are responsible for major conduction delays. Moreover, pacemaker cells appear to have an independent innervation from conducting cells. Vagally induced depression of the junctional region of the atrioventricular node does not

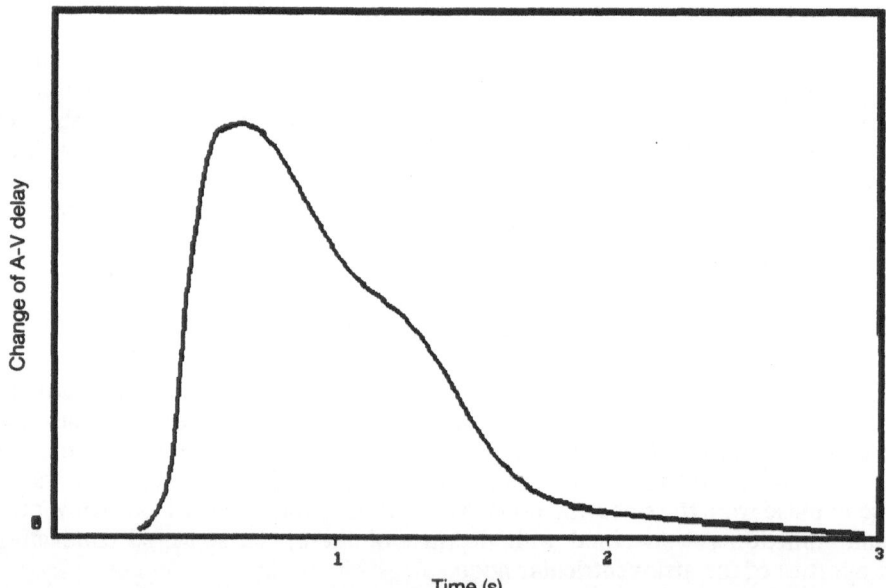

Fig. 1.11. The effect of a short burst of vagal stimulation on conduction in the atrioventricular node.

necessarily influence automaticity and conduction in a similar manner. The response of atrial pacemakers to short bursts of postganglionic vagal stimulation during junctional rhythm resembles that of the sinoatrial node, with acceleration and a triphasic response in some cases, depending on the spontaneous junctional cycle length and on the phase in which the vagal stimulus is delivered. The changes in conduction do not correlate with the changes in automaticity.

The nature of the sympathetic–vagal interaction is different in the atrioventricular node from that in the sinoatrial node. In junctional rhythm the response curves to sympathetic stimulation of both ventricular rate and A-V conduction are substantially parallel, indicating lack of interaction and an algebraic summation of the opposing sympathetic and vagal effects. The conducting properties of the node are particularly sensitive to adrenergic influence but there is no true interaction with the vagal effect (Mazgalev et al. 1988). A short burst of stimulation applied to the stellate ganglion results in a shortened A-V conduction time which takes 20 seconds to return to normal. This is much longer than the second or so it take s the node to recover from vagal stimulation (Spear and Moore 1973).

If sinus rate is prevented from falling by atrial pacing, the depressant effect of vagal stimulation on atrioventricular conduction may be uncovered. Thus, stimulation of baroceptors with a rise in blood pressure may induce atrioventricular block, which is not seen if the sinus rate is allowed to fall (O'Toole et al. 1984).

The parallel effect of the vagus on sinoatrial and atrioventricular nodes during respiratory sinus arrhythmia may be demonstrated in the surface electrocardiogram of healthy subjects. Multiple regression analysis of the P-R intervals in a 3-minute recording showed a significant respiratory variation of the P-R interval of about 1% in 57% of subjects (Rawles et al. 1989)

References

Anderson RH, Ho SY (1990) Accessory atrioventricular connections. In: Touboul P, Waldo AL (eds) Atrial arrhythmias. Current concepts and management. Mosby Year Book, St Louis, pp 9–26
Bleeker WK, Mackay AJC, Masson-Pevet M, Bouman LN, Becker AE (1980) Functional and morphological organization of the rabbit sinus node. Circ Res 46:11–22
Brown GL, Eccles JC (1934) The action of a single vagal volley on the rhythm of the heart beat. J Physiol (Lond) 82:211–41
Cohen RJ, Berger RD, Dushane TE (1983) A quantitative model for the ventricular response during atrial fibrillation. IEEE Trans Biomed Eng 30:769–81
Durrer D, Janse MJ, Lie KI, Van Capelle FJL (1978) Human cardiac electrophysiology. In: Dickinson CJ, Marks J (eds). Developments in cardiovascular medicine. MTP Press, Lancaster
Eckberg DL (1976) Temporal response patterns of the human sinus node to brief carotid baroceptor stimuli. J Physiol (London) 258:769–82
Ewing DJ, Neilson JMM, Travis P (1984) New method for assessing cardiac parasympathetic activity using 24 hour electrocardiograms. Br Heart J 52:396–402
Fisch C (1988) Concealed conduction at the AV nodal level. In: Mazgalev T, Dreifus LS, Michelson EL (eds) Electrophysiology of the sinoatrial and atrioventricular nodes. Alan R Liss: New York, pp 169–86 (Progress in clinical and biological research, volume 275)
Gallagher JJ, Sealy WC, Kasell J, Wallace AG (1976) Multiple accessory pathways in patients with the pre-excitation syndrome. Circulation 54:571–91
Grant RP (1956) The mechanism of A-V arrhythmias with an electronic analogue of the human A-V node. Am J Med 20:334–44
Guevara MR, Glass L (1982) Phase locking, period doubling bifurcations and chaos in a mathematical

model of a periodically driven oscillator: A theory for the entrainment of biological oscillators and the generation of cardiac dysrhythmias. J Math Biol 14:1–23

Harms FMA, Heethar RM, Robles de Medina EO, Meijler FL (1980) Atrio-ventricular nodal "memory" studied by random atrial stimulation. Am J Cardiol 45:459 (abstract)

Jalife J, Moe GK (1979) Phasic effects of vagal stimulation on pacemaker activity of the isolated sinus node of the young cat. Circ Res 45:595–608

Jalife J, Michaels DC, Delmar M (1988) Mechanisms of pacemaker synchronization in the sinus node. In: Mazgalev T, Dreifus LS, Michelson EL (eds) Electrophysiology of the sinoatrial and atrioventricular nodes. Alan R Liss, New York, pp 67–91 (Progress in clinical and biological research, volume 275)

James TN (1963) The connecting pathways between the sinus node and A-V node and between the right and left atrium in the human heart. Am Heart J 66:498–508

James TN (1968) The coronary circulation and conduction system in acute myocardial infarction. Prog Cardiovasc Dis 10:410–49

James TN, Isobe JH, Urthaler F (1979) Correlative electrophysiological and anatomical studies concerning the site of origin of escape rhythm during complete atrioventricular block in the dog. Circ Res 45:109–19

Janse MJ (1969) Influence of the direction of the atrial wave front on A-V nodal transmission in isolated hearts of rabbits. Circ Res 25:439–49

Janse MJ, van Capelle FJL, Anderson RH, Touboul P, Billette J (1976) Electrophysiology and structure of the atrioventricular node of the isolated rabbit heart. In: Wellens HJJ, Lie KI, Janse MJ (eds) The conduction system of the heart. Lea & Febiger, Philadelphia, pp 296–315

Jose AD (1966) Effect of combined sympathetic and parasympathetic blockade on heart rate and cardiac function in man. Am J Cardiol 18:476–8

Josephson ME, Seides SF (1979) Clinical cardiac electrophysiology. Techniques and interpretations. Lea & Febiger, Philadelphia

Kirsh JA, Sahakian AV, Baerman JM, Swiryn S (1988) Ventricular response to atrial fibrillation: role of atrioventricular conduction pathways. J Am Coll Cardiol 12:1265–72

Levy MN, Zieske H (1969) Autonomic control of cardiac pacemaker activity and atrioventricular transmission. J Appl Physiol 27:465–70

Levy MN, Zieske H (1972) Synchronization of the cardiac pacemaker with repetitive stimulation of the carotid sinus nerve in the dog. Circ Res 30:634–41

Levy MN, Iano T, Zieske H (1972) Effects of repetitive bursts of vagal activity on heart rate. Circ Res 30:286–95

Loeb JM, de Tarnowsky JM (1988) Modulation of atrioventricular conduction in vivo: integration of heart rate and autonomic neural input. In: Mazgalev T, Dreifus LS, Michelson EL (eds) Electrophysiology of the sinoatrial and atrioventricular nodes. Alan R Liss, New York, pp 169–86 (Progress in clinical and biological research, volume 275)

Mazgalev T, Dreifus LS, Michelson (1988) Electrophysiological mechanism underlying the effects of brief vagal discharges on atrioventricular conduction. In: Mazgalev T, Dreifus LS, Michelson EL (eds) Electrophysiology of the sinoatrial and atrioventricular nodes. Alan R Liss, New York, pp 133–54 (Progress in clinical and biological research, volume 275)

Meijler FL (1986) Comparative aspects of the dual role of the human atrioventricular node. Br Heart J 55:286–90

Meijler FL, Fisch C (1989) Does the atrioventricular node conduct? Br Heart J 61:309–15

Meijler FL, Janse MJ (1988) Morphology and electrophysiology of the mammalian atrioventricular node. Physiol Rev 68:608–47

Meijler FL, Heethar RM, Harms FMA et al. (1982) Comparative atrioventricular conduction and its consequences for atrial fibrillation in man. In: Kulburtus HE, Olsson SB, Schlepper M (eds) Atrial fibrillation. AB Hassle, Molndal, pp 72–80

Mendez C, Gruhzit CC, Moe GK (1956) Influence of cycle length upon refractory period of auricles, ventricles, and A-V node in the dog. Am J Physiol 184:287–95

Moe GK, Preston JB, Burlington H (1956) Physiologic evidence for a dual A-V transmission system. Circ Res 4:357–75

Olsson SB, Cai N, Dohnal M, Talwar KK (1986) Noninvasive support for and characterization of multiple intranodal pathways in patients with mitral valve disease and atrial fibrillation. Eur Heart J 7:320–33.

O'Toole MF, Wurster RD, Phillips JG, Randall WC (1984) Parallel baroceptor control of sinoatrial rate and atrioventricular conduction. Am J Physiol 246:H149–53

Paes de Carvalho A, de Almeida DF (1960) Spread of activity through the atrioventricular node. Circ Res 8:801–9

Randall WC, Thomas JX, Barber MJ, Rinkema LE (1983) Selective denervation of the heart. Am J Physiol 244:H607–13.

Rawles JM, Pai GR, Reid SR (1989) A method of quantifying sinus arrhythmia: parallel effect of respiration on P-P and P-R intervals. Clin Sci 76:103–8

Robergé FA, Nadeau RA, James TN (1968) The nature of the PR interval. Cardiovasc Res 2:19–30

Robergé FA, Bhereur P, Nadeau RA (1971) A cardiac pacemaker model. Med Biol Eng 9:3–12

Sealy WC, Seaber AV (1979) Cardiac rhythm following exclusion of the sinoatrial node and most of the right atrium from the remainder of the heart. J Thorac Cardiovasc Surg 77:436–47

Simson (1988) A model of conduction through the N region of the AV node. In: Mazgalev T, Dreifus LS, Michelson EL (eds) Electrophysiology of the sinoatrial and atrioventricular nodes. Alan R Liss, New York, pp 97–109 (Progress in clinical and biological research, volume 275)

Smith SA (1982) Reduced sinus arrhythmia in diabetic autonomic neuropathy: diagnostic value of an age-related normal range. Br Med J 285:1599–601

Spear JF, Moore EN (1973) Influence of brief vagal and stellate nerve stimulation on pacemaker activity and conduction within the atrioventricular conduction system of the dog. Circ Res 32:27–41

Van der Pol B, van der Mark J (1928) The heart beat considered as a relaxation oscillator, and an electrical model of the heart. Phil Mag 6:763–75

Wallace AG, Sealy WC, Gallagher JJ, Kasell J (1976) Ventricular excitation in the Wolff–Parkinson–White syndrome. In: Wellens HJJ, Lie KI, Janse MJ (eds) The conduction system of the heart. Lea & Febiger, Philadelphia, pp 613–30

Winfree AT (1980) The geometry of biological time. Springer- Verlag, New York

Winfree AT (1987) When time breaks down. The three-dimensional dynamics of electrochemical waves and cardiac arrhythmias. Princeton University Press, Princeton

Chapter 2

The Pathophysiology of Atrial Fibrillation

Pathology

Histological examination of the heart of a patient who has had chronic atrial fibrillation invariably shows abnormalities of the atria, the sinus node, and the region where they are contiguous (James 1982; Ih and Saitoh 1982). Patchy, focal degeneration and fibrosis of atrial myocardium and the conducting system are characteristic, and predispose to atrial fibrillation by virtue of the resulting inhomogeneity of refractoriness and repolarisation. Focal changes in the atria may result directly from involvement by inflammatory conditions such as rheumatic fever or sarcoidosis, or indirectly by occlusion of small arteries supplying the atria. The atrial myocardium may be patchily infiltrated with amyloid or tumour, or show reactive fibrosis as in haemachromatosis (Hodkinson and Pomerance 1977). Idiopathic focal degeneration of the atria, which may be very extensive, may also be seen, sometimes in several family members (Amat-Y-Leon et al. 1974; Waters et al. 1975).

If the sinoatrial node is normal then sinus rhythm may prevail even in the presence of widespread focal atrial disease. However, the sinoatrial node is invariably abnormal in patients known to have had atrial fibrillation for at least 2 weeks before death (James 1982; Davies and Pomerance 1972). The sinoatrial node, as well as being affected by focal conditions that afflict the atria, is also vulnerable to a variety of specific lesions. These include fibromuscular dysplasia of the sinus node artery, alteration of the node's collagen framework, and circumscriptive perinodal degeneration, in which the sinus node gives the appearance of being separated from the rest of the atrium by fibrous tissue (James et al. 1980). Frequently, in systemic disease, the atrioventricular node is also involved by the same pathological process affecting the sinoatrial node.

In acute myocardial infarction complicated by atrial fibrillation, coronary occlusion proximal to the origin of the sinus node artery is usually seen (James 1961, 1968), but the artery to the left atrium and the atrioventricular node may also be implicated (Hod et al. 1987), and atrial infarction has been documented (Gordon et al. 1984).

Atrial hypertrophy, which is often accompanied by dilatation, is another structural factor predisposing to atrial fibrillation (Henry et al. 1976; Aronow et al. 1987), though it seldom occurs in isolation from pathological conditions that also cause focal fibrosis and cellular degeneration (Thiedemann and Ferrans 1977).

If atrial fibrillation occurs just in the last 2 weeks of life the sinoatrial node is usually normal, but there is frequently evidence of pericarditis or pulmonary embolism (Davies and Pomerance 1972).

There are no histological abnormalities that are specific for atrial fibrillation, and the lesions are an exaggeration of those changes that occur with age (Ih and Saitoh 1982). Similar pathological features to those described above are found in patients with atrial flutter or tachycardia, or indeed in patients with sinus rhythm.

In many cases of atrial fibrillation – or sinoatrial disease in which atrial fibrillation often occurs – histological examination post-mortem shows evidence of cardioneuropathy, with involvement of cardiac nerves in the pathological processes affecting the heart (Demoulin and Kulburtus 1979; James 1980). The significance of these observations is unknown.

The Electrophysiology of Atrial Fibrillation

The surface electrocardiogram of a patient with atrial fibrillation shows a rapid, irregular undulation of the baseline between QRS complexes. Study of several leads simultaneously shows that there is electrical activity going on all the time (Fig. 2.1) ; at the time of an isoelectric voltage in one lead there is a peak or a trough in another lead, the recording of zero voltage in any lead being due to the electrical vector being at right angles to that lead, rather than to electrical activity being absent. The picture is one of a vector, or vectors, continuously turning and twisting in all directions, never still and never absent. This is quite unlike the

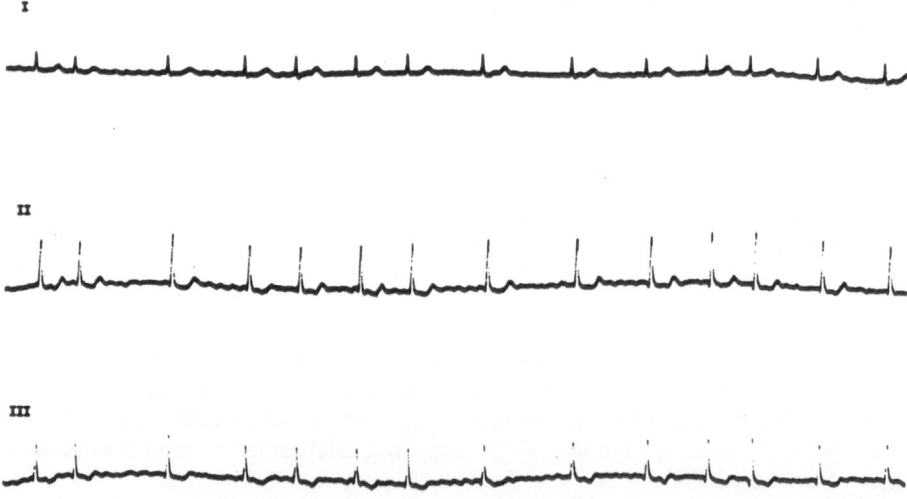

Fig. 2.1. The electrocardiogram in atrial fibrillation. Standard leads I, II, and III are shown.

orderly P-wave vector which sweeps through the same arc at the start of every atrial contraction and then disappears until the next beat.

An intracardiac recording of the electrogram, taken with an electrode located low in the right atrium near to the atrioventricular node, shows a rapid irregular sequence of action potentials (Fig. 2.2). Where do they come from, and where do they go? We owe the answers, more than to anyone else, to Allessie and his colleagues in the University of Limburg, Maastricht (Allessie et al. 1982). These workers have developed a method of mapping the movement of action potentials in the atria, in which fibrillation can be reliably induced by the application of acetylcholine. A pair of hollow egg-shaped shells are inserted into the atria of the isolated perfused dog heart. Each shell bears on its surface 480 tiny silver electrodes that are in contact with the atrial wall. The electrodes are regularly spaced at 3 mm apart, and provide a grid for mapping the passage of action potentials around the atria.

With this mapping technique it has been shown that in atrial fibrillation waves of atrial depolarisation course rapidly, ceaselessly and chaotically this way and that, around natural obstacles, and to and fro between the atria. At any one time there may be several wavelets of depolarisation sweeping through various parts of the atria; at no time is there no wavefront of depolarisation present somewhere. At any point in the atria, for example close to the atrioventricular node, wavelets of depolarisation will arrive frequently from all directions, in a random sequence and at random intervals.

In humans, intracardiac recordings show several types of atrial activity (Gavrilescu and Luca 1975; Puech et al. 1982). Electrical activity may be completely disorganised and chaotic without isoelectric intervals, or may show well-defined though irregular signals with an isoelectric baseline between them, or may exhibit regular flutter-like activity. These different patterns may coexist, so that, for example, one atrium may show flutter-like activity and the other a

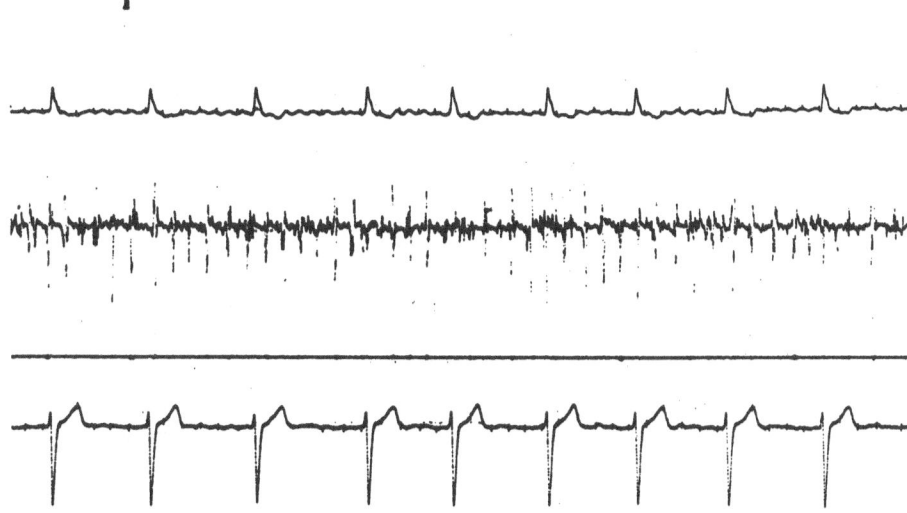

Fig. 2.2. An atrial electrogram in atrial fibrillation. The top channel is standard lead II, the second channel is recorded from low in the right atrium, and the fourth channel is chest lead V_1. (Courtesy of Dr Peter Stafford.)

fibrillatory pattern (Leier and Schaal 1980), or one pattern may merge into another over a short period of time (Wells et al. 1978).

Perpetuation of atrial fibrillation depends on the distance between depolarisation wavefronts, the *wavelength*, being short in relation to the size of the atria. The shorter the wavelength, or the larger the atria, the greater the number of re-entry pathways that can be accommodated. If the wavelength exceeds the available space then atrial fibrillation dies out and is replaced by another rhythm. The wavelength depends on the speed of propagation of the action potential, and the duration of the subsequent refractory period (Smeets et al. 1986). Slow conduction velocity and a brief refractory period mean that atrial myocardium is quickly repolarised and ready to propagate another action potential, so that the distance between depolarisation fronts is short.

Electrophysiological Factors that Predispose to Atrial Fibrillation

Atrial fibrillation in different mammalian species is more likely to occur the larger the size of the animal and its heart (Meijler 1986). In horses atrial fibrillation is a common problem, leading to loss of performance (Stewart et al. 1990). Once initiated in adult cows, atrial fibrillation persists for up to 8 weeks, but only lasts for a few minutes in calves or adult goats, which have small atria (Moore and Spear 1987). Atrial fibrillation is not uncommon in large dogs but is rare in small dogs; it has not been reported in cats (Bohn et al. 1971), and is very difficult to induce experimentally in rats (Meijler 1986a). In humans, whose heart size is intermediate on the mammalian scale, enlargement of the atria is well known to predispose to atrial fibrillation (Henry et al. 1976; Takahashi et al. 1982; Manyari et al. 1990). This is because the larger the atria the greater the number of re-entry circuits that can be accommodated.

The refractory period also affects the length of the re-entry circuit, and hence the number that can exist in atria of a given size. A short refractory period gives rise to a short wavelength and the possibility of more circuits. Stimulation of cholinergic receptors shortens the refractory period, and acetyl-beta- methylcholine or acetylcholine itself may be used to precipitate atrial fibrillation in experimental animals such as the dog (Nahum and Hoff 1940; Burn 1979; Allessie et al. 1982). Vagal activity has long been recognised as a precipitant of atrial fibrillation in animals (Lewis et al. 1921a) and humans (Altschule 1945; Coumel et al. 1978). Acetylcholine released by a vagal stimulus train lowers the atrial fibrillation threshold, determined by delivering extra-stimuli to the right atrium, in the dog (Euler and Scanlon 1987) and lowers the critical mass for initiation of atrial fibrillation in the rabbit (Nilius et al. 1981). Adrenergic mechanisms also participate in the induction of atrial fibrillation by acetylcholine (Hashimoto et al. 1968).

Another factor predisposing to atrial fibrillation is inhomogeneity of refractory periods. This may be brought about by patchy fibrosis, inflammation, infiltration, or by neuropharmacological means. Acetylcholine, besides shortening refractory periods, also causes them to vary in duration because of a non-uniform distribution of its effect in the atria (Alessi et al. 1958; Ninomiya 1966). Thus, vagal stimulation may precipitate atrial fibrillation by the dual action of shortening and temporal dispersion of refractory periods.

The wavelength between depolarisation fronts is determined by conduction velocity as well as the refractory period, slow conduction favouring a short wavelength and the sustenance of fibrillation. Slow inter- and intra-atrial conduction of premature stimuli is an electrophysiological measure that correlates well with a vulnerability to atrial fibrillation (Cosio et al. 1983; Simpson et al. 1988).

The duration and uniformity of the signal-averaged P wave on the surface electrocardiogram may reflect atrial conduction velocities and the dispersion of refractoriness (Ninomiya 1966). In patients with paroxysmal atrial fibrillation the P terminal force in lead V_1 (the product of amplitude and duration of the terminal negative component of the P wave) is greater than normal, suggesting slower intra-atrial conduction velocity (Robitaille and Phillips 1967). The P wave is longer in patients with atrial arrhythmias than in those without this postoperative complication of coronary artery surgery (Buxton and Josephson 1981). Compared with control subjects, patients with paroxysmal atrial fibrillation have P waves that are longer (NS), have more high-frequency content, and higher peak velocity change, suggesting slower and greater dispersion of atrial conduction velocities (Stafford et al. 1990).

The atrial refractory period is positively related to the atrial cycle length; the adaptation to a rapid change of rate is mostly complete within the period of one beat. However, some subjects fail to show rapid adaptation of the refractory period, and also show increased vulnerability to provocation of atrial tachyarrhythmias by extra-stimuli (Attuel et al. 1982). This may be another mechanism leading to dispersion of refractoriness and a predisposition to atrial fibrillation.

Initiation of Atrial Fibrillation

The circumstances in which atrial fibrillation is likely to arise are: an enlarged atrium, patchily involved by a pathological process, and high vagal tone. This combination of conditions results in short refractory periods of varying lengths. The initiating event is often an atrial premature beat (Killip and Gault 1965; Bennett and Pentecost 1970), or a ventricular premature beat that is presumably conducted retrogradely (Csapo 1971), but in many cases atrial fibrillation appears to arise directly from sinus rhythm after passing through a phase of focal atrial tachycardia (Cotoi et al. 1978). Atrial fibrillation is much more likely to be initiated by an extrastimulus falling within the first half of the P-P cycle (Killip and Gault 1965), especially during atrial repolarisation (Reynolds et al. 1967) – "the R on P phenomenon" (Csapo 1971).

A possible sequence of events is that the impulse from a premature beat arrives at a patch of atrial myocardium that is still refractory from the previous beat. The depolarisation wavefront is deflected away from this area, coursing through those irregularly placed patches of atrium that have recovered their excitability. Meanwhile, areas of myocardium that were initially refractory are now recovering at different rates, so new pathways are opening up behind as well as ahead of the wavefront, which may split into several wavelets that later merge and then again diverge. Once fibrillation is established it sets up the circumstances for its own perpetuation since there is then myocardium at all stages of excitation and recovery throughout the atria. Provided there is at all times a pathway linking the excitable areas the erratic passage of the depolarisation wavefront maintains complete desynchronisation of refractoriness. Atrial fibrillation will only die out if the physiological circumstances that give rise to its appearance are changed. Lengthening the refractory period may increase the wavelength so that there is no longer sufficient room for the re-entry circuits. Another solution is to synchronise the refractory periods throughout the atria. This may be achieved by direct current cardioversion which simultaneously depolarises the whole heart with an electric shock so that refractory periods of different areas of the atria are once more in step with each other.

Experimentally, or in electrophysiological studies, atrial fibrillation may be induced by high-frequency stimulation of one or other atria (Haft et al. 1968; Wyndham et al. 1977; Whittington et al. 1979; Brignole et al. 1986); a stimulation frequency of 50 Hz is commonly used. Up to a frequency approaching 10 Hz, atrial myocardium will be depolarised after each stimulus, but above that frequency stimuli will depolarise some fibres but others will be refractory. Turbulent activity will follow, with incoordinate contraction of the atria, lack of effective mechanical activity, and an irregular ventricular response. Similar chaotic behaviour follows the application of aconitine (Scherf 1947). A rapidly firing focus discharges at a frequency which the atria cannot follow. Induced atrial fibrillation may be modified by autonomic action or drugs that affect the refractory period of the atrial myocardium, but clearly no drug will terminate the fibrillation as long as the electrical stimulation is applied or the aconitine focus continues to fire (Moe 1968). Doubts have been expressed about the validity of using induced atrial fibrillation as a model for the real thing (Strackee et al. 1971; Dreifus and Mazgalev 1988).

The Mapping of Atrial Flutter and Fibrillation: The Essential Distinction Between Them

The surface electrocardiogram during atrial flutter shows a regular saw-tooth appearance of the baseline, with about 300 flutter waves per minute. Mapping of flutter in the dog by the same technique used for atrial fibrillation shows that a re-entry circuit is present. In the most common variety of flutter (type 1) the depolarisation wavefront descends in the free wall of the right atrium along the line of the inferior internodal tract and after passing between the orifice of the inferior vena cava and the tricuspid valve ascends in the inter-atrial septum. There is some uncertainty as to whether or not the circuit includes the roof of the right atrium above the superior vena cava, but the left atrium is not included and has a bystander role. There is only a single circuit, and in the less common type 2 flutter there is clockwise rotation compared with counterclockwise rotation in type 1 flutter. In the dog the wavelength between depolarisation fronts is 8–10 cm. This is longer than the wavelength of atrial fibrillation, which is less than 8 cm, but less than that of unsustained atrial tachycardia which is 10–12 cm; the wavelength in normal sinus rhythm is 12–16 cm (Allessie et al. 1990).

In humans two varieties of flutter are also recognised – types 1 and 2 – with respectively negative and positive flutter waves in the precordial leads. The circus movements follow the same route as in the dog, ascending in the septum and posterior wall of the right atrium and descending in the lateral and anterior wall. The path length is about 5 cm. In type 2 flutter the direction of depolarisation is reversed.

The essential difference between atrial fibrillation, atrial flutter, and an unsustainable tachycardia consisting of a few beats, is the distance between depolarisation fronts. When the wavelength is that of flutter the circuit followed is the longest that can be accommodated in the right atrium; a greater wavelength results in the tachycardia being unsustainable, while a shorter wavelength allows multiple pathways to be accommodated.

Lewis et al. (1921b) wrote "The essential difference between flutter and fibrillation is that in the former the gap between the crest and wake of the circulating wave is greater."

The Ventricular Response in Atrial Fibrillation

The Input to the Atrioventricular Node

A moment's consideration of the mechanism of atrial fibrillation – multiple depolarisation wavelets coursing through the atria and impinging on the atrioventricular node – reveals the most likely source of the irregularity of the pulse in atrial fibrillation. The arrival of impulses at the node might be expected to be truly random, with the rate of arrival showing a Poisson distribution (Goldstein and Barnett 1967; Hashida et al. 1973). Such a statistical distribution describes the situation when events happen singly at random, when the timing of one event has no effect on the timing of any other, and the average rate of occurrence of the events is constant over the period of observation. Although these conditions are probably met in atrial fibrillation, there have been few reported studies of the fibrillation rate in the vicinity of the atrioventricular node (Puech et al. 1982; Kirsh et al. 1988). An atrial electrogram is depicted in Fig. 2.2, and Fig. 2.3 (A-A) shows a histogram of the intervals between fibrillation potentials derived from that recording; the mean interval between arrival of action potentials is 165 milliseconds, with a coefficient of variation of 28%. The shortest fibrillation interval is 64 milliseconds; this is because the atrial myocardium in the vicinity of the recording electrode is unable to conduct another wave until after the period of refractoriness is over. One of the conditions for a Poisson distribution is therefore not met: the timing of one event is not independent of the timing of another, as a second event is unable to occur within 64 milliseconds or so of previous depolarisation.

The internodal tracts provide several routes of access to the node, and it could be argued that each fibre within each tract constitutes a route to the node (Goldstein and Barnett 1967), so that access to the node is available at all times and impulses may arrive independently of each other. If that is the case then a different condition for the occurrence of a Poisson distribution is not met, namely that events must happen singly. If multiple routes into the node are conceded then there is the possibility of several stimuli arriving simultaneously. Moreover, wavefronts arriving from different directions may cancel each other out, so that once again the independence of separate events is lost (Janse 1969; Mazgalev et al. 1982).

Nevertheless, from these theoretical considerations, and from a limited number of observations, we conclude that the input to the atrioventricular node is likely to approximate to a Poisson distribution.

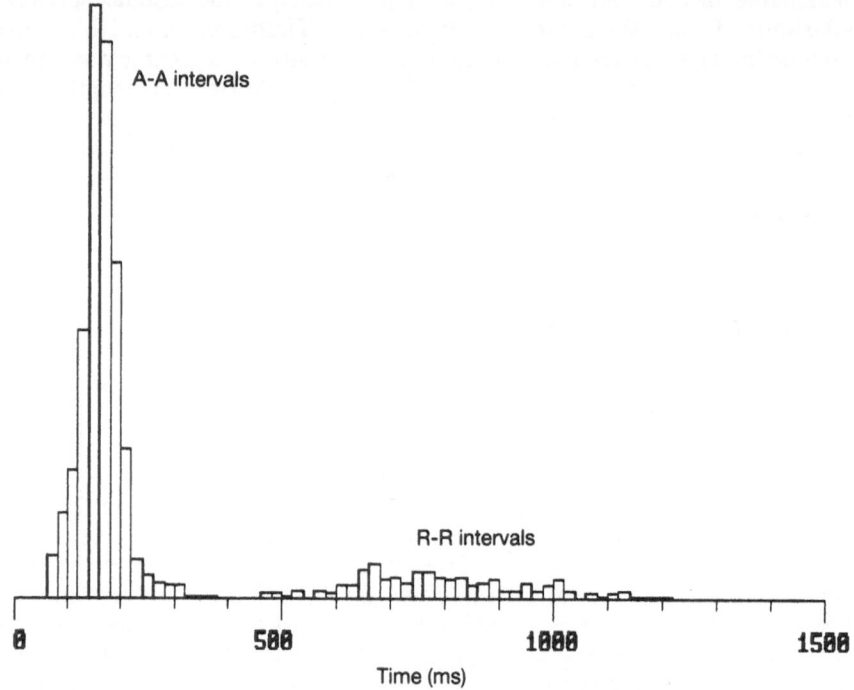

Fig. 2.3. Histogram of intervals between action potentials in the lower right atrium in atrial fibrillation (A- A), and R-R intervals recorded from the surface electrocardiogram (R-R), over the same time period. A-A intervals: mean 165 milliseconds, range 64–312 milliseconds, CV 28%. R-R intervals: mean 809 milliseconds, range 464–1368 milliseconds, CV 22%.

The Ventricular Response in Atrial Fibrillation

Generations of medical students have been taught that the pulse in atrial fibrillation is irregularly irregular, and the nature and cause of the irregularity have received much attention, ever since Lewis (1909) wrote "The irregular pulse is due to fibrillation of the auricle". The title of the first description of atrial fibrillation by Hering (1903) mentioned "pulsus irregularis perpetuus", and Lewis (1909–10) spoke of "the absolute irregularity of the ventricle"; the subject has continued to receive editorial comment (Brody 1970; Meijler 1986b; Dreifus and Mazgalev 1988).

If the atrioventricular node is considered as a through-conductor with a refractory period, then one theoretical possibility is that the ventricular response will be rapid and regular, for as soon as the atrioventricular node has recovered from the passage of one impulse another will arrive and be conducted to the ventricles (Langendorf 1965). That, of course, is not the case, and Fig. 2.4 shows a typical histogram of R-R intervals from a patient with atrial fibrillation with a mean ventricular rate of 75 beats per minute. The distribution of R-R intervals is skewed, with a tail to the right; the coefficient of variation is 23%.

In Fig. 2.3 the histogram of intervals between action potentials in the right atrium (A-A) is contrasted with that of R-R intervals over the same time period (R-R). Compared with A-A intervals, the distribution of R-R intervals is shifted

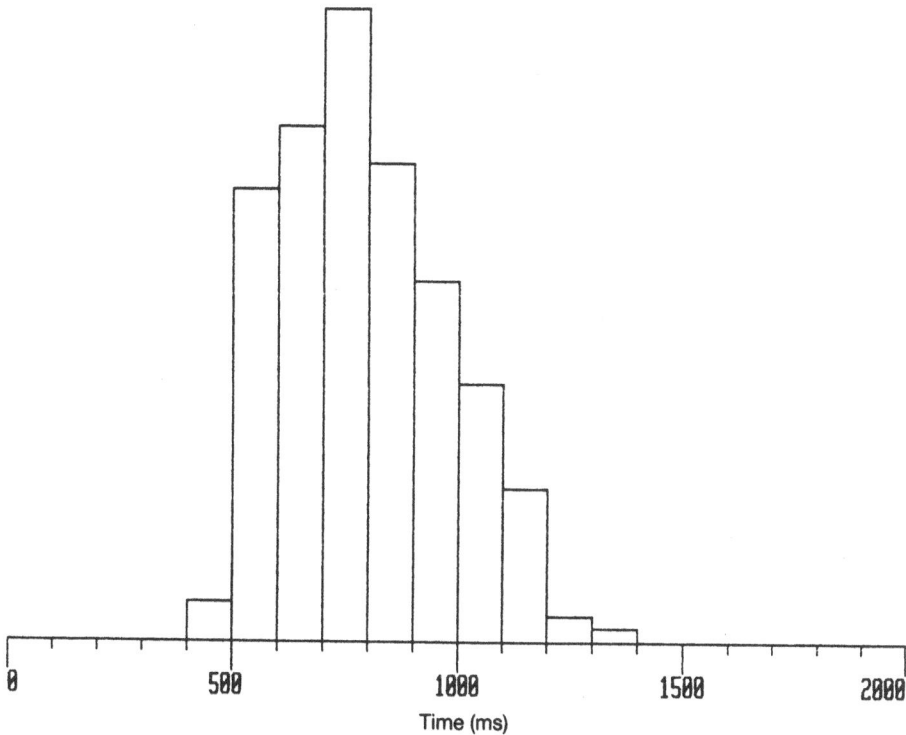

Fig. 2.4. Histogram of R-R intervals in a subject with atrial fibrillation. Mean R-R 799 milliseconds, range 488–1320 milliseconds, modal 700 milliseconds.

so far to the right that there is no overlap; also, there are many fewer R-R than A-A intervals, as indicated by the smaller area of the R-R interval histogram. These two histograms, of fibrillation rates in the atrium and of R-R intervals in the ventricles, embody a central problem in atrial fibrillation: Speaking both mathematically and physiologically, how does the atrioventricular node convert impulses from an atrial to a ventricular distribution? An answer will not be attempted here, but the question is taken up again in the next chapter. Meanwhile, further characteristics of the ventricular response in atrial fibrillation will be described.

In chronic atrial fibrillation a very wide range of average ventricular rates is encountered – from 46 to 183 in one series of patients (Escudero et al. 1976). The mean resting ventricular rate is 90–110 beats per minute (Escudero et al. 1976; Rawles and Rowland 1986; Pai and Rawles 1989). There is no relation between ventricular rate and age, but in patients taking digoxin there is a loose negative relation between the serum digoxin concentration and ventricular rate (Chamberlain et al. 1970; Beasley et al. 1985).

Within subjects the coefficient of variation of R-R intervals is about 20%, and is unrelated to the mean ventricular rate (Escudero et al. 1976), so that although the absolute variability in ventricular rate is less at high than at low ventricular rates, the percentage variation around the mean value is about the same. Roughly speaking, if the mean ventricular rate in a patient with atrial fibrillation is 100, then the range of instantaneous ventricular rates would be from 50–150 beats per minute, a fact we can use in considering the need to reduce ventricular rate (Chapter 6).

In the statistical sense, the pulse in atrial fibrillation is not completely irregular since there are limits to the range of possible R-R intervals that can occur. The shortest R-R interval is about 300 milliseconds, a figure determined by the refractory periods of the atrioventricular node and of ventricular myocardium. At the other extreme, by the time twice the mean R-R interval has elapsed a second R wave will usually have occurred. Holter monitoring has shown pauses of up to 2.8 seconds by day and 4 seconds by night in asymptomatic patients with atrial fibrillation (Pitcher et al. 1986). The occasional occurrence of such long pauses skews the distribution of R-R intervals to the right. In patients with atrial fibrillation and a history of near-syncope, pauses of more than 2 seconds are considered abnormal, and taken as evidence of coexisting heart block (Rebello and Brownlee 1987).

Several authors who have studied the distribution of R-R intervals in atrial fibrillation have noted that the histograms are not always unimodal but may show two or more peaks. Thus, Soderstrom (1950) remarks on the tendency for R-R intervals to congregate around durations of 350, 700, 1000 and 1400 milliseconds. Goldstein and Barnett (1967) reported that modes most commonly occurred at about 450 and 650 milliseconds but were observed anywhere from 350 to 1350 milliseconds. Horan and Kistler (1961) noted a tendency for histograms to have a single peak at high ventricular rates, a double peak at intermediate rates, and a low single peak at low rates. Olsson et al. (1986) found a bi- or tri-modal distribution in 16 of 22 patients with atrial fibrillation, and considered this to be evidence of dual pathways in the atrioventricular node. Urbach et al. (1969) considered peaks on the histogram of R-R intervals in patients with atrial fibrillation to be evidence of accelerated junctional escape rhythm, and possible digitalis toxicity. The significance of these observations will be considered in the next chapter.

If the ventricular response in atrial fibrillation is truly irregular, then not only should the distribution of R-R intervals around the mean be normal, which is only approximately the case, but their distribution in time should be random. This may be tested by the technique of autocorrelation, illustrated in Fig. 2.5. Correlation coefficients are calculated for each R-R interval and its successor (first order autocorrelation coefficient), each interval and its second successor (second order autocorrelation coefficient), third successor (third order autocorrelation coefficient), and so on, up to the tenth successor. If the sequence of R-R intervals is

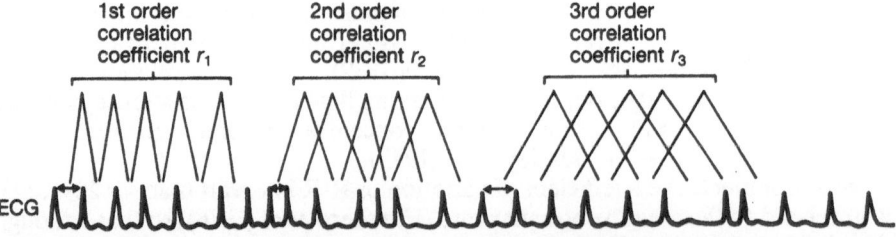

Fig.2.5. The technique of autocorrelation of R-R intervals in atrial fibrillation. If the sequence of R-R intervals is random then r_1–r_n are not significant. (Reproduced by permission of the Editor, *British Heart Journal*.)

random and the duration of a given interval has no influence on those that follow then these autocorrelation coefficients will not differ significantly from zero. On the other hand, if a given R-R interval has an effect on its successors then one or more of the autocorrelation coefficients will achieve statistical significance and may be negative or positive.

In 16 out of 17 patients with atrial fibrillation, Horan and Kistler (1961) found no relationship between consecutive R-R intervals, but Goldstein and Barnett (1967) found evidence of non-random behaviour in half their records. They described "chaining", in which series of beats with almost equal R-R intervals occur, giving rise to a positive relationship between consecutive R-R intervals.

Bootsma et al. (1970) computed autocorrelograms from approximately 2000 beats in 36 cases of atrial fibrillation. In 12 cases (8 positive, 4 negative) the first-order autocorrelation coefficients exceeded the 95% confidence limit, and in 9 cases the 99% limit. In 23 cases the autocorrelograms were repeated after exercise, when 13 positive and 1 negative first-order coefficients were significant ($p<0.05$), compared with 3 positive and 3 negative before exercise; there was a positive shift in the mean value of the first-order coefficients.

Moe and Abildskov (1964) showed in experimentally induced atrial fibrillation in the dog that stellectomy, which reduced the ventricular rate, led to abolition of a positive first-order autocorrelation.

In a study of 58 patients with atrial fibrillation, Honzicova et al. (1973) found a random ventricular response in 51, but 7 had alternation of cycle lengths and a negative first-order autocorrelation coefficient.

Rawles and Rowland (1986) studied 74 patients with atrial fibrillation, of whom 22 (30%) had non-random rhythm as evidenced by significant departure of first-order autocorrelation coefficients from zero; there were equal numbers of those with significant negative or positive coefficients.

In summary, these studies show that while the majority of cases of atrial fibrillation exhibit a random ventricular response, non-random rhythms with either positive or negative first-order autocorrelation coefficients are present in a substantial minority. The magnitude of statistically significant first-order correlation coefficients is low, especially when the number of beats is large. Thus, in our series, the coefficients ranged from -0.423 to $+0.407$. Less than 20% of the variance of R-R intervals even in significant cases can be explained in terms of the correlation between consecutive beats. Thus, from a clinical point of view, the dogma that the pulse in atrial fibrillation is completely irregular can be upheld (Meijler 1986b). However, in trying to understand how the ventricular response comes about, these minor departures from complete irregularity are of considerable interest.

The Role of the Autonomic Nervous System in Atrial Fibrillation

Mackenzie (1922) recognised that in atrial fibrillation there is loss of control of heart rate by the sinoatrial node, control being taken up by the atrioventricular node; indeed, he used the term "nodal rhythm" for atrial fibrillation.

Although involvement of cardiac nerves by the pathological processes underlying atrial fibrillation has been reported (Demoulin and Kulburtus 1979; James 1980), there is no compelling evidence of widespread cardiac autonomic neuropathy. The ventricular rate in atrial fibrillation increases with exercise (Aberg et al. 1972; Beasley et al. 1985) or with atropine (Horan and Kistler 1961; Graybiel 1964), suggesting that at rest it is restrained by a measure of autonomic tone. On the other hand, the intrinsic heart rate in atrial fibrillation is said to be no different from that before autonomic blockade (Jose 1966), which might indicate that the autonomic nervous system does not exercise any control over ventricular rate, or that the balance between vagal and sympathetic tone is different from that in sinus rhythm, when vagal tone predominates.

Evidence of anatomical integrity of the baroreflex arc in atrial fibrillation is given in a case report of a hypertensive patient with an implanted carotid sinus nerve stimulator (Borst and Meijler 1984). Stimulation resulted in slowing of ventricular rate after a delay of about 1.5 seconds. Because of the unphysiological nature of the stimulus, no conclusions can be drawn from this single case about the normal role of the baroreflexes in atrial fibrillation.

Meijler et al. (1984), while maintaining that the rhythm of atrial fibrillation in humans is random, showed that it may be non-random in the horse, in which atrial fibrillation is characterised by a skewed distribution of R-R intervals, a few lasting as long as 5 seconds. These exceptionally long intervals were associated with fluctuations of blood pressure thought to result in baroreflex activity which modified atrioventricular nodal function, imposing a pattern on the otherwise random rhythm. The explanation for the alleged lack of pattern in atrial fibrillation in humans was said to be the absence of such long ventricular pauses as occur in the horse.

In atrial fibrillation the baroceptor/heart rate reflex, if it functions at all, has to act through the atrioventricular node rather than the sinoatrial node as in sinus rhythm. The extent to which its operation is thereby impaired will be discussed in a later chapter. It should be noted here that if the baroceptor/heart rate reflex operates in atrial fibrillation, even in a modified manner (such as by entrainment, as described in Chapter 1), some imposition of order on the otherwise random ventricular response might be expected.

Autocorrelation and the Baroreflex

Imagine that an exceptionally long R-R interval occurs in a sequence of beats of roughly average duration in atrial fibrillation (Fig. 2.6). The beat terminating this long interval, during which the ventricle fills more than average, will generate higher than average systolic pressure. About half a second after the acute rise of blood pressure in systole the efferent vagal discharge rate will increase and this might be expected to increase the atrioventricular delay and postpone the occurrence of the next beat, prolonging the next R-R interval. This next interval would therefore also be longer than average. By a similar argument a shorter than average R-R interval will lead to lower than average systolic blood pressure, the withdrawal of vagal tone, less AV delay, and another short R-R interval. This, then, could be the mechanism of "chaining", in which there is a tendency for consecutive R-R intervals to be more alike than would be expected by chance. It

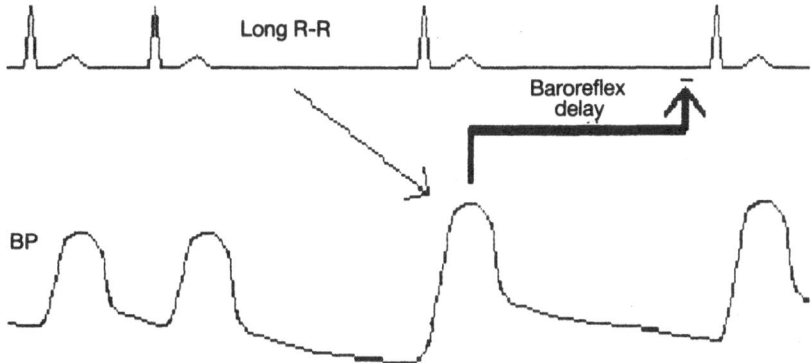

Fig. 2.6. A possible mechanism of chaining, involving operation of the baroreflex through the atrioventricular node in atrial fibrillation.

will be recalled that this occurrence is marked by a statistically significant positive first-order autocorrelation coefficient.

Consider now the possibility that the time taken for the baroreflex to operate is long in relation to the mean R-R interval, which is halved compared with the example just given. An exceptionally long R-R interval would then result in prolongation not of the next, but of the next-but-one R-R interval (Fig. 2.7). The beat terminating the exceptionally long interval will be followed by an interval of average duration but will generate higher than average systolic pressure. This will result in the next-but-one interval being longer than normal. Thereafter, there would be a self- perpetuating tendency for long and short intervals to alternate, giving a negative first-order autocorrelation coefficient. Thus, both chaining and alternation of R-R intervals may be explained by operation of the baroreflex, the sign of the correlation between consecutive intervals depending on the relative speed of operation of the reflex and the average duration of cardiac intervals. On this analysis some indication of the operation of the baroreflex might be gleaned from examination of the autocorrelogram, a negative first-order autocorrelation

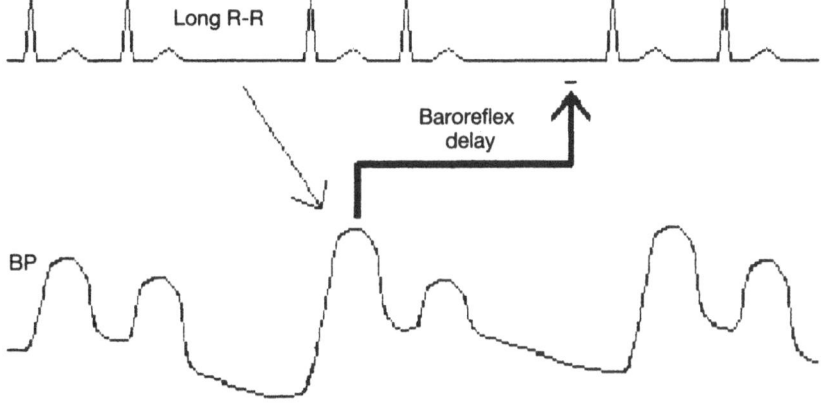

Fig. 2.7. A possible mechanism of alternation of cycle lengths in atrial fibrillation.

coefficient being expected with high ventricular rates, and a positive coefficient with slower ventricular rates. However, when first-order autocorrelation coefficients are plotted against mean R-R intervals no significant relationship is seen (Fig. 2.8).

There is, in fact, a serious flaw in using the autocorrelogram to investigate the randomness of the rhythm in atrial fibrillation. Autocorrelation is a technique borrowed from time- series analysis, often used for economic forecasting. Here, information on, say, exports is available at regular intervals. From a knowledge of the correlation between consecutive monthly export totals an attempt is made to predict next month's figures from this month's. In the case of atrial fibrillation what is being measured is time itself – the interval between beats – and this is sampled at irregular intervals determined by the time of occurrence of each beat. The series of beat-to-beat intervals has only a spurious similarity to a time-series, and autocorrelation can only be properly applied if the relationship between the duration of one interval and the next is not time- related. For example, suppose intervals are classified as long or short, depending on whether they are of greater or less than average duration. Suppose that the occurence of a long beat results in a switch being thrown into the "long" position, which renders the next interval longer than it would otherwise be, by an amount that is not changed with the passage of time. The switch is not reset until the next beat occurs, however much later that may be. Under these circumstances it would be appropriate to test for the presence of such a mechanism by means of the autocorrelogram, which might be called an ordinal autocorrelogram because it is the order in which beats occur

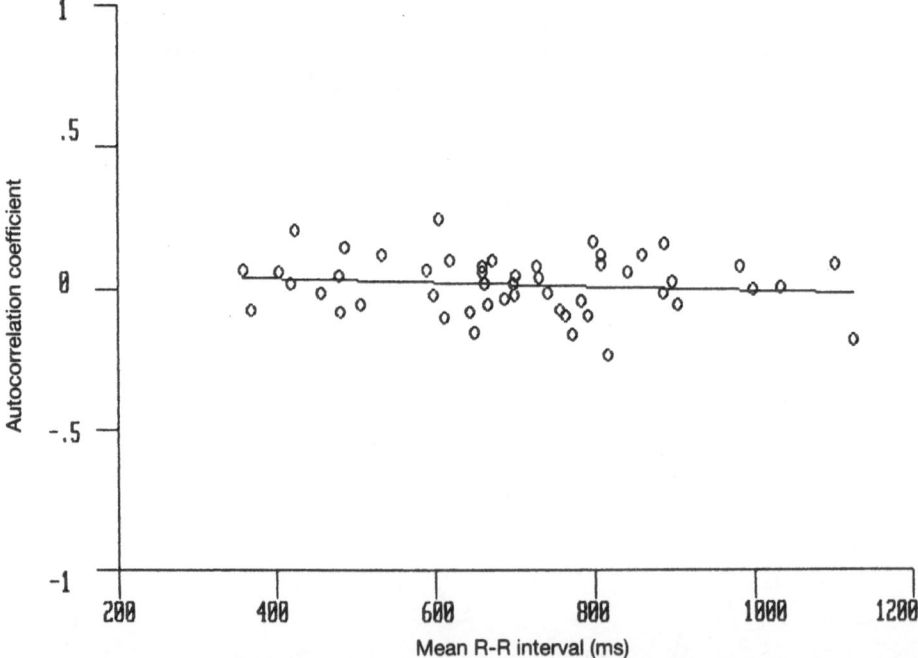

Fig.2.8. A plot of first order autocorrelation coefficients against mean R-R intervals in 50 patients with atrial fibrillation.$r= -0.13$, p not significant.

that is under investigation. However, such a mechanism is very improbable, and it is much more likely that any influence that one beat has on the timing of the next decays rapidly with the passage of time.

Fig. 2.9 takes as an example atrial fibrillation with an average R-R interval of 800 milliseconds and a coefficient of variation of R-R intervals of 20%. Starting with the occurrence of an index beat, the probability of the occurrence of the next beat is plotted against time. For a period of time equal to the refractory period of the atrioventricular node there is no chance of another beat occurring, but thereafter the probability increases to a maximum, corresponding to the modal R-R interval; after that the probability declines until there is virtual certainty that the beat has already occurred, and therefore negligible chance that it is yet to occur at any instant. The cumulative area under the curve is the probability that a beat has occurred by that time. By the time 1600 milliseconds have elapsed it is certain that the beat will have occurred, and the area under the curve is 1. What has been drawn is the probability density function, which has exactly the same shape and time-base as the distribution of a large number of R-R intervals. For the purposes of the illustration this has been represented as a normal distribution with the familiar bell- shaped contour.

Let us now draw the probability density functions for the first and subsequent beats up to the tenth; 99% confidence intervals are shown (Fig. 2.10). The curves indicating the probability of the occurrence of the second to tenth beats are wider and lower than that for the first beat, though the area under each curve is the same and equal to 1, i.e. the total probability that each of the second to tenth beats will occur. The upper tail of the probability density function of the first beat is overlapped by the lower tail of that of the second beat, and the degree of overlap in the possible time of occurrence of consecutive beats becomes progressively

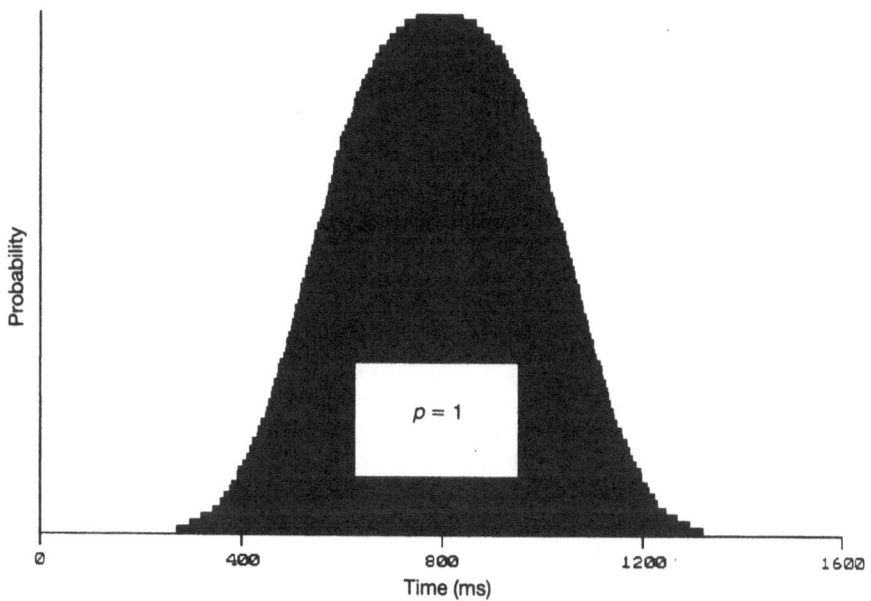

Fig. 2.9. The probability of occurrence of the first beat following an index beat in atrial fibrillation. The mean R-R interval is 800 milliseconds, and the coefficient of variation of R-R intervals is 20%.

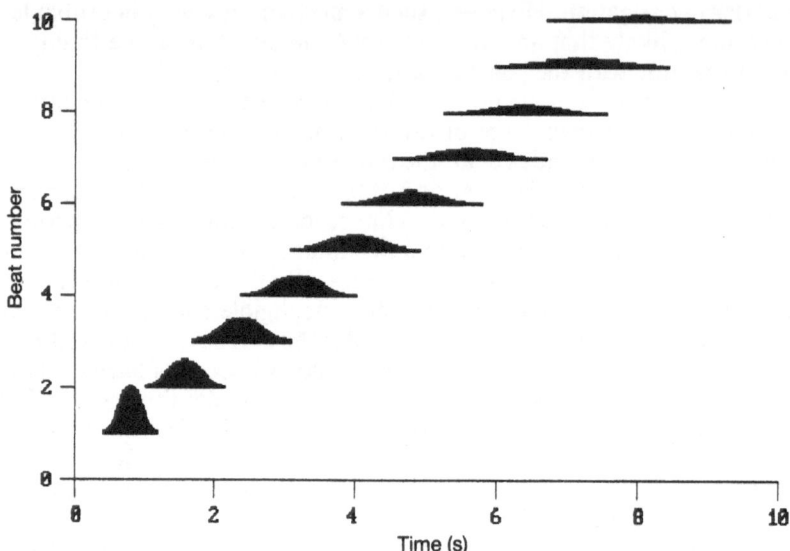

Fig. 2.10 Probability density functions for beats 1–10 after an index beat in atrial fibrillation.

higher as the beat number increases; the probability density curve becomes lower as well as wider.

Five seconds after the index beat four beats will certainly have occurred, with a high possibility that the fifth has taken place. There is rather more than a 50% chance that the sixth beat has occurred, and a small chance that the seventh has taken place too; there is only an outside chance that the eighth beat has occurred ($p<.01$). The probability of *any* beat occurring at any particular point in time after an index beat is calculable from the sum of all the probabilities of the overlapping density functions of individual beats. As these get flatter and wider their combined height evens out at a constant value depending on the average heart rate and the duration of the time-sample under consideration.

In an ordinal autocorrelogram the index beat is correlated with the nth following beat even though that beat may occur over a wide range of times, times that overlap with those of the $(n+1)$th and $(n-1)$th and several subsequent and preceding beats. The ordinal autocorrelogram therefore blurs and distorts any association that may be present between the occurrence of a heart beat and subsequent events that occur after a fixed time delay – such as the time it takes for the baroreflex to operate.

A type of autocorrelogram that is methodologically correct has been applied to atrial fibrillation by Braunstein and Franke (1961). To distinguish it from the ordinal autocorrelogram that we have employed hitherto we might call it a temporal autocorrelogram. The electrocardiogram is scanned at 200 millisecond intervals for a total of 10 seconds after each beat, and the results of the scans summated. The number of subsequent beats falling in each time-sample is noted. An example is shown in Fig. 2.11. The continuous line shows the number of "hits" expressed as a percentage of the total number of beats. The dashed lines indicate the 99% confidence limits for the number of hits that would be expected just by chance, assuming a completely random distribution of beats in time. Thus,

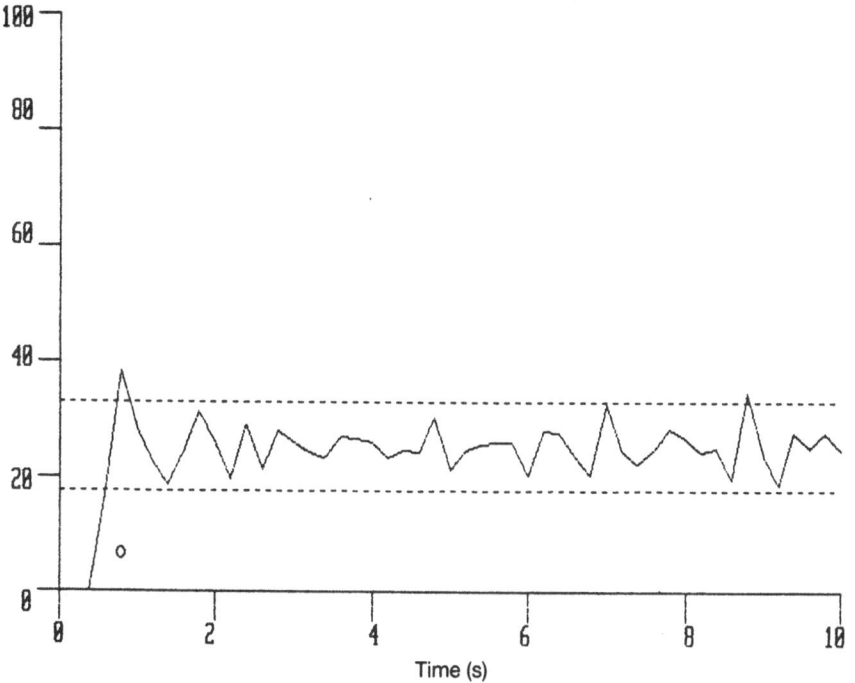

Fig.2.11. A temporal autocorrelogram for a patient with atrial fibrillation. The mean R-R interval is shown by a *small circle*; the *dashed lines* indicate the 99% confidence intervals for the occurrence of beats.

if the average R-R interval is 800 milliseconds and the time- sample is 200 milliseconds then there is a 25% chance that any sample will contain a beat, and the 99% confidence limits for this proportion are 18%–32% for a total of 225 beats in the sequence.

By 4 seconds after the index beat the number of hits is exactly what would be expected by chance, and corresponds to the sum of the probabilities of the first to nth beats occurring at that time. We can say that the occurrence of heart beats 4 seconds after any given beat appears to be random, and by this time any influence on the timing of subsequent beats by the index beat has completely waned. Immediately after the index beat, however, there are far fewer beats than expected by chance, this being the refractory period of the atrioventricular node and ventricles. There are more beats than would be expected by chance at a time corresponding to the mean R-R interval and often, though not in this example, the second and third multiples of it. These multiples of the average cardiac cycle length cover the 1- -2 second period during which the baroreflex might be expected to act. Because of the tendency for beats to cluster in time around the average R-R interval, an indirect consequence of the existence of a refractory period, it is not possible to recognise any influence of the baroreflex itself on the otherwise random temporal distribution of heart beats in atrial fibrillation. We may postulate that the operation of the baroreflex in atrial fibrillation may impose some pattern on the otherwise near-random occurrence of heart beats in atrial fibrillation. However, the ordinal autocorrelogram is a flawed method for detecting any such pattern, and the temporal autocorrelogram, while theoretically preferable, cannot distinguish any pattern imposed by the baroreflex from the non-random distribution of beats indirectly caused by the existence of a refractory period.

We may conclude that any short-term effect of the baroreflex on heart rate in atrial fibrillation is likely to be small and difficult to recognise, and further study of the baroreflex through spontaneous changes of heart rate in atrial fibrillation is unlikely to be fruitful.

Fibrillatory Waves in Atrial Fibrillation

Action potentials from the fibrillating right atrium may be recorded using an intra-atrial suction electrode (Caceres et al. 1965; Olsson et al. 1971). Such monophasic action potentials occur with a frequency of 2–700 per minute and have a widely variable duration inversely related to their frequency, though with high frequencies demarcation of consecutive action potentials is difficult or impossible. The characteristics of the fibrillatory waves recorded in this way are no different in patients with paroxysmal or chronic atrial fibrillation, or in those with underlying rheumatic or coronary heart disease (Cotoi et al. 1979). Patients with longer and less frequent action potentials are more likely to convert to, and remain in, sinus rhythm than those with more rapid fibrillation associated with shorter action potentials (Gavrilescu et al. 1976). The monophasic action potential recorded immediately after cardioversion also has prognostic significance, a longer duration indicating greater refractoriness and increased likelihood of remaining in sinus rhythm (Olsson et al. 1971).

The surface electrocardiogram also shows fibrillatory waves, or f waves, during atrial fibrillation, and these correspond to the organised irregular activities recorded with endocardial electrodes. However, the fine, continuous, disorganised activity seen on an atrial electrogram is not seen in the surface electrocardiogram (Puech et al. 1982).

Aysha and Hassan (1988) claim that f wave amplitude on the surface electrocardiogram correlates strongly with left atrial size, endorsing the findings of Peter et al. (1968). The reverse view is expressed by Morganroth et al. (1979) who state that there is no relationship between fibrillatory wave amplitude and left atrial size or aetiology of fibrillation – a conclusion reached by Aberg (1969) and Garber et al. (1976). The latter suggest that f wave size correlates with the age of the patient, fibrillation becoming finer with age. The average f wave amplitude may be greater in chronic than in paroxysmal atrial fibrillation (Takahashi et al. 1983; Aysha and Hassan 1988).

The Brody Phenomenon in Atrial Fibrillation

Brody (1956) showed theoretically that the magnitude of the QRS vector on the surface electrocardiogram is related to the intracavitary blood mass. Electrical

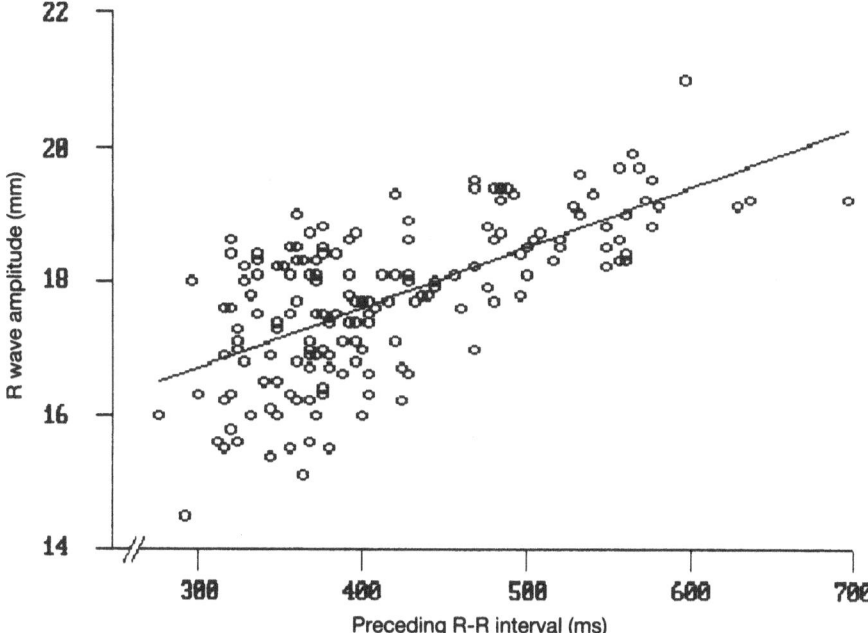

Fig. 2.12. Correlation between amplitude of R waves and preceding R-R intervals in a patient with atrial fibrillation. $r = 0.65$, $p<0.001$.

conductivity is 10 times higher in blood than in myocardium so an increased end-diastolic volume would be expected to augment the radial and attenuate the tangential components of myocardial dipoles. This prediction has been confirmed in healthy subjects (Feldman et al. 1985), and an increase in P wave or R wave voltage may be seen in acute heart failure with atrial or ventricular dilatation.

In atrial fibrillation there is beat-to-beat variation of R-R intervals, diastolic filling periods and hence ventricular end- diastolic volumes. The R wave voltage might therefore be expected to vary with the preceding R-R interval. In a series of 50 patients with atrial fibrillation studied by Dr. G.R. Pai (unpublished), 35 (70%) patients showed a correlation between the voltage of the R wave and the preceding R-R interval (Fig. 2.12). This phenomenon was less common in older patients and those on digoxin, perhaps because the elderly and those in cardiac failure have lower ventricular compliance. In them, an increasing length of diastole would not lead to such a large increase in diastolic volume as in younger, fitter patients. Opposite results have been reported by Teichmann et al. (1987).

The QT Interval in Atrial Fibrillation

The QT interval of the surface electrocardiogram corresponds to electrical systole, during which ventricular muscle is depolarised, as may be demonstrated by unipolar electrodes placed within the heart. If there is a regular ventricular rate and an extrastimulus is applied immediately after depolarisation there is no response, and the ventricle is said to be refractory. If the stimulus is applied later then depolarisation occurs, but its duration depends on the interval between the previous beat and the stimulus. The curve depicting the relationship between the delay in the stimulus and the duration of depolarisation is called the restitution curve. The shape and position of the restitution curve is affected by the basic ventricular rate, being depressed at higher heart rates. With a change of ventricular rate the duration of depolarisation, and the QT interval, changes immediately and then more gradually over the ensuing few minutes (Seed et al. 1987). Atrial fibrillation is characterised by marked beat-to-beat changes of heart rate, with a wide range of mean heart rates in different subjects. It therefore provides an opportunity to examine the QT interval as a function of both the instantaneous and the mean heart rate. In atrial fibrillation each beat may be considered as a depolarising stimulus, occurring a variable time after the preceding beat. The duration of the subsequent QT interval will vary according to the restitution curve at that particular mean ventricular rate. The within-subject QT/R-R interval relationship corresponds to the restitution curve except that, in contrast to the experimental situation and to sinus rhythm, the rhythm of ventricular contraction is irregular. In atrial fibrillation, therefore, both instantaneous and mean ventricular rates would be expected to determine the duration of the QT interval, and this is what was seen in a group of 50 patients with atrial fibrillation (Pai and Rawles 1989). The influence of the mean ventricular rate exceeded that of the instantaneous ventricular rate, the slope of the QT/R-R interval relationship between subjects being 21%, compared with an average of 7% within subjects (Fig. 2.13). With increasing mean ventricular rate both the position and the slope of the within-subject QT/R-R interval relationship were altered, the position being depressed and the slope increasing.

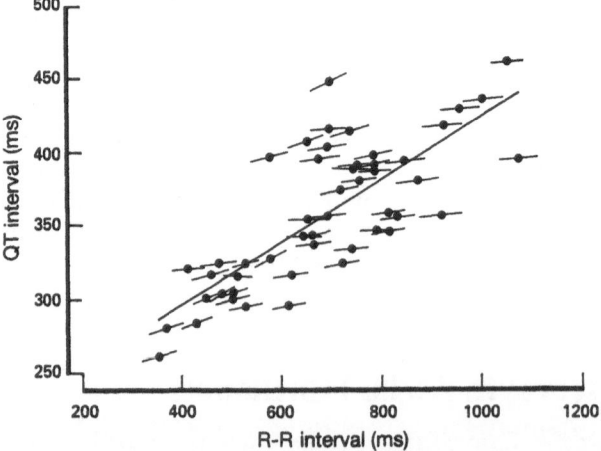

Fig.2.13. The relationship between QT and R-R intervals within and between 50 subjects with atrial fibrillation. (Reproduced by permission of the Editor, *British Heart Journal.*)

Is the QT Interval in Atrial Fibrillation Different from that in Sinus Rhythm?

The occurrence of atrial fibrillation implies some underlying cardiac pathology, which was overt in many cases in the series. It is therefore difficult to be sure whether any difference from normal of QT intervals is due to the arrhythmia, or underlying pathology. Ischaemia and inflammation are important causes of QT prolongation (Lepeshkin 1951), but no patient was known to have any cardiac inflammatory process, and none of the patients with ischaemic heart disease had ischaemia at rest or had recently suffered myocardial infarction.

A scatter plot of mean uncorrected QT and R-R intervals, while mostly in the upper half of the normal distribution, showed only 8 results outside the normal limits reported by Lepeschkin (1951, p. 180), based on 6000 healthy subjects in sinus rhythm. However, the regression line of QT against R-R intervals in atrial fibrillation is above that reported by Schlamowitz (1946) from 495 subjects in sinus rhythm, and that of Ahnve (1985) from 152 patients with ischaemic heart disease on digitalis. The regression line for Simonson et al.'s (1962) data from 960 healthy subjects has a lower slope and intercepts with our own at its midpoint.

We may tentatively conclude that in atrial fibrillation uncorrected QT intervals are generally longer than in sinus rhythm for the same ventricular rate, but that only a minority of mean QT intervals fall outside the wide range encountered in the largest published series. Comparison with data from patients with ischaemic heart disease suggests that the prolongation of QT intervals in atrial fibrillation may result from the arrhythmia rather than any underlying pathology.

Atrial Fibrillation and the Wolff–Parkinson–White Syndrome

The accessory atrioventricular connections present in the Wolff–Parkinson–White syndrome allow the atrioventricular node to be bypassed in sinus rhythm, leading to ventricular pre- excitation and the appearance of a delta wave at the commencement of the QRS complex of the electrocardiogram. The accessory pathway also makes it possible for re-entry tachycardia to occur, the antegrade route being through the atrioventricular node, with retrograde conduction via the accessory pathway. The maximum ventricular rate during re-entry tachycardia is determined by the refractory periods of the atrioventricular node and the accessory pathway (Castellanos et al. 1975; Milstein et al. 1986), but the mean ventricular rate during atrial fibrillation is not accurately predicted by any electrophysiological measurements (Brugada et al. 1984).

Atrial fibrillation seems to occur rather more commonly in the Wolff–Parkinson–White syndrome than might be expected, particularly since most subjects with this syndrome have no overt heart disease (Gressard et al. 1981). Of 100 patients with accessory pathways studied by Campbell et al. (1977), 16 had atrial fibrillation during electrophysiological study, and 32 had atrial fibrillation documented at some time. Reciprocating tachycardia may be the trigger for atrial fibrillation, or intrinsic atrial abnormalities that predispose to atrial fibrillation may be present (Fujimura et al. 1990). The presence of the accessory pathway itself may favour atrial fibrillation; its section leads to cure of paroxysmal atrial fibrillation as well as reciprocating tachycardia (Sharma et al. 1985; Waspe et al. 1986). The accessory pathway may provide a route for the initiation of atrial fibrillation by retrogradely conducted ventricular premature beats which may fall in the atrial vulnerable period, or merge with the sinus beat causing uneven refractoriness (Shen and Sung 1982). But it is not clear why some patients with the Wolff–Parkinson–White syndrome have paroxysmal atrial fibrillation while others do not (Bauernfeind et al. 1981).

During atrial fibrillation, conduction to the ventricles may occur through the atrioventricular node, the accessory pathway, or both. The electrocardiogram during atrial fibrillation may show bizarre broadened complexes due to pre-excitation, aberration, or runs of either alternately; ventricular tachycardia may be misdiagnosed. Conduction predominantly over the bypass tract may lead to a very rapid ventricular response, degenerating into ventricular fibrillation (Farre et al. 1979). The ventricular rate in atrial fibrillation is determined by the shorter of the refractory periods of the accessory pathway and the atrioventricular node (Castellanos et al. 1973; Wellens and Durrer 1974). A short refractory period of the accessory pathway marks a high risk of life-threatening tachycardia following onset of atrial fibrillation (Fromer et al. 1985). However, refractory periods determined during electrophysiological examination may alter absolutely and relative to each other following induction of reciprocating tachycardia (Tonkin et al. 1975), during exercise, or with isoprenaline (Levy et al. 1979; Wellens et al. 1982; German et al. 1983; Crick et al. 1985). Thus, in the Wolff- -Parkinson–White syndrome the risk of ventricular fibrillation following the onset of atrial fibrillation during normal activities cannot be accurately predicted by electrophysiological testing in the laboratory.

Aberration in Atrial Fibrillation

Unless there is coexisting bundle branch block, the QRS complex in atrial fibrillation is similar to that in sinus rhythm. However, under certain conditions functional bundle branch block may occur producing broad complexes that may be misinterpreted as ventricular ectopics. Usually the right bundle branch is affected rather than the left, and this is helpful diagnostically (Gulamhusein et al. 1985). A second point of distinction is that the coupling interval to the preceding beat is variable in aberration, but generally fixed in ventricular ectopic beats. Ashman (Gouaux and Ashman 1947) described how aberration occurred when a short R-R interval followed a long one, and the phenomenon has been named after him. The explanation is that the refractory period of the conducting His–Purkinje system is positively related to the preceding cycle length, so an impulse occurring a short interval after a long interval is blocked in the right bundle branch. Aberration is unlikely to occur when the ventricular rate is very slow or very fast (Heimonas et al. 1982). This pattern of behaviour of refractory periods in relation to cycle lengths is unique to the His– Purkinje system.

Schamroth (1969) describes two cases of atrial fibrillation with aberration.

However, some early impulses, in which aberration would have been expected, were conducted normally, a phenomenon attributed to supernormal conduction. This term is a misnomer since conduction is not supernormal but merely a temporary improvement in a depressed state of conduction.

Akiyama et al. (1989) have reported two cases of atrial fibrillation exhibiting "Ashman phenomenon of the T wave". In the absence of aberrancy of QRS complexes there was T wave inversion after short preceding intervals, accentuated by a long pre- preceding interval.

References

Aberg H (1969) Atrial fibrillation. II. A study of fibrillatory wave size on the regular scalar electrocardiogram. Acta Med Scand 185:381–5

Aberg H, Strom G, Werner I (1972) Heart rate during exercise in patients with atrial fibrillation. Acta Med Scand 191:315–20

Ahnve S (1985) Correction of the QT interval for heart rate: Review of different formulas and the use of Bazett's formula in myocardial infarction. Am Heart J 109:568–74

Akiyama T, Richeson JF, Faillace RT, Lockhart J, Scherer JC (1989) Ashman phenomenon of the T wave. Am J Cardiol 63:886– 90

Alessi R, Nusynowitz M, Abildskov JA, Moe GK (1958) Nonuniform distribution of vagal effects on the atrial refractory period. Am J Physiol 194:406–10

Allessie MA, Lammers W, Smeets J, Bonke F, Hollen J (1982) Total mapping of atrial excitation during acetylcholine-induced atrial flutter and fibrillation in the isolated canine heart. In: Kulbertus HE, Olsson SB, Schlepper M (eds) Atrial fibrillation. AB Hassle, Molndal, pp 44–61

Allessie MA, Rensma W, Brugada J, Smeets JLRM, Penn O, Kirchhof CJHJ (1990) Modes of atrial re-entry. In: Touboul P, Waldo AL (eds) Atrial arrhythmias. Current concepts and management. Mosby Year Book, St Louis, pp 112–30

Altschule MD (1945) The relation between vagal activity and auricular fibrillation in various clinical conditions. New Eng J Med 233:265–7

Amat-Y-Leon F, Racki AJ, Denes P et al. (1974) Familial atrial dysrhythmia with A-V block. Intracellular microelectrode, clinical electrophysiologic, and morphologic observations. Circulation 50:1097–1104

Aronow WS, Schwartz KS, Koenigsberg M (1987). Prevalence of enlarged left atrial dimension by echocardiography and its correlation with atrial fibrillation and an abnormal P terminal force in lead V1 of the electrocardiogram in 588 elderly persons. Am J Cardiol 59:1003–4

Attuel P, Childers R, Cauchemez B, Poveda J, Mugica J, Coumel P (1982) Failure in the rate adaptation of the atrial refractory period: its relationship to vulnerability. Int J Cardiol 2:179– 97

Aysha MH, Hassan AS (1988) Diagnostic importance of fibrillatory wave amplitude: a clue to echocardiographic left atrial size and etiology of atrial fibrillation. J Electrocardiol 21:247–51

Bauernfeind RA, Wyndham CR, Swiryn SP, et al. (1981) Paroxysmal atrial fibrillation in the Wolff–Parkinson–White syndrome. Am J Cardiol 47:562–9

Beasley R, Smith DA, McHaffie DJ (1985) Exercise heart rate at different serum digoxin concentrations in patients with atrial fibrillation. Br Med J 290:9–11

Bennett MA, Pentecost BL (1970) The pattern of onset and spontaneous cessation of atrial fibrillation in man. Circulation 41:981–8

Bohn FK, Patterson DF, Pyle RL (1971) Atrial fibrillation in dogs. Br Vet J 127:485–96

Bootsma BK, Hoelen AJ, Strackee J Meijler FL (1970) Analysis of R-R intervals in patients with atrial fibrillation at rest and during exercise. Circulation 41:783–94

Borst C, Meijler FL (1984) Baroreflex modulation of ventricular rhythm in atrial fibrillation. Eur Heart J 5:870–5

Braunstein JR, Franke (1961) Autocorrelation of ventricular response in atrial fibrillation. Circ Res 9:300–4

Brignole M, Menozzi C, Sartore B, Barra M, Monducci I (1986) The use of atrial pacing to induce atrial fibrillation and flutter. Int J Cardiol 12:45–54

Brody DA (1956) A theoretical analysis of intracavitary blood mass influence on the heart-lead relationship. Circ Res 4:731– 8

Brody DA (1970) Ventricular rate patterns in atrial fibrillation. Circulation 41:733–5

Brugada P, Roy D, Weiss J, Dassen WR, Wellens HJ (1984) Dual atrio-ventricular nodal pathways and atrial fibrillation. PACE 7:240–7

Burn JH (1979) Excitatory actions of acetylcholine in the heart. Br J Pharmacol 67:3–12

Buxton AE, Josephson ME (1981) The role of P wave duration as a predictor of postoperative atrial arrhythmias. Chest 80:68– 73

Caceres CA, Calatayud JB, Nutter D, Kelser GA (1965) The intracavitary electrocardiogram of atrial fibrillation. Angiology 16:584–93

Campbell RW, Smith RA, Gallagher JJ, Pritchett EL, Wallace AG (1977) Atrial fibrillation in the pre-excitation syndrome. Am J Cardiol 40:514–20

Castellanos A Jr, Myerburg RJ, Craparo K, Befeler B, Agha AS (1973) Factors regulating ventricular rates during atrial flutter and fibrillation in pre-excitation (Wolff–Parkinson–White) syndrome. Br Heart J 35:811–16

Castellanos A, Levy S, Befeler B, Myerburg RJ (1975) Mechanisms determining the ventricular rate in Wolff–Parkinson–White arrhythmias. Adv Cardiol 14:221–32

Chamberlain DA, White RJ, Howard MR, Smith TW (1970) Plasma digoxin concentrations in patients with atrial fibrillation. Br Med J iii:429–32

Cosio FG, Palacios J, Vidal JM, Cocina EG, Gomez-Sanchez MA, Tamargo L (1983) Electrophysiologic studies in atrial fibrillation. Slow conduction of premature impulses: a possible manifestation of the background for re-entry. Am J Cardiol 51:122–30

Cotoi S, Carasca E, Georgescu C, Lazar P (1978) Two varieties of the onset of atrial fibrillation studied by monophasic action potential recording. Jpn Heart J 19:479–84

Cotoi S, Georgescu C, Kifor I (1979) Analysis of human atrial fibrillatory waves using monophasic action potential technique. Am Heart J 98:465–71

Coumel P, Attuel P, Lavallee J, Flammang D, Leclercq JF, Slama R (1978) [The atrial arrhythmia syndrome of vagal origin] Arch Mal Coeur 71:645–56

Crick JC, Davies DW, Holt P, Curry PV, Sowton E (1985) Effect of exercise on ventricular response to atrial fibrillation in Wolff- -Parkinson–White syndrome. Br Heart J 54:80–5

Csapo G (1971) Role of ventricular premature beats in initiation and termination of atrial arrhythmias. Br Heart J 33:105–10

Davies MJ, Pomerance A (1972) Pathology of atrial fibrillation in man. Br Heart J 34:520–5

Demoulin J-C, Kulbertus HE (1979) Histopathological correlates of sinoatrial disease. Br Heart J 40:1384–9

Dreifus LS, Mazgalev T (1988) "Atrial paralysis": does it explain the irregular ventricular rate during atrial fibrillation? J Am Coll Cardiol 11:546–7

Escudero EM, Iveli CA, Moreyra AE, Lorente H, Cingolani HE (1976) The pulse in patients with atrial fibrillation: its irregularity and inequality. Eur J Cardiol 4:31–8

Euler DE, Scanlon PJ (1987) Acetylcholine release by a stimulus train lowers atrial fibrillation threshold. Am J Physiol 253:H863–8

Farre J, Ross DL, Wiener I, Bar FW, Vanagt E, Wellens HJ (1979) Electrophysiological studies in patients with the Wolff- Parkinson–White syndrome. Herz 4:38–46

Feldman T, Borow KM, Neumann A, Lang RM, Childers RW (1985) Relation of electrocardiographic R-wave amplitude to changes in left ventricular chamber size and position in normal subjects. Am J Cardiol 55:1168–74

Fromer M, Gloor H, Kappenberger L (1985) [Effect of heart rate on the refractory period of the accessory atrioventricular pathway in Wolff–Parkinson–White syndrome]. Schweiz Med Wochenschr 115:1545–51

Fujimura O, Klein GJ, Yee R, Sharma AD, Boahene KA (1990) Atrial fibrillation in the Wolff–Parkinson–White syndrome. In: Touboul P, Waldo AL (eds) Atrial arrhythmias. Current concepts and management. Mosby Year Book, St Louis, pp 262–9

Garber EB, Morgan MG, Glasser SP (1976) Left atrial size in patients with atrial fibrillation: an echocardiographic study. Am J Med Sci 272:57–64

Gavrilescu S, Luca C (1975) Right atrium monophasic action potentials during atrial flutter and fibrillation in man. Am Heart J 90:199–205

Gavrilescu S, Dragulescu SI, Streian C, Luca C (1976) Monophasic action potentials of the right atrium during atrial fibrillation in man. Cor Vasa 18:264–70

German LD, Gallagher JJ, Broughton A, Guarnieri T, Trantham JL (1983) Effects of exercise and isoproterenol during atrial fibrillation in patients with Wolff–Parkinson–White syndrome. Am J Cardiol 51:1203–6

Goldstein RE, Barnett GO (1967) A statistical study of the ventricular irregularity of atrial fibrillation. Comp Biomed Res 1:146–61

Gordon S, Finck DR, Perera RD, Levine J, Barnes SJ (1984) Atrial infarction complicating an acute inferior myocardial infarction. Arch Intern Med 144:193

Gouaux JL, Ashman R (1947) Auricular fibrillation with aberration simulating ventricular paroxysmal tachycardia. Am Heart J 34:366- -73

Graybiel A (1964) Auricular fibrillation in an asymptomatic young man. Effects of exercise, digitalization, atropinization and the restoration of normal rhythm. Am J Cardiol 14:828–36

Gressard A, Atallah G, Chatelain MT, Touboul P (1981) [Genesis of auricular fibrillation in the Wolff–Parkinson–White syndrome]. Arch Mal Coeur 74:1277–82

Gulamhusein S, Yee R, Ko PT, Klein GJ (1985) Electrocardiographic criteria for differentiating aberrancy and ventricular extrasystole in chronic atrial fibrillation: validation by intracardiac recordings. J Electrocardiol 18:41–50

Haft JI, Lau SH, Stein E, Kosowsky BD, Damato AN (1968) Atrial fibrillation produced by atrial stimulation. Circulation 37:70- -4EP Hashida E, Inoue T, Tasaki T (1973) A probabilistic study of the irregularity of R-R interval durations in atrial fibrillation - a preliminary report. Jpn Circ J 37:1423–6

Hashimoto K, Chiba S, Tanaka S, Hirata M, Suzuki Y (1968) Adrenergic mechanism participating in induction of atrial fibrillation by ACh. Am J Physiol 215:1183–91

Heimonas ET, Cokkinos DV, Hatzivasiloglou C, Papoulis S, Ioannou N, Chaniotakis M (1982) A study of the factors influencing the appearance of the Ashman phenomenon. Acta Cardiol 37:175–81

Henry WL, Morganroth J, Pearlman AS et al. (1976) Relation between echocardiographically determined left atrial size and atrial fibrillation. Circulation 53:273–9

Hering HE (1903) Analyse des pulsus irregularis perpetuus. Prag Med Wochenschr 28:377–81

Hod H, Lew AS, Keltai M et al. (1987) Early atrial fibrillation during evolving myocardial infarction: a consequence of impaired left atrial perfusion. Circulation 75:146–50

Hodkinson HM, Pomerance A (1977) The clinical significance of senile cardiac amyloidosis: a prospective clinico-pathological study. Q J Med 46:381–7

Honzicova N, Fiser B, Semrad B (1973) Ventricular function in patients with atrial fibrillation. Cor Vasa 4:257–64

Horan LG, Kistler JC (1961) Study of ventricular response in atrial fibrillation. Circ Res 9:305–11

Ih S, Saitoh S (1982) The histopathological substratum for atrial fibrillation in man. Acta Pathol Jpn 32:183–91

James TN (1961) Myocardial infarction and atrial arrhythmias. Circulation 24:761–76

James TN (1968) The coronary circulation and conduction system in acute myocardial infarction. Prog Cardiovasc Dis 10:410- 49

James TN (1980) Neural control of the heart in health and disease. In: Stollerman GH (ed) Advances in internal medicine, vol 26. Year Book Medical Publishers, Chicago, pp 317–45

James TN (1982) Diversity of histopathologic correlates of atrial fibrillation. In: Kulburtus HE, Olsson SB, Schlepper M (eds) Atrial fibrillation. AB Hassle, Molndal, pp 13–32

James TN, Pearce WN, Givhan EG (1980) Sudden death while driving. Role of sinus perinodal degeneration and cardiac neural degeneration and ganglionitis. Am J Cardiol 45:1095–102

Janse MJ (1969) Influence of the direction of the atrial wave front on A-V nodal transmission in isolated hearts of rabbits. Circ Res 25:439–49

Jose AD (1966) Effect of combined sympathetic and parasympathetic blockade on heart rate and cardiac function in man. Am J Cardiol 18:476–78

Killip T, Gault JH (1965) Mode of onset of atrial fibrillation in man. Am Heart J 70:172–9

Kirsh JA, Sahakian AV, Baerman JM, Swiryn S (1988) Ventricular response to atrial fibrillation: role of atrioventricular conduction pathways. J Am Coll Cardiol 12:1265–72

Langendorf R, Pick A, Katz LN (1965) Ventricular response in atrial fibrillation. Role of concealed conduction in the AV junction. Circulation 32:69–75

Leier CV, Schaal SF (1980) Biatrial electrograms during coarse atrial fibrillation and flutter-fibrillation. Am Heart J 99:331- 41

Lepeschkin E (1951) Modern electrocardiography, vol 1. Williams & Wilkins, Baltimore

Levy S, Broustet JP, Clementy J, Vircoulon B, Guern P, Bricaud H (1979) [Wolff–Parkinson–White syndrome. Correlation between the results of electrophysiological investigation and exercise tolerance testing on the electrical aspect of pre-excitation] Arch Mal Coeur 72:634–40

Lewis T (1909) Auricular fibrillation: A common clinical condition. Br Med J ii:1528

Lewis T (1909-10) Auricular fibrillation and its relationship to clinical irregularity of the heart. Heart 1:306–72

Lewis T, Drury AN, Bulger HA (1921a) Observations upon flutter and fibrillation. VII. The effects of vagal stimulation. Heart 8:141–70

Lewis T, Drury AN, Wedd AM, Iliescu CC (1921b) Observations upon the action of certain drugs upon fibrillation of the auricles. Heart 9:207–67

Mackenzie J (1922) Observations on the process which results in auricular fibrillation. Br Med J ii:71–3

Manyari DE, Patterson C, Johnson D, Melendez L, Kostuk WJ, Cape RDT (1990) Atrial and ventricular arrhythmias in asymptomatic active elderly subjects: Correlation with left atrial size and left ventricular mass. Am Heart J 119:1069–76

Mazgalev T, Dreifus LS, Bianchi J, Michelson EL (1982) Atrioventricular nodal conduction during atrial fibrillation in rabbit heart. Am J Physiol 243:H754–60

Meijler FL (1986a) Comparative aspects of the dual role of the human atrioventricular node. Br Heart J 55:286–90

Meijler FL (1986b) Editorial: the pulse in atrial fibrillation. Br Heart J 56:1–3

Meijler FL, Kroneman J, Tweel I van der, Herbscleb JN, Heethar RM, Borst C (1984) Nonrandom ventricular rhythm in horses with atrial fibrillation and its significance for patients. J Am Coll Cardiol 4:316–23

Milstein S, Sharma AD, Klein GJ (1986) Electrophysiologic profile of asymptomatic Wolff–Parkinson–White pattern. Am J Cardiol 57:1097–100

Moe GK (1968) A conceptual model of atrial fibrillation. J Electrocardiol 1:145–6

Moe GK, Abildskov JA (1964) Observations on the ventricular dysrhythmia associated with atrial fibrillation in the dog heart. Circ Res 14:447–60

Moore EN, Spear JF (1987) Electrophysiological studies on atrial fibrillation. Heart Vessels [Suppl] 2:32–9

Morganroth J, Horowitz LN, Josephson ME, Kastor JA (1979) Relationship of atrial fibrillatory wave amplitude to left atrial size and etiology of heart disease. An old generalization re- examined. Am Heart J 97:184–6

Nahum LH, Hoff HE (1940) Production of auricular fibrillation by application of acetyl-beta-methyl-choline chloride to localized region on the auricular surface. Am J Physiol 129:P428

Nilius B, Schuttler K, Boldt W (1981) Parasympathetic influence on electrical vulnerability in the atrial myocardium of the rabbit. Acta Biol Med Ger 40:821–9

Ninomiya I (1966) Direct evidence of nonuniform distribution of vagal effects on dog atria. Circ Res 19:576–83

Olsson SB, Cotoi S, Varnauskas E (1971) Monophasic action potential and sinus rhythm stability after conversion of atrial fibrillation. Acta Med Scand 190:381–7

Olsson SB, Cai N, Dohnal M, Talwar KK (1986) Noninvasive support for and characterization of multiple intranodal pathways in patients with mitral valve disease and atrial fibrillation. Eur Heart J 7:320–33

Pai GR, Rawles JM (1989) The QT interval in atrial fibrillation. Br Heart J 61:510–3

Peter RH, Gracey JG, Beach TB (1968) Significance of fibrillatory waves and the P terminal force in idiopathic atrial fibrillation. Ann Intern Med 68:1296–300

Pitcher D, Papouchado M, James MA, Rees JR (1986) Twenty four hour ambulatory electrocardiography in patients with chronic atrial fibrillation. Br Med J 292:594

Puech P, Grolleau R, Rebuffat G (1982) Intra-atrial mapping of atrial fibrillation in man. In: Kulburtus HE, Olsson SB, Schlepper M (eds) Atrial fibrillation. AB Hassle, Molndal, pp 94–107

Rawles JM, Rowland E (1986) Is the pulse in atrial fibrillation irregularly irregular? Br Heart J 56:4–11

Rebello R, Brownlee WC (1987) Intermittent ventricular standstill during chronic atrial fibrillation in patients with dizziness or syncope. PACE 10:1271–6

Reynolds EW, MacDonald WJ, Greenfield BM, Semion AA (1967) Mechanisms of onset and termination of abnormal cardiac rhythm studied by constant monitoring. Am Heart J 74:473–81

Robitaille GA, Phillips JH (1967) An analysis of the P wave in patients with transient benign atrial fibrillation. Dis Chest 52:806–12

Schamroth L (1969) The supernormal phase of intraventricular conduction. Br Heart J 31:337–42

Scherf D (1947) Studies on auricular tachycardia caused by aconitine administration. Proc Soc Exp Bol Med 64:233–9

Schlamowitz I (1946) An analysis of the time relationships within the cardiac cycle in electrocardiograms of normal men. I. The duration of the Q-T interval and its relationship to the cycle length (R-R interval). Am Heart J 31:329–42

Seed WA, Noble MIM, Oldershaw P et al.(1987) Relation of human cardiac action potential duration to the interval between beats: implications for the validity of rate corrected QT interval (QTc). Br Heart J 57:32–7

Sharma AD, Klein GJ, Guiraudon GM, Milstein S (1985) Atrial fibrillation in patients with Wolff–Parkinson–White syndrome: incidence after surgical ablation of the accessory pathway. Circulation 72:161–9

Shen EN, Sung RJ (1982) Initiation of atrial fibrillation by spontaneous ventricular premature beats in concealed Wolff– Parkinson–White syndrome. Am Heart J 103:911–2

Simonson E, Cady LD, Woodbury M (1962) The normal Q-T interval. Am Heart J 63:747–53

Simpson RJ Jr, Amara I, Foster JR, Woelfel A, Gettes LS (1988) Thresholds, refractory periods, and conduction times of the normal and diseased human atrium. Am Heart J 116:1080–90

Smeets JLRM, Allessie MA, Lammers WJEP, Bonke FIM, Hollen J (1986) The wavelength of the cardiac impulse and re-entrant arrhythmias in isolated rabbit atrium. The role of heart rate, autonomic transmitters, temperature, and potassium. Circ Res 58:96–108

Soderstrom N (1950) What is the reason for the ventricular arrhythmia in cases of auricular fibrillation? Am Heart J 40:212–23

Stafford P, Turner I, Vincent R (1990) Quantitative analysis of the P wave in paroxysmal atrial fibrillation. Br Heart J 64:87 (abstract)

Stewart GA, Fulton LJ, McKellar CD (1990) Idiopathic atrial fibrillation in a champion Standardbred racehorse. Aust Vet J 67:187–91

Strackee J, Hoelen AJ, Zimmerman ANE, Meijler FL (1971) Artificial atrial fibrillation in the dog. An artifact? Circ Res 28:441–5

Takahashi N, Imataka K, Seki A, Fujii J (1982) Left atrial enlargement in patients with paroxysmal atrial fibrillation. Jpn Heart J 23:677–83

Takahashi N, Seki A, Imataka K, Fujii J (1983) Fibrillatory wave size in paroxysmal atrial fibrillation. Jpn Heart J 24:309– 14

Teichmann W, Wanke R, Hecht P (1987) [Effect of ventricular filling volume on the R-amplitude in the ECG]. Z Gesamte Inn Med 42:366–8

Thiedemann KU, Ferrans VJ (1977) Left atrial ultrastructure in mitral valvular disease. Am J Pathol 89:575–604

Tonkin AM, Miller HC, Svenson RH, Wallace AG, Gallagher JJ (1975) Refractory periods of the accessory pathway in the Wolff- Parkinson-White syndrome. Circulation 52:563–9

Urbach JR, Grauman JJ, Straus SH (1969) Quantititive methods for the recognition of atrioventricular junctional rhythms in atrial fibrillation. Circulation 39:803–17

Waspe LE, Brodman R, Kim SG, Fisher JD (1986) Susceptibility to atrial fibrillation and ventricular tachyarrhythmia in the Wolff- -Parkinson–White syndrome: role of the accessory pathway. Am Heart J 112:1141–52

Waters DD, Nutter DO, Hopkins LC, Dorney ER (1975) Cardiac features of an unusual x-linked humeroperoneal neuromuscular disease. N Engl J Med 293:1017–22

Wellens HJ, Durrer D (1974) Wolff–Parkinson–White syndrome and atrial fibrillation. Relation between refractory period of accessory pathway and ventricular rate during atrial fibrillation. Am J Cardiol 34:777–82

Wellens HJ, Brugada P, Roy D, Weiss J, Bar FW (1982) Effect of isoproterenol on the anterograde refractory period of the accessory pathway in patients with the Wolff–Parkinson–White syndrome. Am J Cardiol 50:180–4

Wells JL, Karp RB, Kouchoukos NT, MacLean WA, James TN, Waldo AL (1978) Characterization of atrial fibrillation in man: studies following open heart surgery. PACE 1:426–38

Whittington JR, Cross MR, Raftery EB (1979) An effective conscious animal model of atrial fibrillation. Cardiovasc Res 13:105–12

Wyndham CRC, Amat-y-Leon F, Wu D et al. (1977) Effects of cycle length on atrial vulnerability. Circulation 55:260–7

Chapter 3

Models of the Atrioventricular Node

The mechanism of the ventricular response in atrial fibrillation has long been a puzzle, the answer to which lies in the working of the atrioventricular node. Any attempted solution must be consistent with, and preferably also illuminate, the known behaviour of the atrioventricular node in sinus rhythm. Essentially, the problem is mathematical: how to convert the distribution and sequence of impulses arriving at the atrioventricular node to those of the impulses leaving the node. However, the mathematical solution has to correspond to the known electrophysiological properties of the conducting system. The main features that need to be accommodated in a model are as follows: reduction of the number of impulses entering the node to the number leaving the node; an essentially random sequence of R-R intervals with the distribution properties found in atrial fibrillation, including the presence of narrow peaks on the histogram of R-R intervals; a relatively long atrioventricular conduction interval in sinus rhythm; rapid change of atrioventricular delay with change of ventricular rate; and a negative relation between cycle length of conducted impulses and the effective refractory period of the atrioventricular node.

The Electrophysiological Model of the Atrioventricular Node

The orthodox electrophysiological model of the atrioventricular node is represented by the ladder diagram, which shows the arrival of an impulse from the atria, the delayed passage through the node, and the emergent stimulus to ventricular contraction (Fig. 3.1). A cross bar indicates block of the passage of an impulse – for example in the Wenckebach phenomenon. A cross bar also indicates concealed conduction, where an impulse which penetrates the node, while not propagated itself, alters the node's behaviour to the next stimulus. The atrioventricular node may be penetrated by a stimulus arising in a ventricle and conducted retrogradely into the atrioventricular node. The response of the node to retrograde stimulation is qualitatively similar to that for antegrade stimulation.

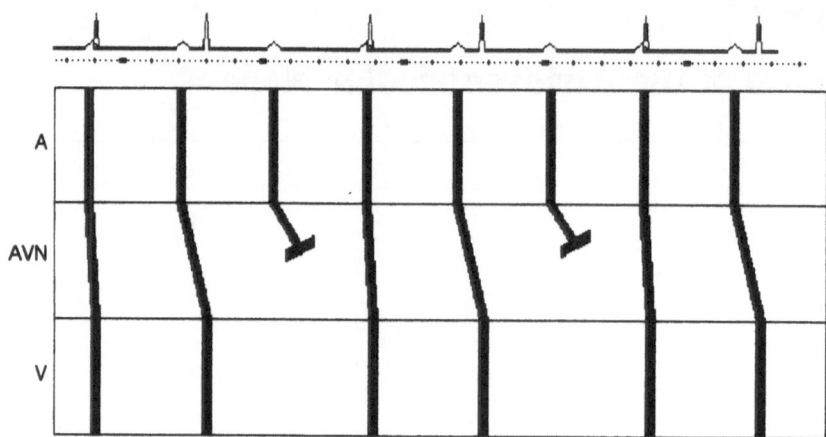

Fig. 3.1. Upper. Stylised electrocardiogram (ECG) showing 3:2 Wenckebach atrioventricular block. T waves are omitted for clarity. The smallest intervals on the time scale are 40 milliseconds, corresponding to those normally found on an electrocardiogram recorded at 25 mm s^{-1}. *Below:* Ladder diagram showing the passage of impulses from atria (A), through the atrioventricular node (AVN), to the ventricles (V). Two impulses are blocked in the atrioventricular node.

In the dog, the atrial rhythm during ventricular fibrillation, though slower, is very similar to the ventricular rhythm during atrial fibrillation (Scher et al 1976). Any model, therefore, has to allow for passage of impulses in either direction through the node.

In atrial fibrillation numerous stimuli arrive at the atrioventricular node at random intervals, from many directions, and with varying power to penetrate the node. Many stimuli penetrate into the proximal part of the atrioventricular node where they can be detected with microelectrodes (Moore 1967). In the ladder diagram representing atrial fibrillation, the mechanism that determines the number of stimuli emerging from the node is not depicted, but most impulses that enter the node are shown to be blocked, and some stimuli that emerge from the

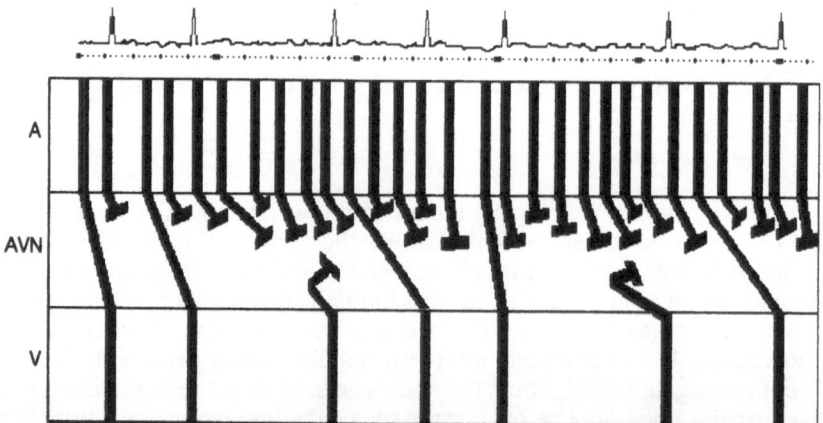

Fig. 3.2. ECG and ladder diagram of atrial fibrillation. Most impulses entering the atrioventricular node are blocked. Two ventricular impulses originate as junctional escape beats.

node are considered to arise from a junctional pacemaker (Langendorf et al. 1965) (Fig. 3.2).

Analogue Models of the Atrioventricular Node

Van der Pol and van der Mark (1928) described an electronic model of the conducting system of the heart that comprised three interconnected oscillating circuits, representing the sinus node, the atria and the ventricles. Each circuit was a simple combination of battery, resistance, capacitance and neon tube that discharged periodically, depending on the time constant of the resistance–capacity network. Such a circuit, described as a relaxation oscillator, may be represented by an ordinary differential equation and has the following general properties. It has periodicity, determined by the time constant or relaxation time; it can be readily entrained to run at other than its fundamental frequency; the amplitude is hardly influenced by the frequency of oscillation; and the waveform deviates considerably from that of a sine-wave, so many higher harmonics of the fundamental frequency are present. The electronic model described by van der Pol and van der Mark (1928) simulated sinus rhythm, ectopic beats and various heart blocks, but not the Wenckebach phenomenon.

A mathematical model of the conduction system of the heart as two weakly coupled relaxation oscillators, based on van der Pol's differential equation, was described nearly 50 years later by Katholi et al. (1977).

Grant (1956) replicated van der Pol and van der Mark's electrical model and, like the original authors, was unable to simulate the Wenckebach phenomenon. This was only achieved after the waveforms of both relaxation oscillators were considerably modified in an unphysiological manner. A more up-to-date version of the van der Pol model using semiconductors was described by Robergé et al. (1971).

The advent of the digital computer rendered analogue models obsolete as research tools, and greatly extended the possibilities for modelling electrophysiological systems.

Mathematical Models of the Atrioventricular Node in Atrial Fibrillation

Moe et al. (1964) developed a computer model of atrial fibrillation which simulated multiple re-entry in a two- dimensional multi-element structure that represented myocardium. Their work has been replicated (Osher and Cairns 1967). More recently Malik and Camm (1989) have modelled fibrillation-like chaotic behaviour in a one-dimensional computer simulation, thereby showing that the topology of myocardium is unimportant in the development of fibrillation.

The conceptual model of the atrioventricular node proposed by Moe and Abildskov (1964) depends for its operation in atrial fibrillation on "concealed conduction" in two hypothetical compartments of the node. This model requires the behaviour of the node in atrial fibrillation to be rather different from that in sinus rhythm. Moreover, the model predicts a bimodal histogram with broad peaks, and cannot account for histograms with multiple or narrow peaks (Cohen et al. 1983).

A variation of Moe and Abildskov's model is presented by Honzicova et al. (1973). This model predicts a series of evenly spaced peaks in the histogram of R-R intervals, and an abrupt step to the shortest interval, which always constitutes the highest peak. This predicted distribution is unlike any that is observed in patients with atrial fibrillation.

In order to explain the difference between the number of stimuli arriving at the node and emerging from it, Hashida et al. (1978) invoke summation of excitatory stimuli until a threshold value is reached, when onward transmission of an impulse occurs. Evidence suggesting that summation is a feature of atrioventricular conduction has been presented by Zipes et al. (1973).

Several mathematical models have been proposed that are purely descriptive, and in which minimal attempts are made to explain the mode of operation in terms of electrophysiological mechanisms. One such is the retarded excitation model of ten Hoopen (1966), which predicts unimodal histograms. Heethar et al. (1973a,b) stimulated the right atrial appendage of the isolated rat heart with a variety of pulse sequences and observed the resulting conduction times. An empirical model was then constructed which accurately predicted the P-R interval in vivo from a knowledge of the four or five preceding P-P intervals. The model also considered the behaviour of atrial stimuli which were blocked. This investigation, though descriptive, is helpful in demonstrating that the time constant for change of the A-V conduction time with a change of heart rate is short, most of the change being executed within the duration of one beat. If the time constant for that relationship were to be prolonged over several beats it might have the effect of regularising the ventricular rhythm in atrial fibrillation (Meijler et al. 1982).

Van der Tweel et al. (1983) studied the ventricular response to random atrial stimulation in three patients. They then computed transfer functions by time-series analysis. The cross correlogram of stimulus–stimulus to stimulus–response intervals showed a negative correlation coefficient at lag 0, and one or two less negative but significant coefficients at lags 1 and 2, thereby showing that the behaviour of the human atrioventricular node is similar to that of the rat and the dog (Billette 1981). The short time constant of atrioventricular delay in a variety of species has been remarked by Meijler et al. (1982).

Other descriptive models of atrioventricular conduction are those of Dorveaux et al. (1988) and Simson (1988). The latter author explains the slow conduction through the N region of the atrioventricular node in terms of a high intercellular resistance and a low space constant. A voltage threshold that decreases with time is described; this could account for the increased conduction time with an abrupt reduction of cycle length.

The Model of Cohen et al. (1983)

Cohen et al. (1983) have developed a simple, quantitative model for the genesis of R-R intervals, based on the well-known electrophysiological properties of the

specialised conducting system. The atrioventricular node's behaviour is considered to be equivalent to that of a single atrioventricular junctional cell. This
equivalent cell is bombarded by impulses from the fibrillating atria that have the
time distribution characteristic of a Poisson process. During phases 0–3 of the
action potential of the junctional cell equivalent it is refractory and unresponsive
to any atrial input. During phase 4 there is spontaneous depolarisation with a rise
in the transmembrane potential. Each time an atrial impulse arrives the
transmembrane potential increases by a discrete amount, and when the potential
reaches threshold by any combination of spontaneous depolarisation and atrial
increments the junctional cell fires, and there is a new action potential (Fig. 3.3).
Cohen et al. (1983) are at pains to point out that the atrioventricular junctional
equivalent cell is a hypothetical lumped structure that is used to represent the
overall behaviour of the various regions of the atrioventricular node. The effect
of an atrial impulse on an individual cell may be different from the behaviour
depicted by the model, yet the model may accurately describe the overall
response of the atrioventricular node.

   Cohen et al.'s model is characterised by four parameters: the mean rate of
arrival of atrial impulses, the amplitude of the atrial impulse, the rate of phase 4
depolarisation, and the refractory period of the atrioventricular junctional
equivalent cell. The form of the histogram of R-R intervals depends on the
relative magnitude of spontaneous phase 4 depolarisation and the rise of
membrane potential due to the atrial increments. If the rate of spontaneous
depolarisation is negligible compared with the atrial increments then the
histogram will have a broad, smooth outline. If atrial increments are negligible by
comparison with phase 4 depolarisation then the histogram will have a single
narrow peak. When both factors contribute to reaching threshold, so that
sometimes the threshold is crossed by spontaneous depolarisation and sometimes
with a final atrial increment, then a broad histogram with narrow peaks is

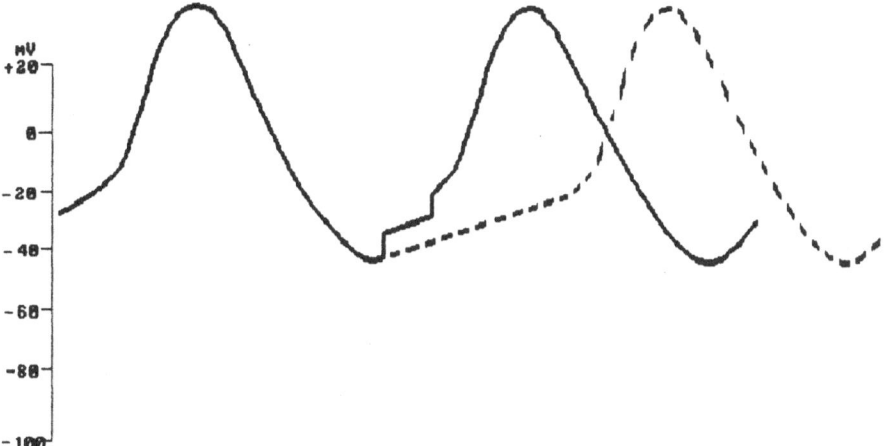

Fig. 3.3. Cohen et al.'s model of the atrioventricular node. The action potential of the atrioventricular
junctional equivalent cell is shown; phase 4 depolarisation is accelerated by the arrival of two atrial
stimuli. The *dashed line* indicates the action potential that would have occurred in the absence of the
stimuli.

obtained. Thus, this model is capable of generating a variety of histograms that match those that are found in patients.

In their paper Cohen et al. (1983) go on to modify the model slightly so that the refractory period of the junctional equivalent cell is not fixed but is related to the previous R-R interval (Mendez et al. 1956). This modification introduces a weak positive correlation between consecutive R-R intervals – the "chaining" reported by Goldstein and Barnett (1967). The authors discuss how their model could be further modified to render its functioning more lifelike by the introduction of an element of "noise", which, in physiological terms, would be small, short-term variations in the parameters of the model. Thus, the rate of diastolic depolarisation, or the magnitude of the increment of transmembrane potential associated with the arrival of an atrial impulse, as well as the rate of their arrival, might all fluctuate to some extent. An additional source of variability is autonomic tone, which could alter the rate of arrival of impulses from the atria, as well as the intrinsic rate of the atrioventricular junctional equivalent cell, expressed as the rate of phase 4 depolarisation. It is possible that some atrial stimuli, depending on the time in the node's cycle when they arrive, and the direction from which they come, result in hyperpolarisation rather than depolarisation of the junctional cell membrane. An adjustment of Cohen et al's model to incorporate this phenomenon is suggested by Meijler and Fisch (1989).

Cohen et al. (1983) address the problem of the ventricular response in atrial fibrillation, and they are conspicuously successful in their solution. Their model explains the occurrence of a minimum R-R interval, the greater number of stimuli arriving at the node than leaving it, and the form of the histogram of R-R intervals. The occurrence in some patients of bi- or tri-modal histograms is seen to arise from the pattern of diastolic depolarisation rather than to be evidence of dual conduction pathways, as has been suggested by Olsson et al. (1986). Cohen et al. did not, however, deal with other aspects of atrioventricular node function that need to be explained, namely the long atrioventricular delay and its rapid adaptation to a change in ventricular rate.

The Atrioventricular Node as a Biological Oscillator

The concept of the atrioventricular node as an oscillator is not new, it dates back to 1928 (van der Pol and van der Mark), and has reappeared from time to time since then in analogue form. More recently, a model of the dog atrioventricular node as a periodically perturbed biological oscillator was reported (van der Tweel et al. 1986). The authors stimulated the right atrial appendage of anaesthetised dogs under autonomic blockade. Atrial stimuli that were premature with respect to the basic stimulation cycle length were applied and the conduction times to the left ventricle were observed. The results are presented as latency– phase curves, where phase is the time between a QRS complex and an atrial stimulus (R-S interval), and latency the time between this stimulus and the following QRS complex (S-R interval). An exponential relationship was obtained between S-R and R-S intervals, for which the regression equation was obtained in each animal. Application of the equation enabled prediction of the response to a change in atrial stimulation frequency. Following an abrupt change in stimulation rate, a new steady state was reached faster in the models than in the dogs from which the models had been derived, and the Wenckebach phenomenon was predicted but was not observed. Although the presentation of the results was in the form of a latency–phase curve, which is appropriate when considering the node as an oscillator, this model was essentially descriptive, and the accuracy of the predictions was not contingent on any underlying assumptions about the mode of operation of the atrioventricular node.

The Model of Guevara and Glass (1982)

Guevara and Glass (1982) explore the consequences of considering the sinoatrial and atrioventricular nodes as two coupled oscillators, a notion first mooted by van der Pol and van der Mark (1928). However, instead of using the van der Pol differential equation, they consider the output of each oscillator as a sine-wave, which is a reduction of the pacemaker action potential to one of the simplest forms possible (Fig. 3.4). A sine-wave is not completely unphysiological, as van der Tweel et al. (1973) have shown that it is possible to entrain both sinoatrial and atrioventricular nodes with a sinusoidal stimulus.

The sine-wave output is generated by a radius, length 1 unit, rotating in an anticlockwise direction (Fig. 3.5). The cycle starts with the radius in the 3 o'clock position, when its phase angle, φ, is 0°. The output of the oscillator is cos φ. The sinusoidal output of this simple geometrical model is an extremely crude simplification of the complex cycle of voltage changes observed in a cardiac pacemaker. However, while retaining the essential behaviour of excitable cardiac cells, the model has an advantage over other, more complex versions in that its behaviour under a wide range of conditions may be generalised and represented in a simple diagram: the Poincaré map. It could be refined to give a more lifelike representation of intracellular voltage changes without altering its basic proper-

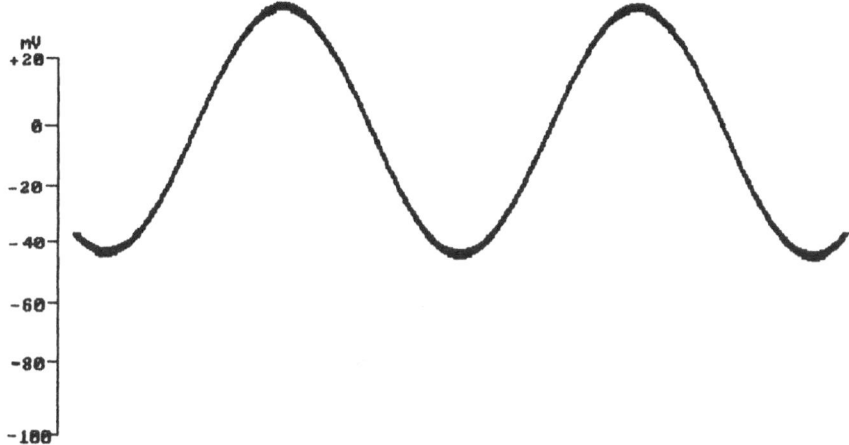

Fig. 3.4. The action potential of a pacemaker cell considered as a sine wave.

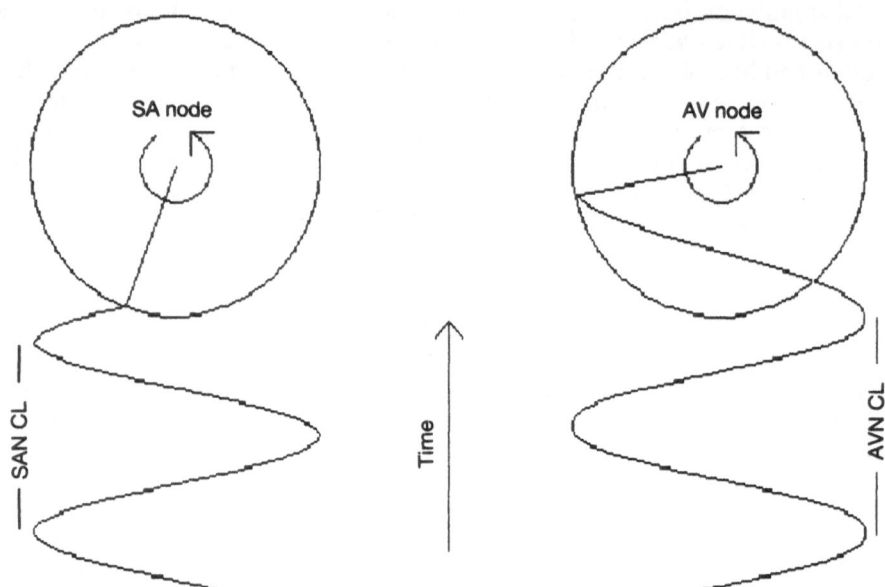

Fig. 3.5. Sinoatrial (SAN) and atrioventricular (AVN) nodes as sine wave oscillators generating a sinusoidal membrane potential. The cycle length (CL) of the SAN is shorter than that of the AVN.

ties. Simple as it is, Guevara and Glass's model can give us deep insight into the working of the atrioventricular node.

Resetting the Oscillator

Guevara and Glass (1982) consider both nodes as sine-wave oscillators, each discharging when its radius passes the starting position, when φ equals 0° (Fig. 3.5). Discharge of sinoatrial and atrioventricular oscillators corresponds to atrial and ventricular depolarisation respectively. The two oscillators are coupled by means of a vector that arises from the sinoatrial node every time it discharges. It corresponds, of course, to the stimulus that enters the atrioventricular node following atrial depolarisation. The vector runs horizontally, parallel to the x axis. Its length is variable, depending on the degree of coupling between oscillators, and the unit of length is 1 radius. "Coupling" represents the ability of the depolarisation wavefront arriving at the atrioventricular node to penetrate it; it will depend on the number of atrial myocardial cells adjacent to the node that are depolarised at the same time (Sealy and Seaber 1979), the direction from which the wavefront arrives (Janse 1969), and whether several wavefronts arrive more or less simultaneously (Janse 1990). The power of the stimulus to penetrate the atrioventricular node ("coupling") will probably be greater in sinus rhythm than in atrial fibrillation. The coupling vector resets the atrioventricular oscillator by determining the new position of the radius and therefore the new phase angle (Fig. 3.6). Thus, during the first half of the cycle, arrival of the vector retards the phase, and during the second half the phase of the oscillator is advanced. In electrophysiological terms the effect of a positive shift of the transmembrane

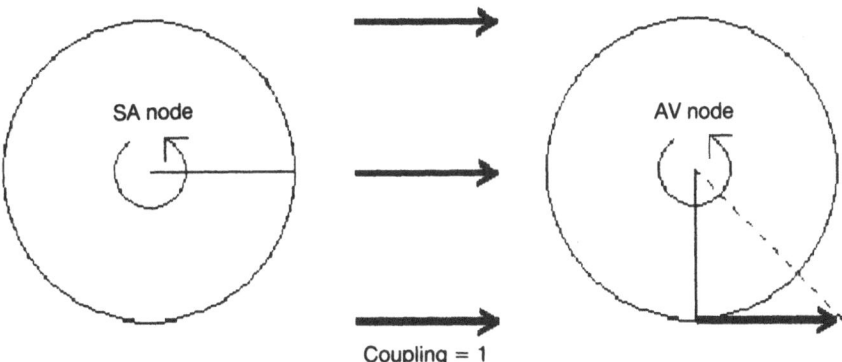

Fig. 3.6. Resetting the AV oscillator. The unit of coupling is 1 radius. The new phase of the AV oscillator after being advanced by 1 radius is indicated by the *dashed line*.

potential when the potential is becoming more negative is to retard the progress of the oscillator; the effect of a positive shift of the transmembrane potential when the potential is becoming more positive is to advance it. Advancement of the oscillator by arrival of a stimulus during diastole is illustrated in Fig. 3.7. The symmetry of operation of the model is unrealistic, likewise the absence of any refractory period. However, its simplicity allows ready calculation of the effect of periodic perturbation of the atrioventricular oscillator by discharge of the sinoatrial oscillator, over a wide range of relative cycle lengths and coupling strengths.

Guevara and Glass (1982) show that when the coupling strength is less than 1 (radius), the phase reset curve is that of a weak stimulus, with an average slope of 1, but when the coupling strength is greater than 1 it is that of a strong stimulus, with an average slope of 0 (Fig. 3.8). Phase reset curves were described in Chapter 1, and we will return to them later in this chapter.

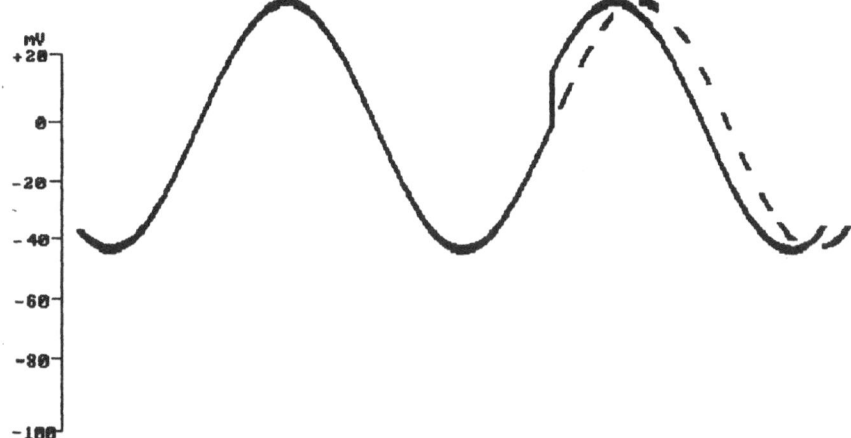

Fig. 3.7. Guevara and Glass's model of the atrioventricular node. Phase 4 depolarisation of the action potential is accelerated by the arrival of an atrial stimulus. The *dashed line* indicates the action potential that would have occurred in the absence of the stimulus.

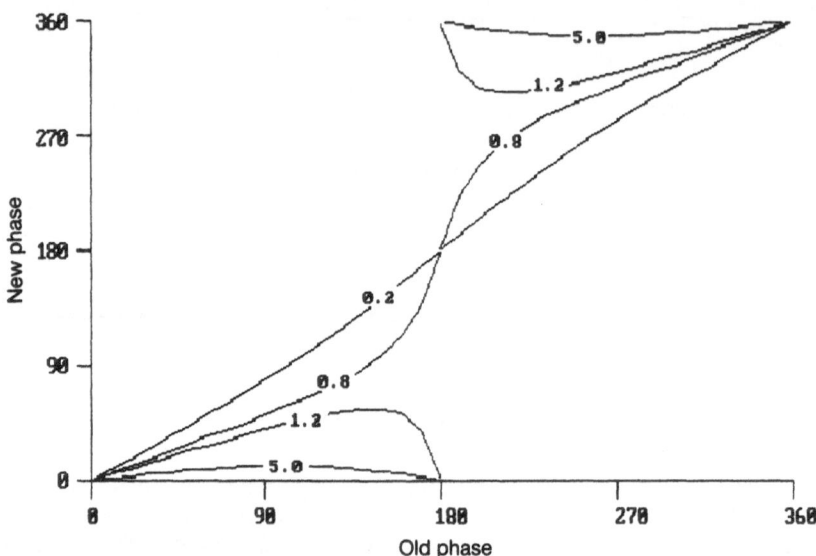

Fig. 3.8. Four phase reset curves with stimulus strengths of 0.2 to 5.0 in Guevara and Glass's model of the atrioventricular node. Weak stimuli are less than 1 radius, strong stimuli greater than 1 radius and their reset curves are in two parts.

When the coupling strength exceeds 1, and the cycle length of the sinoatrial oscillator is 75% or more of that of the atrioventricular oscillator, then the atrioventricular oscillator is locked 1 : 1 to the frequency of the sinoatrial oscillator. For example, with a coupling strength of 1.0 and sinoatrial and atrioventricular node cycle lengths of 800 and 1000 milliseconds respectively, the atrioventricular cycle length after it has been advanced by the sinoatrial stimulus three quarters of the way through its intrinsic cycle becomes the same as that of the sinoatrial node (Fig. 3.9, top). The atrioventricular oscillator has been entrained by the sinoatrial oscillator so that they both run at the same frequency, though the atrioventricular oscillator lags behind the sinoatrial oscillator by a delay that corresponds to the PR interval on the simulated surface electrocardiogram.

These events may be represented in a more elaborate ladder diagram that shows the state of the atrioventricular oscillator (Fig. 3.9, bottom). The atrioventricular node is now represented by three rungs of the ladder, the topmost rung showing the state of the oscillator at the time of arrival of the stimulus at the atrioventricular node, the middle rung the state of the oscillator when it has been reset, and the bottom rung the state of the oscillator when the stimulus passes to the ventricles. The state of the oscillator is indicated by the position of the radius, which rotates anticlockwise starting from the 3 o'clock position, following the same trigonometrical convention used in Fig. 3.5. At the 3 o'clock starting position the atrioventricular pacemaker discharges. Resetting is taken to be instantaneous, so that the line joining the top two rungs is vertical; after resetting completion of the pacemaker cycle takes an appreciable time indicated by the obliquity of the line joining the second and third rungs of the ladder diagram.

In Fig. 3.9, the atrioventricular pacemaker is 223° through its cycle when an

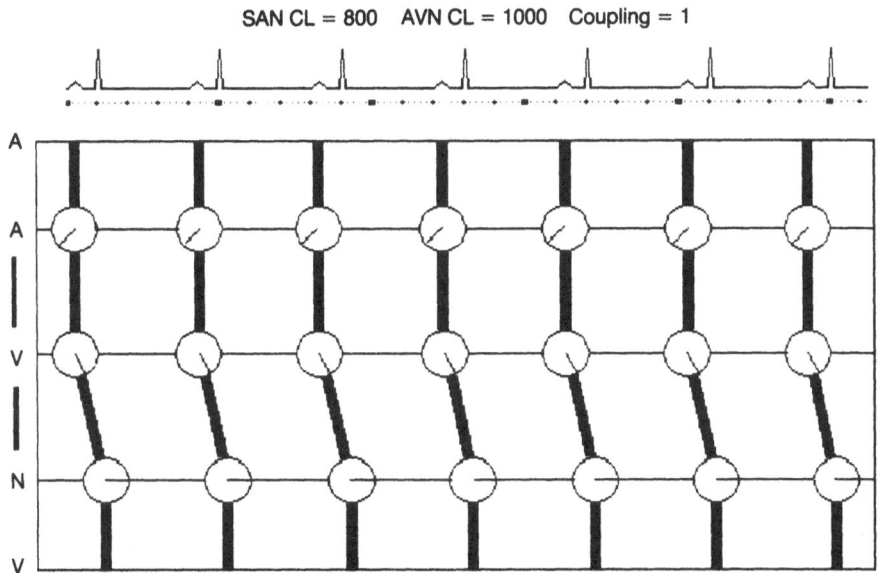

Fig. 3.9. In this and subsequent figures, ECG rhythm strips generated by Guevara and Glass's model at various settings of sinoatrial node (SAN) cycle length (CL) and coupling strength are shown. Atrioventricular cycle length is fixed at 1000 milliseconds. The *circles* inserted in the ladder diagram show the angular position of the atrioventricular oscillator at the time of arrival of the atrial stimulus (*top row*), after being reset (*middle row*), and at the time of discharge to the ventricles (*bottom row*); 0° is in the 3 o'clock position. Sinus rhythm; the atrioventricular delay (P-R interval) is 180 milliseconds.

atrial stimulus arrives (top row). The pacemaker is instantly reset to 295° (middle row), and takes 180 milliseconds to progress through 65° to reach the starting position of 0°/360° (bottom row), when it discharges and the ventricles are stimulated. In this example the atrioventricular delay is 180 milliseconds, this being the time taken for the atrioventricular oscillator to complete its cycle after arrival of the entraining stimulus from the sinoatrial oscillator.

If, while keeping the coupling strength at 1, the cycle length of the sinoatrial oscillator is shortened relative to that of the atrioventricular oscillator, 650 : 1000 milliseconds, then second degree heart block with a 3 : 2 Wenckebach cycle is produced, as illustrated in Fig. 3.10. The first atrial stimulus resets the pacemaker, from 311° to 330°, and it discharges 83 milliseconds later after completing a further 30° travel; a ventricular beat ensues. The second atrial stimulus finds the pacemaker just over halfway through its cycle; it is advanced to 285° and discharges after a delay of 208 ms. The third atrial stimulus finds the pacemaker less than halfway through its cycle, and it is retarded to 77°; it has progressed to 311° by the time of arrival of the fourth atrial stimulus, and then the pattern is repeated. In this sequence successive atrioventricular delays increase until a beat is dropped, a pattern that characterises second degree atrioventricular "block" of the Wenckebach variety.

Guevara and Glass (1982) published a diagram describing in general terms the behaviour of the two oscillators with all combinations of coupling strength and relative cycle length. This "Poincaré map" (named after the mathematician who studied the behaviour of coupled pendulum oscillators) delimits the areas of 1 : 1,

SAN CL = 650   AVN CL = 1000   Coupling = 1

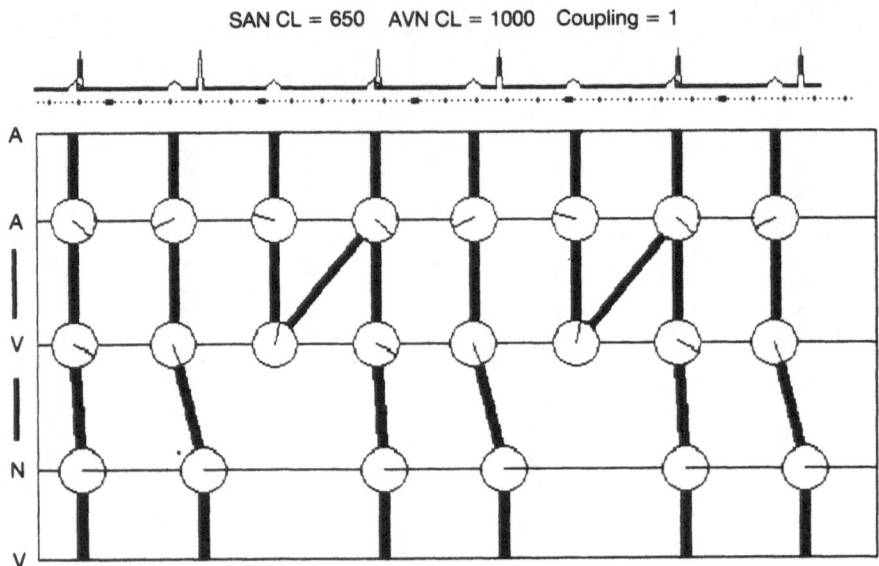

Fig. 3.10. Second degree heart block with 3 : 2 Wenckebach cycles.

SAN CL = 600   AVN CL = 1000   Coupling = 0.4

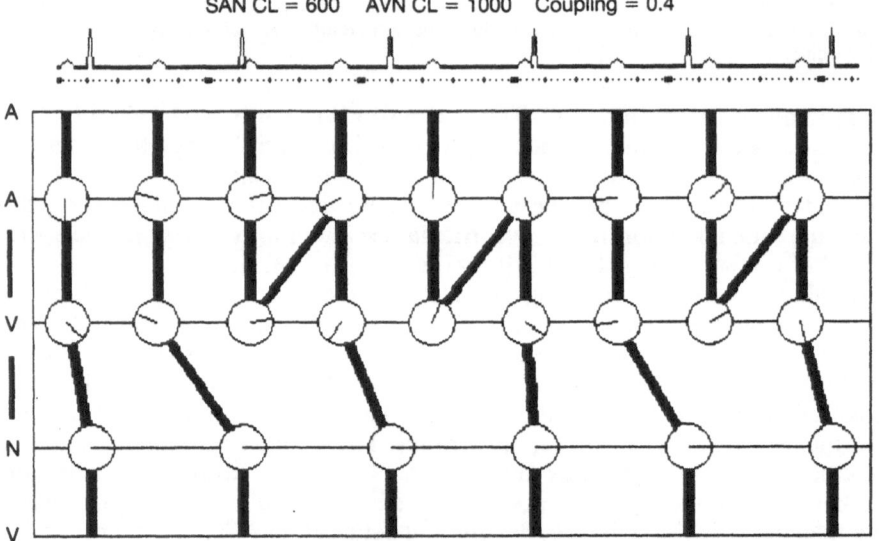

Fig. 3.11. Irregular Wenckebach cycles.

3 : 2, and 2 : 1 phase locking, and also indicates areas which contain phase-locked and non-phase-locked dynamics. An example of the latter is shown in Fig. 3.11. With a coupling of 0.4 and sinoatrial and atrioventricular cycle lengths of 600 and 1000 milliseconds respectively, an irregular rhythm consisting of variable Wenckebach cycles is generated. Another example is shown in Fig. 3.12. Here the

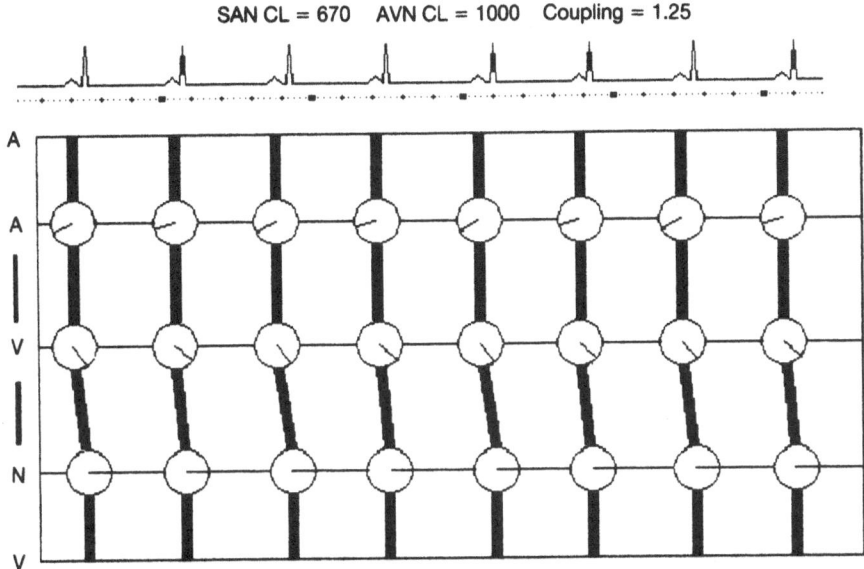

Fig. 3.12. Alternating long and short P-R intervals, so- called 2 : 2 block.

atrioventricular delays are alternately long and short, this being an example of so-called 2 : 2 block.

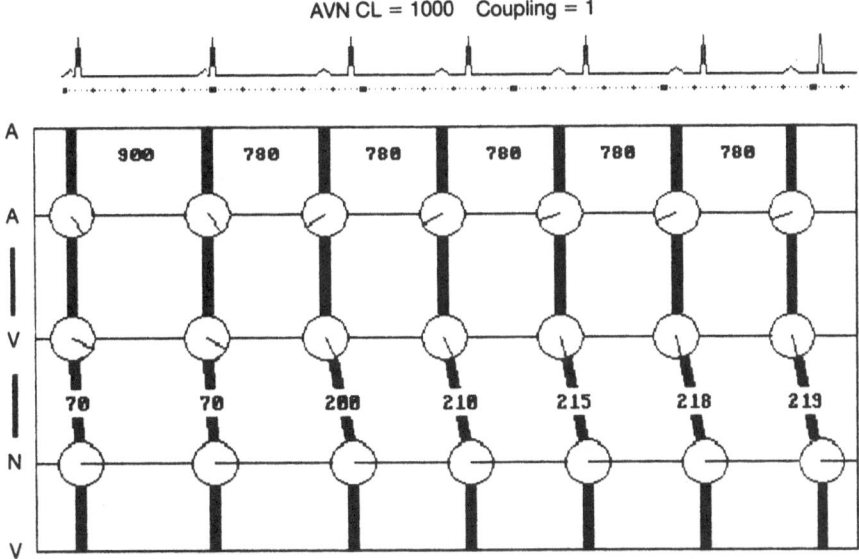

Fig. 3.13. Rapid upward adjustment of atrioventricular delay with reduction of atrial cycle length from 900 to 780 milliseconds.

These examples serve to illustrate that this simple model of the cardiac conduction system is capable of simulating very complex rhythm disturbances, some of which are aperiodic, that is, have no repeating pattern, in spite of their origin from the interaction of two periodic phenomena (Glass et al. 1986). The varieties of heart block illustrated all have their clinical counterparts, which are very difficult indeed to explain in traditional electrophysiological terms, but which are predictable consequences of the coupling of two oscillators (Gold-berger et al. 1985).

If the rate of the sinoatrial oscillator is abruptly increased then the stimulus it gives rise to when it fires arrives at an earlier phase of the atrioventricular oscillator's cycle (Fig. 3.13). The atrioventricular delay will therefore increase immediately and the adjustment will be mostly complete within one cycle. Thus, the model also simulates two further aspects of behaviour that are characteristic of the atrioventricular node - - a relatively long atrioventricular delay in sinus rhythm, and a negative relationship between atrioventricular delay and sinoatrial cycle length, with rapid adjustment of the former with change of the latter.

The Response of a Sine-Wave Oscillator to a Rapid Random Input

Guevara and Glass (1982) did not consider how their model might behave in atrial fibrillation. We can see, however, that although there are some differences, in general the sine-wave model is similar to that of Cohen et al. (1983), which so successfully explains the generation of R-R intervals in atrial fibrillation. The main difference between the models is that Cohen et al.'s has a refractory period when it is completely unresponsive to the atrial input. In Guevara and Glass's model, stimuli arriving in the first half of the cycle, corresponding to phases 2 and 3 when the membrane potential is becoming more negative, cause a positive shift of potential by electrotonic modulation. The phase of the pacemaker is therefore retarded. This produces a characteristic phase resetting curve, which has been demonstrated electrophysiologically for cardiac tissue (Jalife and Antzelevitch 1979; Glass et al. 1984).

If the atrioventricular sine-wave oscillator is bombarded with impulses with a Poisson distribution at a high frequency, it discharges irregularly, being randomly accelerated or retarded. Interestingly, when the model is run on a microcomputer the average rate of discharge of the atrioventricular oscillator is lower than its intrinsic rate, the accelerating effect of impulses arriving in the second half of the cycle not compensating fully for the retarding effect of impulses arriving in the first half of the cycle. Indeed, if the coupling strength is sufficient the cycle is repeatedly retarded to such an extent that the oscillator never gets past midcycle and never discharges. With loose coupling of say 0.1 or 0.2, the atrioventricular oscillator's discharge rate is negatively related to the fibrillation rate. The discharge rate is also negatively related to the coupling strength, but positively related to the intrinsic frequency of the atrioventricular oscillator. Thus the higher the strength and rate of atrial input the lower the rate of discharge.

In the dog the intrinsic rate of the atrioventricular node is about two-thirds that of the sinoatrial node (Urthaler et al. 1973). Applied to humans this would give an intrinsic rate for the atrioventricular node of about 50, which is approximately what is seen in heart block with junctional escape rhythm. If the nodal rate were to be depressed by impulses arriving in the first half of the atrioventricular node's cycle, this would lead to a mean ventricular rate of less than 50, a rate that is seen only exceptionally in atrial fibrillation. A major problem with the sine-wave model is, therefore, how to account for the high ventricular rates that are so characteristic of atrial fibrillation.

In Cohen et al.'s (1983) model the ventricular rate is positively related to the rate of arrival of atrial stimuli at the atrioventricular node; in Guevara and Glass's (1982) model the ventricular rate is negatively related to the rate of arrival of atrial stimuli once the intrinsic rate of the atrioventricular oscillator is sufficiently exceeded.

Direct evidence on the relation between atrial and ventricular rates is scanty. Exercise is not associated with any change of fibrillation rate, the increase in ventricular rate being the result of altered behaviour of the atrioventricular node (Aberg and Furberg 1975). Neither is the fall in ventricular rate with digitalisation associated with any change in the fibrillation rate (Aberg and Nordgren 1970). The ventricular response in both humans and dogs is greater in atrial tachycardia than in atrial fibrillation (Cohen et al. 1970), and acceleration of ventricular rate with a change to atrial flutter from fibrillation has been reported (Langendorf et al. 1965). In dogs with atrial fibrillation induced by vagal stimulation, lidocaine is effective in restoring sinus rhythm. During cardioversion by this means there is slowing of atrial activity and this is associated with ventricular acceleration (David et al. 1990). Electrograms recorded low in the right atrium showed flutter during periods of fast ventricular rates and fibrillation during periods of slower response in patients studied by Leier et al. (1979). Knowlton and Falk (1990) reported an increase in ventricular rate before conversion to sinus rhythm in patients with recent-onset atrial fibrillation, and this may well be associated with a falling fibrillation rate. However, it may be incorrect to equate the fibrillation rate derived from a surface electrocardiogram, which summates the electrical activity of multiple re-entry wavelets, with the rate of atrial input to the atrioventricular node. But also, a change in ventricular rate occurring during pharmacological cardioversion, or associated with the transition of fibrillation to flutter, or fibrillation to atrial tachycardia, could well result from a change of power in depolarisation wavefronts rather than a change in their rate of arrival at the atrioventricular node.

In 12 patients with atrial fibrillation and normal atrioventricular conduction reported by Kirsh et al. (1988) there was a strong positive correlation between mean A-A and R-R intervals. This is the best evidence we have that the ventricular rate in atrial fibrillation is positively related to the rate of arrival of stimuli at the atrioventricular node.

Modification of the sine-wave oscillator model so that it is relatively refractory and unresponsive to weak resetting stimuli during the first half of its cycle goes some way towards explaining the high ventricular rates in atrial fibrillation. A period of refractoriness renders the model more life-like – the atrioventricular node does have a refractory period – and it then behaves in a manner which is intuitively correct. The ventricular rate is positively related to the rate of atrial input, the coupling strength, and the intrinsic rate of the atrioventricular oscillator. The main remaining differences from the model of Cohen et al. is that the form of the action potential is sinusoidal, and the magnitude of the increment in membrane potential varies according to the point in the cycle when the stimulus arrives. This would not materially affect the functioning of the model, however, and both models would be expected to respond very similarly to a rapid random input. Fig. 3.14 shows a histogram of R-R intervals generated by the modified version of Guevara and Glass's model.

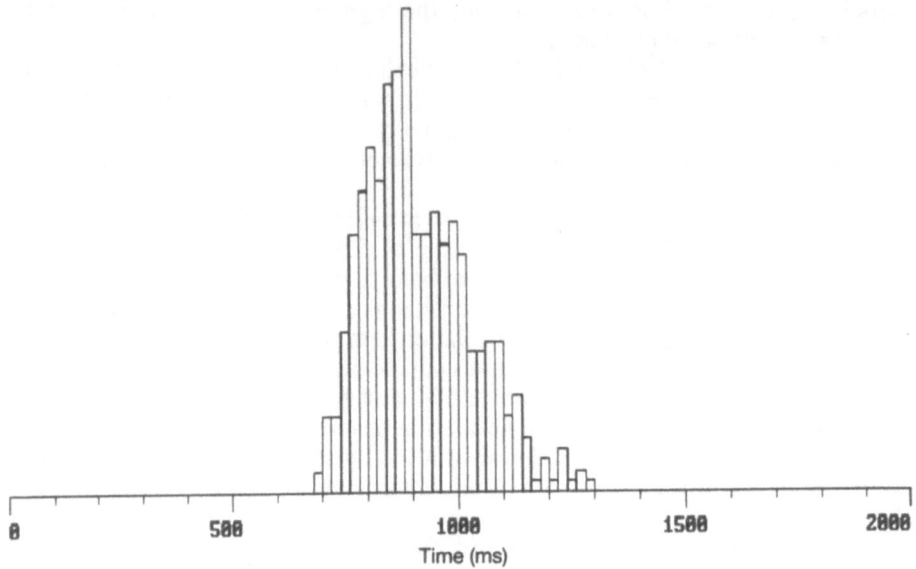

Fig. 3.14. Histogram of R-R intervals generated by Guavera and Glass's model in atrial fibrillation. Mean A-A interval 150 milliseconds, AVN CL 1200 milliseconds, coupling = 1. Mean R-R interval 917 milliseconds, range 660–1290 milliseconds, modal 880 milliseconds.

The Effect of a Refractory Period on the Behaviour of the Sine-Wave Model of the Atrioventricular Node

In order to explain high ventricular rates in atrial fibrillation by the sine-wave model, it is necessary to make the assumption that the atrioventricular oscillator is refractory to any weak input during the first half of its cycle. Otherwise, if the frequency of atrial input is more than twice that of the atrioventricular oscillator, the latter is repeatedly retarded in the first half of its cycle and never discharges at all.

The sine-wave model, modified in this way, still exhibits similar features to the unmodified version, which has no refractory period. The duration of the atrioventricular delay during 1 : 1 phase locking – equivalent to normal sinus rhythm – is unchanged. The Poincaré map (Fig. 3.15), though no longer symmetrical, still has areas of phase-locked and non-phase-locked behaviour, indicating that the modified model is still capable of generating the same variety of rhythm disturbances as before.

If we make the assumption that the sine-wave model of the atrioventricular node is completely refractory even to strong non-physiological stimuli for a time at the beginning of its cycle, let us say for the first quarter, then the effective refractory period of the model node is negatively related to the atrial cycle length, as it is in reality (Josephson and Seides 1979, p 44). This is illustrated in Fig. 3.16. With an atrial cycle length of 780 milliseconds and an atrioventricular nodal cycle length of 1000 milliseconds (right), the atrioventricular pacemaker will be unresponsive to stimuli for a period of 250 milliseconds after its discharge, which takes place 220 milliseconds after arrival of the atrial stimulus. The effective refractory period – the longest interval after a normal atrial stimulus where a

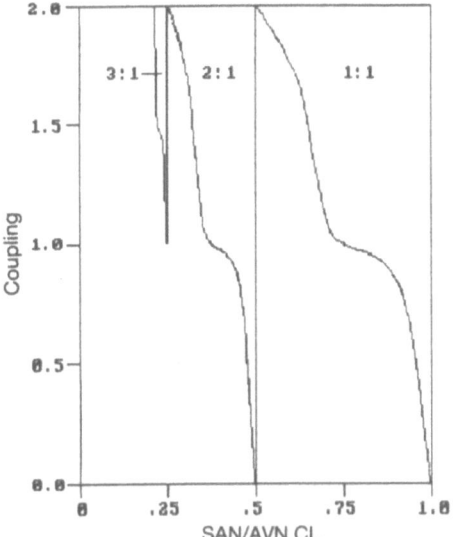

Fig. 3.15. Poincaré map of modified Guevara and Glass model of the atrioventricular node. Zones of combination of coupling strength and ratios between sinoatrial and atrioventricular node cycle lengths (SAN/AVN CL) that give rise to 1 : 1, 2 : 1 and 3 : 1 phase-locked behaviour are indicated. In the unmarked zones phase-locked and aperiodic dynamics occur.

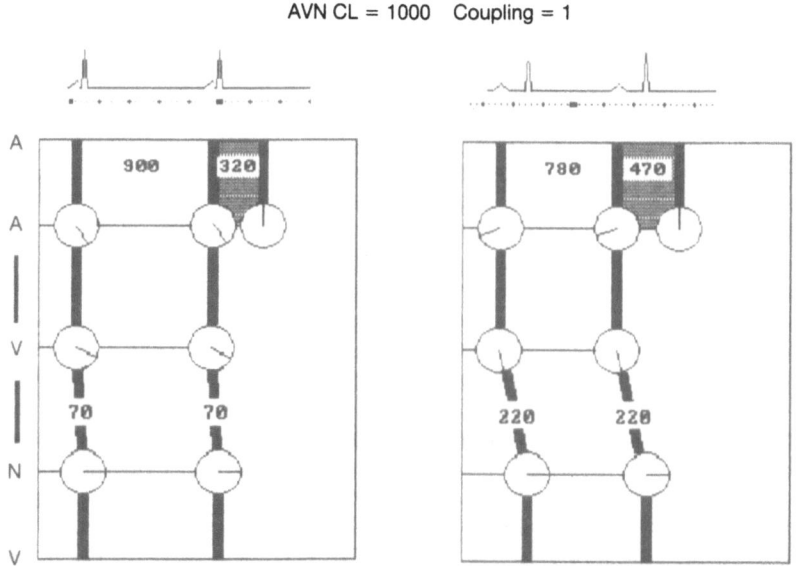

Fig. 3.16. The refractory period of the atrioventricular node (*shaded portion*) is shorter with an atrial cycle length of 900 milliseconds (*left*) than with an atrial cycle length of 780 milliseconds (*right*).

premature extrastimulus is not followed by a ventricular response – is 470 milliseconds. When the atrial cycle length is 900 milliseconds the atrioventricular delay is 70 milliseconds, and the effective refractory period is 320 milliseconds. The effective refractory period of the atrioventricular node is therefore negatively related to the driving cycle length. These effective refractory periods are unrealistically prolonged because the sine-wave model distorts the duration and shape of the action potential so that it is one- quarter of the cycle length of the atrioventricular oscillator. In reality it is a variable proportion of the cycle length of the oscillator, and would usually be less than one quarter. Nevertheless, the same principles governing the duration of these electrophysiological intervals would pertain, and the sine-wave model will be retained for simplicity and the ease of demonstrating the general principles of operation. The model runs on a microcomputer, and the simple program is given in the Appendix.

Further development of the model to show its interaction with the autonomic nervous system is described in Chapter 8.

What Values Should Be Assigned to the Sine-Wave Model's Parameters?

Before running the model, three parameters have to be specified: the mean cycle length of the sinoatrial oscillator or, if atrial fibrillation is to be simulated, the mean interval between arrival of impulses at the atrioventricular node; the mean cycle length of the atrioventricular oscillator; and the coupling strength.

In 18 patients with atrial fibrillation the average duration of A-A intervals recorded from atrial electrograms was 185 milliseconds, equivalent to an atrial input of 324 impulses per minute with a range of 210–571 impulses per minute (Kirsh et al. 1988). Monophasic action potentials have the same frequency when recorded from different atrial sites, with a reported range of 311–578 impulses per minute and a mean of 485 impulses per minute in lone atrial fibrillation and 364 impulses per minute in ischaemic heart disease (Gavrilescu et al. 1976). The mean A-A interval in the model could therefore realistically be between 100–200 milliseconds.

The cycle length of the atrioventricular oscillator assigned in the preceding examples has been 1000 milliseconds, or 60 impulses per minute. However, like the sinoatrial node, the atrioventricular node is under autonomic control, and its rate is likely to vary according to the prevailing autonomic conditions. The discharge rate for the sinoatrial node varies over at least a threefold range, and there is no reason to assume otherwise for the atrioventricular node. The intrinsic rate of the atrioventricular node is about 50 per minute so its rate could be as high as 150 per minute (cycle length 400 milliseconds) under conditions of high sympathetic drive and vagal withdrawal, such as exercise. The model's atrioventricular nodal cycle length could therefore be between 400 and 1200 milliseconds.

The appropriate coupling strength to employ in the model is unknown, and can only be guessed. We do know, however, that the atrial pacing rate at which 1 : 1 atrioventricular conduction fails and Wenckebach block occurs is about 140 beats per minute or a cycle length of 428 milliseconds, about half that of the atrioventricular node (Curry 1975). On the Poincafe map of the modified Guevara and Glass model (Fig. 3.15) the boundary of 1 : 1 conduction with a sinoatrial/atrioventricular cycle length ratio of approximately 0.5 has a coupling strength of 2 or more. It would be reasonable, therefore, to assume that in sinus rhythm the atrial input constitutes a strong resetting stimulus with a coupling strength of 2 or more; in atrial fibrillation this would be reduced to between 1 and 2, or perhaps to that of a weak stimulus with a coupling strength of less than 1.

More Electrophysiological Phenomena: Ventricular Extrasystoles

Fig. 3.17 shows a conventional ladder diagram of a ventricular extrasystole occurring in sinus rhythm. It is not conducted into the atria and does not reset the sinoatrial pacemaker, which therefore continues to fire at a steady rate, although the P wave following the extrasystole is obscured. Retrograde conduction into the atrioventricular node does take place, however, and this renders the node refractory to the arrival of the next atrial stimulus. The R'-R interval following the extrasystole is prolonged beyond the duration of the regular R-R interval by the extent to which the extrasystole is premature: the compensatory pause.

In the extended ladder diagram (Fig. 3.18) the extrasystole is shown conducted retrogradely into the atrioventricular node, where it acts as a strong resetting stimulus, advancing the atrioventricular oscillator so that it discharges soon afterwards. The first atrial stimulus after the extrastimulus finds the atrioventricular oscillator refractory, but the following one advances the oscillator which then discharges in the usual way. The atrioventricular delay after the extrasystole may be slightly different from normal.

In explaining the compensatory pause, both the electrophysiological and the oscillator models assume retrograde penetration of the atrioventricular node by a stimulus arising from the ventricular extrasystole, with modification of the node's behaviour towards the subsequent atrial impulse. Both models therefore utilise the concept of concealed conduction, but differ in the way they explain its effect

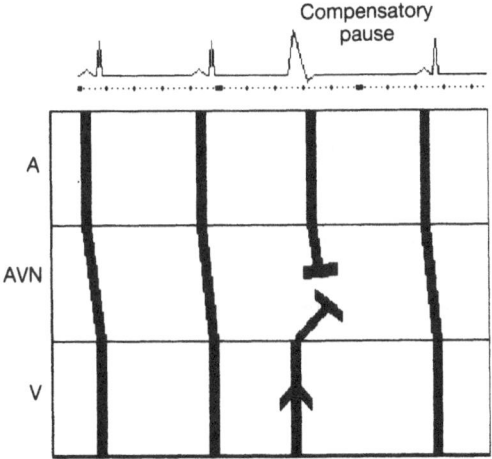

Fig. 3.17. Ladder diagram showing the compensatory pause after a ventricular extrasystole.

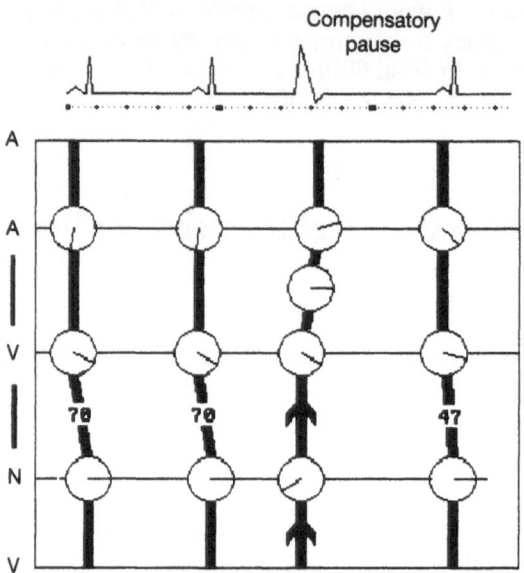

Fig. 3.18. Extended ladder diagram showing the the state of the atrioventricular oscillator when sinus rhythm is interrupted by a ventricular extrasystole. The atrial beats and the extrasystole are strong resetting stimuli with a coupling strength of 2.

on subsequent events. The conventional electrophysiological explanation is in terms of altered refractoriness, while in the oscillator model the interpretation of the same phenomenon is in terms of resetting the atrioventricular pacemaker.

The "Compensatory Pause" in Atrial Fibrillation

When a ventricular extrasystole occurs in atrial fibrillation the interval between the extrasystole and the next normal beat is often longer than the average R-R interval, reminiscent of the compensatory pause in sinus rhythm. In sinus rhythm the compensatory pause after a premature extrasystole allows the ventricular rhythm to get in step once more with the sinoatrial rhythm. The use of the term in atrial fibrillation is therefore misleading, since there is no underlying regularity with which the ventricular rhythm is out of step.

In atrial fibrillation post-extrasystolic intervals are as variable in duration as the intervals between normal beats, so the existence of a compensatory pause can only be determined by careful statistical analysis of a number of extrasystoles, and not by selection of examples of ventricular ectopic beats that are followed by a particularly long interval. Pritchett et al. (1980) induced atrial fibrillation in 5 patients by rapid atrial pacing. Premature ventricular contractions were induced by coupling a right ventricular extrastimulus to every eighth or tenth beat, changing the coupling interval until the entire cardiac cycle was scanned. In each patient the mean interval after the induced extrasystole was longer than the mean control interval (by 107–136 milliseconds), confirming the existence of a compensatory pause.

Figs. 3.19 and 3.20 show how this observation may be explained in terms of the oscillator model, using an extended ladder diagram which includes an additional rung to indicate the bundle of His. After the first ventricular beat (Fig. 3.19) two atrial stimuli arrive when the atrioventricular oscillator is relatively refractory, so its setting is unchanged. The third atrial stimulus advances the oscillator which discharges to give the second ventricular beat. Antegrade conduction down the His–Purkinje network takes 40 milliseconds. The next two atrial stimuli arrive in quick succession and advance the oscillator so that the next ventricular beat occurs 700 milliseconds after the previous one. In Fig. 3.20 the same sequence of atrial stimuli is depicted, but a ventricular extrasystole replaces the second beat of Fig. 3.19. In order that the timing of events after the extrasystole should be the same as after the normal ventricular beat, the extrasystole has to occur earlier than the normal beat by an amount equal to the conduction time from ventricular myocardium to atrioventricular node plus that from atrioventricular node to myocardium. Retrograde conduction in the His–Purkinje system is slower than antegrade (Curry 1975), and the combined conduction times in both directions (40 + 80 = 120 milliseconds in Fig. 3.20) are of the same order as the compensatory pause found by Pritchett et al. (1980) and Morady et al. (1986).

The duration of any interbeat interval in atrial fibrillation is determined by the length of the atrioventricular node's pacemaker cycle, beginning and ending with its discharge, and modulated by the random arrival of atrial impulses. If the discharge at the beginning of a cycle is the result of spontaneous depolarisation, perhaps hastened by arrival of atrial stimuli, it will be followed by a ventricular beat. If, on the other hand, the discharge is brought about by a strong resetting stimulus conducted retrogradely from a ventricular extrasystole, it will be preceded by a ventricular beat, and the post-extrasystolic interval will be a little longer on average than the intervals between normal beats.

The existence of a compensatory pause has been adduced as evidence for retrograde concealed conduction of ventricular extrasystoles in atrial fibrillation

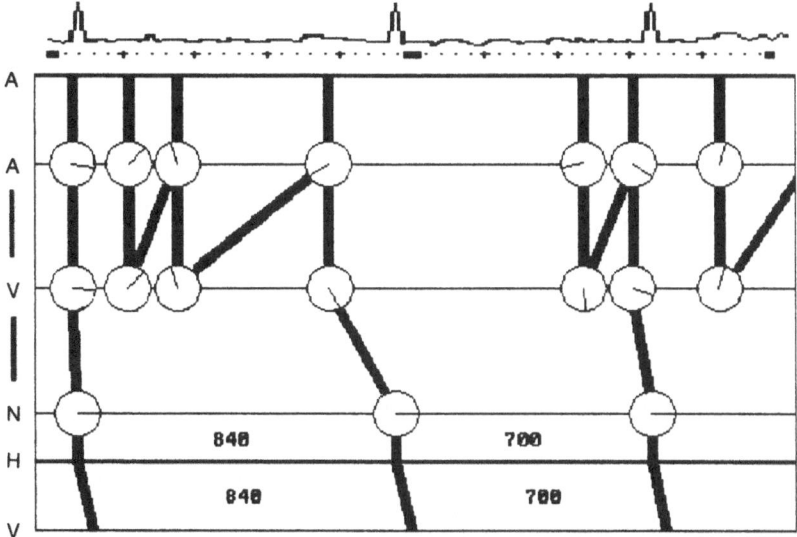

Fig. 3.19. Extended ladder diagram with an additional rung for the bundle of His (H), showing three conducted beats in atrial fibrillation. Note that the time scale has been expanded.

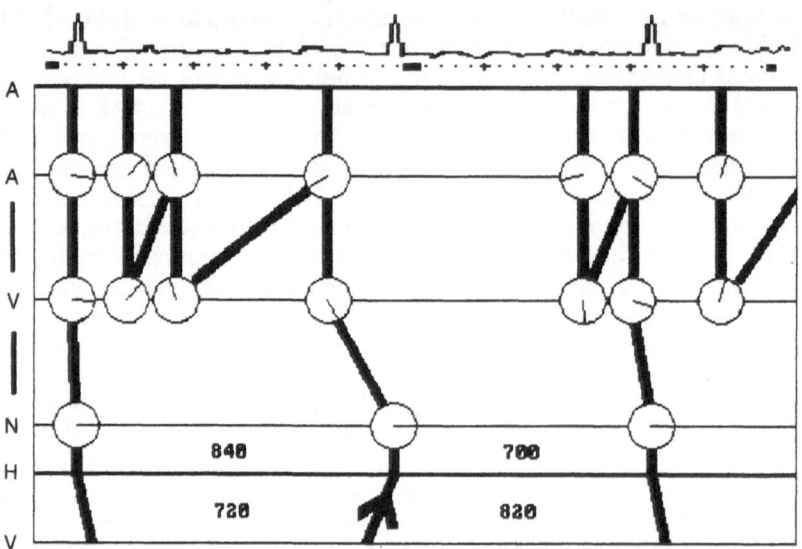

Fig. 3.20. As Fig. 3.19 except that a ventricular extrasystole replaces the second beat.

(Langendorf et al. 1965). The oscillator model also infers concealed retrograde conduction, but explains explicitly how the compensatory pause in atrial fibrillation comes about.

Peeling Back Refractoriness

A ventricular extrasystole, under certain circumstances, may appear to improve conductivity of the atrioventricular node – a phenomenon that has been described as "peeling back refractoriness" (Moe et al. 1968; Moore and Spear 1971). The left-hand side of Fig. 3.21 depicts an atrial extrastimulus delivered at the effective refractory period; in the right-hand side of the figure the second conducted ventricular beat is replaced by a ventricular extrastimulus. The atrial extrastimulus, delivered at the same time in relation to the basic atrial cycle, is now followed by a ventricular beat because it falls outside the effective refractory period; the effective refractory period occurs earlier because the atrioventricular oscillator is discharged prematurely by the retrogradely conducted ventricular extrastimulus.

The Effect of Right Ventricular Pacing in Atrial Fibrillation

Wittkampf et al. (1988) described the effect of right ventricular pacing on the distribution of R-R intervals in 13 patients with atrial fibrillation. A demand pacemaker, as expected, eliminated R-R intervals that were longer than the pacing interval. However, rather surprisingly, in each patient a pacing rate was found that led to the elimination of R-R intervals that were shorter than the pacing interval. The pacing interval at which more than 95% of the spontaneous beats were eliminated was usually a little shorter than the mean R-R interval in

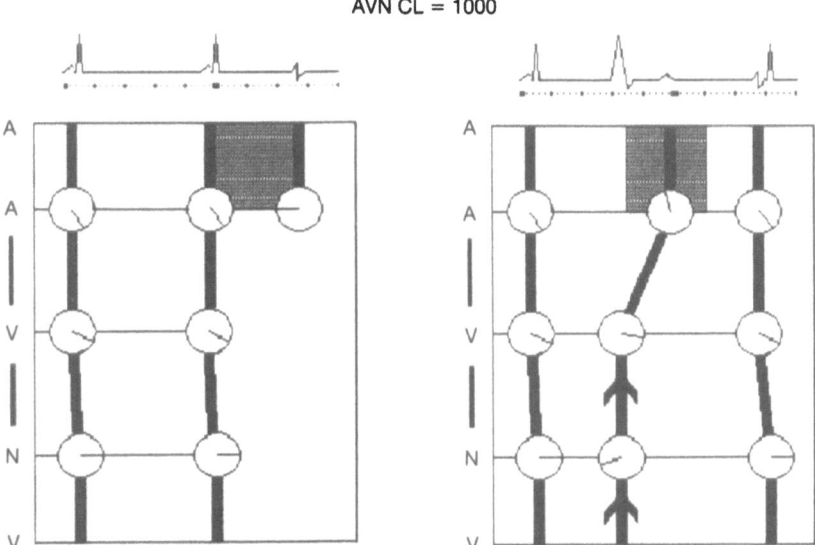

Fig. 3.21. *Left.* Two beats of sinus rhythm followed by an atrial extrastimulus at the effective refractory period of the atrioventricular node. *Right.* The second ventricular beat is replaced by a ventricular extrasystole. The atrial extrastimulus is now conducted.

unpaced atrial fibrillation, and about 60% of the duration of the longest intervals. The authors point out that this phenomenon is not explicable by concealed retrograde conduction and alteration of refractoriness, but could be explained by overdrive suppression of the atrioventricular pacemaker by retrograde depolarisation from the right ventricle. Overdrive suppression would necessitate a pacing cycle shorter than the intrinsic cycle of the atrioventricular pacemaker (Lange 1965), and the authors appear to suggest that the pacing interval at which more than 95% of the shorter intervals are eliminated is only a little less than the atrioventricular pacemaker interval. It is then difficult to explain the occurrence of intervals in unpaced atrial fibrillation which are as much as twice the pacing cycle length (Dreifus and Mazgalev 1988).

A better interpretation of the findings may be that the longest interval in unpaced atrial fibrillation represents the cycle length of the atrioventricular pacemaker. The length of the pacing cycle that eliminates more than 95% of spontaneous beats is on average about 60% of the longest cycle.

Wittkampf et al. (1988) make the distinction between depolarisation and electrotonic modulation of the atrioventricular pacemaker. Antegrade spread of the impulse through the atrioventricular node is normally by electrotonic modulation, depolarisation only occurring retrogradely from the ventricle, either after a paced beat or after a ventricular extrasystole. Depolarisation results in pacemaker suppression by reducing or eliminating phase 4 depolarisation in the pacemaker cells.

In our model of the atrioventricular node the distinction between electrotonic modulation and depolarisation of the atrioventricular oscillator is not so clear cut. The atrial input in atrial fibrillation would be described as a weak resetting stimulus (coupling strength less than 1), and that in sinus rhythm as a strong resetting stimulus (coupling strength more than 1), but all gradations of coupling

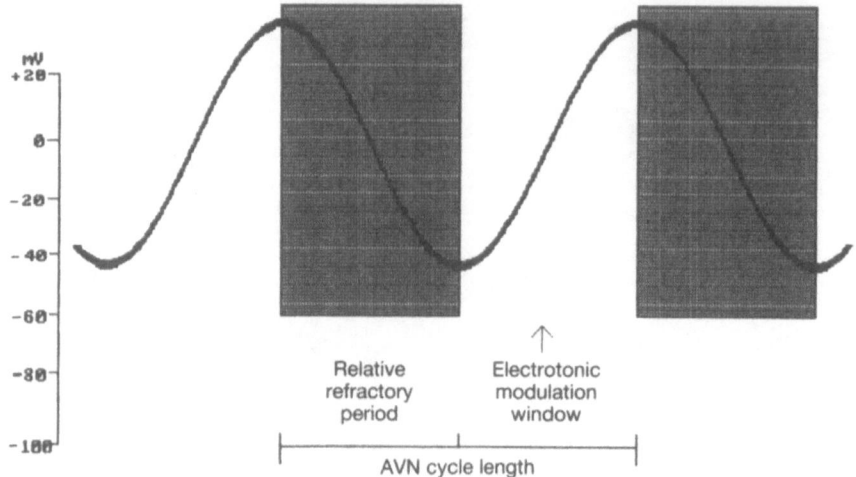

Fig. 3.22. Sine-wave membrane potential generated by the atrioventricular oscillator. Electrotonic modulation may take place during half of the cycle when the potential is rising.

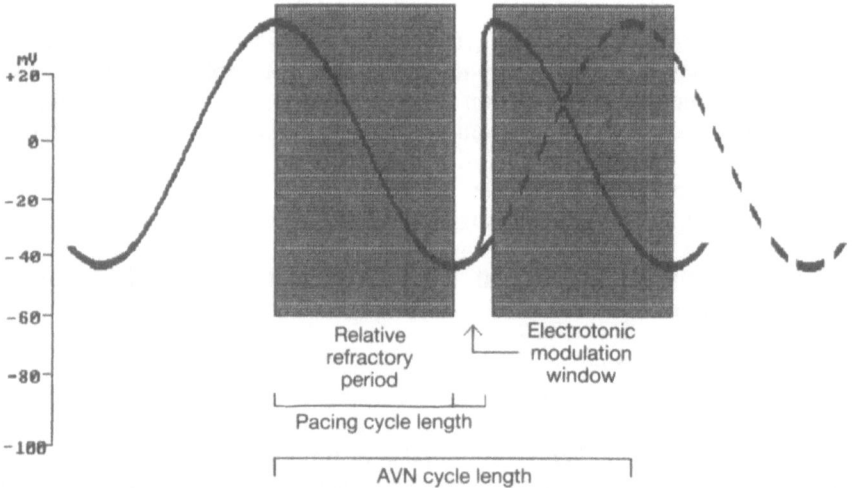

Fig. 3.23. As Fig. 3.22, during ventricular pacing. The window for electrotonic modulation is much smaller than in unpaced atrial fibrillation.

strengths are possible. A paced ventricular beat, or a ventricular extrasystole, conducted retrogradely to the atrioventricular node would be considered a strong resetting stimulus.

Fig. 3.22 shows the sinusoidal transmembrane potential of the atrioventricular oscillator. When the potential is falling the oscillator is relatively refractory, and is unresponsive to weak incoming atrial stimuli. While the potential is rising the oscillator is susceptible to electrotonic modulation by weak, random stimuli from the fibrillating atria, or by the strong stimulus of a retrogradely conducted paced beat. When the right ventricle is paced with a cycle length less than that of the atrioventricular oscillator the time window for electrotonic modulation by atrial stimuli is shortened (Fig. 3.23). When the pacing cycle length is reduced to about 60% of that of the atrioventricular oscillator, all but 5% of R-R intervals shorter than the pacing cycle are eliminated. The occasional R-R interval that is shorter than the pacing interval arises when, by chance, a sufficient number of atrial stimuli to discharge the oscillator fall in the small time window for electrotonic modulation.

The Wolff–Parkinson–White Syndrome

The survival value that results from the atrioventricular node not being a through-conductor is illustrated by the Wolff- Parkinson–White syndrome, where the node is bypassed by one or more accessory tracts. In the event of atrial fibrillation, the bypass tract may conduct impulses to the ventricles at such a rapid rate that syncope or ventricular fibrillation may follow. The ventricular rate in atrial fibrillation has generally been found to be related to the refractory period of the atrioventricular node or the accessory pathway, depending on the predominant route of conduction (Billette et al. 1974; Wellens and Durrer 1974; Rowland et al. 1981; Fillette et al. 1983). However, the refractory period of the accessory pathway does not always differ sufficiently from that of the atrioventricular node to explain the difference in ventricular response between these two atrioventricular routes.

Repetitive concealed conduction within the node has been invoked to explain the slower ventricular rate with nodal conduction (Meijler et al. 1985), but concealed conduction in the accessory pathway has also been adduced to explain the irregularity of ventricular response when pre-excitation occurs (Klein et al. 1984; Milstein et al. 1987).

However, the protection against ventricular fibrillation afforded by the atrioventricular node in atrial fibrillation (which is bypassed in the Wolff–Parkinson–White syndrome) is due not to its conduction properties but rather to its being a biological oscillator, and not a through-conductor at all.

Conclusions

The Guevara and Glass sine-wave oscillator model of the atrioventricular node, when modified to include a relative refractory period, is very similar to that of Cohen et al, and explains the phenomena of the compensatory pause, peeling back refractoriness, and the abolition of shorter R-R intervals with right ventricular pacing. It also explains most of the other puzzling features of the atrioventricular node: the long atrioventricular delay in sinus rhythm, the rapid variation in the delay with change of heart rate, the negative relation between cycle length and effective refractory period, the differential between the number of impulses entering the node and those leaving it in heart block, and the peculiar distribution of R-R intervals in atrial fibrillation. In addition, it allows for a wide range of ratios between input and output frequencies of the atrioventricular node in atrial fibrillation.

Nevertheless, the ability to construct a model that describes all these phenomena is not in itself sufficient proof that the atrioventricular node works in such a way. That the sinoatrial node is an oscillator is not disputed (James 1973), and the obvious parallels between sinoatrial and atrioventricular nodes in structure and function lend support to the notion that the atrioventricular node too is an oscillator. But the power of the oscillator model of the atrioventricular node lies in its ability not to replace, but to embrace classical electrophysiological mechanisms, such as block, or concealed or decremental conduction, and to impart a deeper level of meaning to them. The model gains strength too from its parsimony, the same mechanisms being invoked in sinus rhythm, partial heart block and atrial fibrillation. However, it is difficult to think of a crucial experiment that would confirm or refute the oscillator hypothesis as an explanation for the atrioventricular node's behaviour. It might be objected that the behaviour of the model does not correspond to that of any of the cellular elements that make up the atrioventricular node (Dreifus and Mazgalev 1988). The rejoinder to this criticism is that electrotonic modulation of depolarisation and repolarisation of a community of functionally disparate cell types may lead to total behaviour that is much more than the sum of its parts (Joyner 1986).

Phase Resetting: An Achilles Heel

Using the unmodified Guevara and Glass model of sinoatrial and atrioventricular nodes we will explore further what happens when the regular operation of the model is interrupted by a stimulus, equivalent in the heart to an extrasystole or an external electrical impulse. A stimulus of zero strength does not affect the oscillator, but a weak stimulus may advance or retard the phase of the oscillator, depending on its timing. If a weak stimulus (less than 1 radius) is applied in the first half of the cycle then the oscillator will be retarded: if it is applied in the second half it will be advanced. With a strong stimulus of, say, 5 units, the post-stimulus phase is much the same and close to 0 °or 360° whatever the phase of the cycle in which the stimulus is delivered (Fig. 3.8).

Fig. 3.24 shows what happens when stimuli of different strengths are applied at different phases of the oscillator cycle. Contour lines are drawn to join all post-stimulus beats that have the same phase. For low intensity stimuli the contours run upwards and are almost parallel. As the stimulus strength approaches 1 the contours converge on a point which marks the delivery of a stimulus of strength 1 at 180° or at the midpoint of the oscillator's cycle. What does this point represent? If we return to the geometrical model we see that if a stimulus of 1 radius strength is applied in midcycle then the centrifugal end of the radius is displaced exactly to the centre of the circle. It is therefore meaningless to speak of the phase of that point, the phase being its angular position with respect to the centre. Mathematically, this point represents a singularity, a black hole, where the outcome of the particular combination of stimulus strength and timing is unknown and cannot be predicted. At this point the contour lines run into each other just as they would on a physical map representing a vertical hole.

Although the model is grossly simplified the depiction of a singularity is true to life. A phase/stimulus-strength map of the sinoatrial node would not be so regular

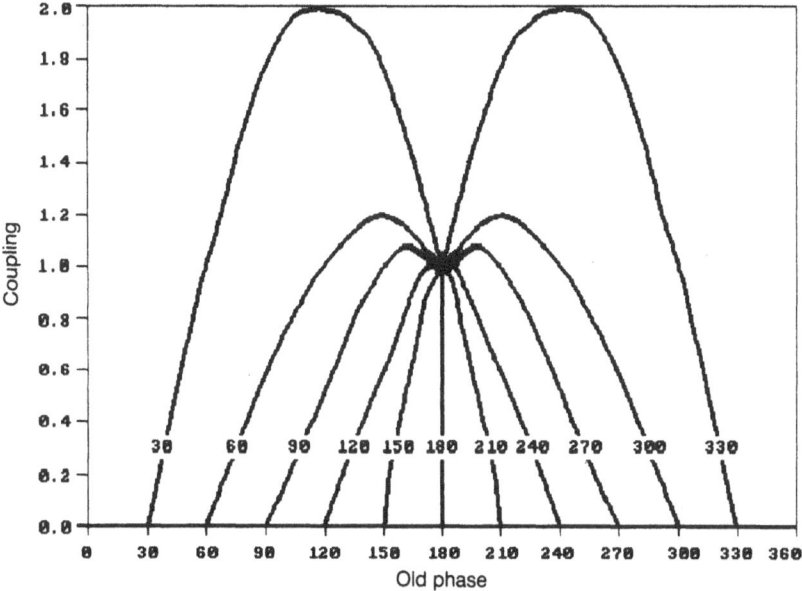

Fig. 3.24. Phase–stimulus strength diagram of the Guevera and Glass model of an oscillator. The abscissa shows the phase before a resetting stimulus of strength indicated on the ordinate is applied. The contour lines join points of the same phase after resetting.

nor be symmetrical. However, the sinoatrial node shows a similar resetting behaviour to the model, with a sudden transition of response from a weak to a strong stimulus, and it is this that determines the topological relationship of the contours and makes the existence of a singularity inescapable. Readers wishing a fuller exposition of this subject should refer to Winfree (1987), who emphasises the inevitable association of singularities with the mode of operation of cardiac pacemaker cells. In the atria, and in the ventricles too, there is an oscillating system consisting of a pacemaker and a mass of excitable cells that are normally driven by the pacemaker. The existence of a singularity means that a stimulus with a certain combination of strength and timing in the cardiac cycle will inevitably result in disruption of the normal operation of the system, so-called annihilation of the pacemaker (Jalife and Antzelevitch 1979). The essential nature of a singularity is that the consequence of the singular combination of conditions is unpredictable. If the conditions for singularity are fulfilled normal rhythm is suspended. What follows, whether a fatal dysrhythmia or the restoration of sinus rhythm, depends on the underlying cardiac state, but the suspension of normal rhythm represents a life-threatening Achilles heel. This is highly relevant to the occurrence of fibrillation, not only of the atria, which concerns us here, but also of the ventricles, which is such an important cause of sudden death.

The singularity may be a point, as in Fig. 3.24, or it may be an area, that is, there may be a range of values for intensity and timing of the stimulus where the outcome is unpredictable. If the underlying conditions for the occurrence of atrial fibrillation are present then atrial fibrillation may become established once normal rhythm is disrupted. Otherwise, after a period of instability, sinus rhythm may be restored (Jalife and Antzelevitch 1980; Reiner and Antzelevitch 1985).

References

Aberg H, Furberg B (1975) Atrial activity during exercise in patients with atrial flutter or atrial fibrillation. Ups J Med Sci 80:20–3

Aberg H, Nordgren L (1970) The effect of digitalis on the atrial activity in atrial fibrillation. Acta Soc Med Ups 75:13–18

Billette J (1981) Short time constant for rate-dependent changes of atrioventricular conduction in dogs. Am J Physiol 241:H28–33

Billette J, Nadeau RA, Roberge F (1974) Relation between the minimum RR interval during atrial fibrillation and the functional refractory period of the AV junction. Cardiovasc Res 8:347–51

Cohen RJ, Berger RD, Dushane TE (1983) A quantitative model for the ventricular response during atrial fibrillation. IEEE Trans Biomed Eng 30:769–81

Cohen SI, Lau SH, Berkowitz WD, Damato AN (1970) Concealed conduction during atrial fibrillation. Am J Cardiol 25:416– 19

Curry PVL (1975) Fundamentals of arrhythmias: Modern methods of investigation. In: Krikler DM, Goodwin JF (eds) Cardiac arrhythmias. The modern electrophysiological approach. Saunders, Philadelphia, pp 39–80

David D, Lang RM, Neumann A, Borrow KM, Akselrod S, Mor-Avi V (1990) Parasympathetically modulated anti-arrhythmic action of lidocaine in atrial fibrillation. Am Heart J 119:1061–8

Dorveaux L, Twidale N, Tonkin A (1988) Direct identification of parameters in a mathematical model describing conduction through the atrioventricular node. Int J Biomed Comp 23:69–76

Dreifus LS, Mazgalev T (1988) "Atrial paralysis": does it explain the irregular ventricular rate during atrial fibrillation? J Am Coll Cardiol 11:546–7

Fillette F, Fontaine G, Frank R, Grosgogeat Y (1983) Ventricular response in atrial fibrillation : normal conduction, consequences of conduction disorders and pre-excitation syndromes. Therapeutic implications. Ann Cardiol Angeiol (Paris) 32:7–19

Gavrilescu S, Dragulescu SI, Streian C, Luca C (1976) Monophasic action potentials of the right atrium during atrial fibrillation in man. Cor Vasa 18:264–70

Glass L, Guevara MR, Belair J, Shrier A (1984) Global bifurcations of a periodically forced biological oscillator. Physiol Rev 29:1348–57

Glass L, Shrier A, Belair J (1986) Chaotic cardiac rhythms. In: Holden AV (ed) Chaos. Manchester University Press, Manchester, pp 237–56

Goldberger AL, Bhargava V, West BJ, Mandell AJ (1985) Nonlinear dynamics of the heartbeat. II Subharmonic bifurcations of the cardiac interbeat interval in sinus node disease. Physica 17D:207–14

Goldstein RE, Barnett GO (1967) A statistical study of the ventricular irregularity of atrial fibrillation. Comp Biomed Res 1:146–61

Grant RP (1956) The mechanism of A-V arrhythmias with an electronic analogue of the human A-V node. Am J Med 20:334–44

Guevara MR, Glass L (1982) Phase locking, period doubling bifurcations and chaos in a mathematical model of a periodically driven oscillator: A theory for the entrainment of biological oscillators and the generation of cardiac dysrhythmias. J Math Biol 14:1–23

Hashida E, Yoshitani N, Tasaki T (1978) A study on the irregularity of the sequence of R-R intervals in chronic atrial fibrillation in man based on the time series analysis and information theory. Jpn Heart J 19:839–51.

Heethaar RM, van der Gon JJD, Meijler FL (1973a) Mathematical model of A-V conduction in the rat heart. Cardiovasc Res 7:105– 14

Heethaar RM, de vos Burchart RM, van der Gon JJD, Meijler FL (1973b) A mathematical model of A-V conduction in the rat heart. II. Quantification of concealed conduction. Cardiovasc Res 7:542–56

Honzicova N, Fiser B, Semrad B (1973) Ventricular function in patients with atrial fibrillation. Cor Vasa 4:257–64

Jalife J, Antzelevitch C (1979) Phase resetting and annihilation of pacemaker activity in cardiac tissue. Science 206:695–7 James TN (1973). The sinus node as a servo mechanism. Circulat Res 32:307–11

Jalife J, Antzelevitch C (1980) Pacemaker annihilation: diagnostic and therapeutic implications. Am Heart J 100:128–30

Janse MJ (1969) Influence of the direction of the atrial wave front on A-V nodal transmission in isolated hearts of rabbits. Circ Res 25:439–49

Janse MJ (1990) Propagation of atrial impulses through the atrioventricular node. In: Touboul P,

Waldo AL (eds) Atrial arrhythmias. Current concepts and management. Mosby Year Book, St Louis, pp 141–52

Josephson ME, Seides SF (1979) Clinical cardiac electrophysiology. Techniques and interpretations. Lea & Febiger, Philadelphia

Joyner RW (1986) Modulation of repolarization by electrotonic interactions. Jpn Heart J 27[Suppl I]:167–83

Katholi CR, Urthaler F, Macy J, James TN (1977) A mathematical model of automaticity in the sinus node and AV junction based on weakly coupled relaxation oscillations. Comp Biomed Res 10:529–43EP

Kirsh JA, Sahakian AV, Baerman JM, Swiryn S (1988) Ventricular response to atrial fibrillation: role of atrioventricular conduction pathways. J Am Coll Cardiol 12:1265–72

Klein GJ, Yee R, Sharma AD (1984) Concealed conduction in accessory atrioventricular pathways: an important determinant of the expression of arrhythmias in patients with Wolff–Parkinson–White syndrome. Circulation 70:402–11

Knowlton AA, Falk RH (1990) Paradoxical increase in heart rate before conversion to sinus rhythm in patients with recent-onset atrial fibrillation. Am J Cardiol 65:930–2

Lange G (1965) Action of driving stimuli from intrinsic and extrinsic sources on in situ cardiac pacemaker tissues. Circ Res 17:449–59

Langendorf R, Pick A, Katz LN (1965) Ventricular response in atrial fibrillation. Role of concealed conduction in the AV junction. Circulation 32:69–75

Leier CV, Johnson TM, Lewis RP (1979) Uncontrolled ventricular rate in atrial fibrillation. A manifestation of dissimilar atrial rhythms. Br Heart J 42:106–9

Malik M, Camm AJ (1989) Computer simulation of myocardial fibrillation using a one-dimensional model of excitation and recovery processes. Cardiovasc Res 23:132–44

Meijler FL, Fisch C (1989) Does the atrioventricular node conduct? Br Heart J 61:309–15

Meijler FL, Heethar RM, Harms FMA et al. (1982) Comparative atrioventricular conduction and its consequences for atrial fibrillation in man. In: Kulburtus HE, Olsson SB, Schlepper M (eds) Atrial fibrillation. AB Hassle, Molndal, pp 72–80

Meijler FL, van der Tweel I, Herbschleb JN, Hauer RN, Robles de Medina EO (1985) Role of atrial fibrillation and atrioventricular conduction (including Wolff–Parkinson–White syndrome) in sudden death. J Am Coll Cardiol 5:17B–22B

Mendez C, Gruhzit CC, Moe GK (1956) Influence of cycle length upon refractory period of auricles, ventricles, and A-V node in the dog. Am J Physiol 184:287–95

Milstein S, Klein GJ, Rattes MF, Sharma AD, Yee R (1987) Comparison of the ventricular response during atrial fibrillation in patients with enhanced atrioventricular node conduction and Wolff–Parkinson–White syndrome. J Am Coll Cardiol 10:1244–8

Moe GK, Abildskov JA (1964) Observations on the ventricular dysrhythmia associated with atrial fibrillation in the dog heart. Circ Res 14:447–60

Moe GK, Rheinboldt WC, Abildskov JA (1964) A computer model of atrial fibrillation. Am Heart J 67:200–20

Moe GK, Childers RW, Meredith J (1968) An appraisal of "Supernormal" A-V conduction. Circulation 38:5–28

Moore EN (1967) Observations on concealed conduction in atrial fibrillation. Circ Res 21:201–8

Moore EN, Spear JF (1971) Experimental studies on the facilitation of AV conduction by ectopic beats in dogs and rabbits. Circ Res 29:29–39

Morady F, Dicarlo LA, Krol RB, de Buitleir M, Baerman JM (1986) An analysis of post-pacing R-R intervals during atrial fibrillation. PACE 9:411–16

Olsson SB, Cai N, Dohnal M, Talwar KK (1986) Noninvasive support for and characterization of multiple intranodal pathways in patients with mitral valve disease and atrial fibrillation. Eur Heart J 7:320–33

Osher WJ, Cairns TW (1967) Computer simulation in the study of atrial fibrillation. J Okla State Med Assoc 60:6–10

Pritchett EL, Smith WM, Klein GJ, Hammill SC, Gallagher JJ (1980) The "compensatory pause" of atrial fibrillation. Circulation 62:1021–5

Reiner VS, Antzelevitch C (1985) Phase resetting and annihilation in a mathematical model of the sinus node. Am J Physiol 249:H1143–53

Robergé FA, Bhereur P, Nadeau RA (1971) A cardiac pacemaker model. Med Biol Eng 9:3–12

Rowland E, Curry P, Fox K, Krikler D (1981) Relation between atrioventricular pathways and ventricular response during atrial fibrillation and flutter. Br Heart J 45:83–7

Scher AM, Heethaar RM, Zimmerman ANE, Meijler FL (1976) Atrial rhythm during ventricular fibrillation in the dog. Circ Res 38:41–5

Sealy WC, Seaber AV (1979) Cardiac rhythm following exclusion of the sinoatrial node and most of the right atrium from the remainder of the heart. J Thorac Cardiovasc Surg 77:436–47

Simson (1988) A model of conduction through the N region of the AV node. In: Mazgalev T, Dreifus LS, Michelson EL (eds) Electrophysiology of the sinoatrial and atrioventricular nodes. Alan R Liss, New York, pp 97–109 (Progress in clinical and biological research, volume 275)

Ten Hoopen M (1966) Ventricular response in atrial fibrillation. A model based on retarded excitation. Circ Res 19:911–6

Urthaler F, Katholi CR, Macy J, James TN (1973) Mathematical relationship between automaticity of the sinus node and the AV junction. Am Heart J 86:189–95

Van der Pol B, van der Mark J (1928) The heartbeat considered as a relaxation oscillator, and an electrical model of the heart. Phil Mag 6:763–75

Van der Tweel LH, Meijler FL, Van Capelle FJL (1973) Synchronization of the heart. J Appl Physiol 34:283–87

Van der Tweel I, Herbshleb JN, Meijler FL (1983) A transfer function model for AV conduction in the human heart. Automedica 4:251-256

Van der Tweel I, Herbschleb JN, Borst C, Meijler FL (1986) Deterministic model of the canine atrioventricular node as a periodically perturbed, biological oscillator. J Appl Cardiol 1:157–73

Wellens HJ, Durrer D (1974) Wolff–Parkinson–White syndrome and atrial fibrillation. Relation between refractory period of accessory pathway and ventricular rate during atrial fibrillation. Am J Cardiol 34:777–82

Winfree AT (1987) When time breaks down. The three-dimensional dynamics of electrochemical waves and cardiac arrhythmias. Princeton University Press, Princeton

Wittkampf FHM, de Jongste MJL, Lie HI, Meijler FL (1988) Effect of right-ventricular pacing on ventricular rhythm during atrial fibrillation. J Am Coll Cardiol 11:539–45

Zipes DP, Mendez C, Moe GK (1973) Evidence for summation and voltage dependency in rabbit atrioventricular nodal fibers. Circ Res 32:170–7

Chapter 4

The Measurement of Cardiac Output in Sinus Rhythm and Atrial Fibrillation

Assessment of Cardiac Function

The Frank–Starling cardiac function curve is of central importance in understanding how cardiac function is assessed (Fig. 4.1). As cardiac function declines the curve relating stroke output and left ventricular end-diastolic pressure shifts to the right and has a lower slope, so that stroke output is reduced for the same filling pressure or filling pressure is increased for the same stroke output. Complete definition of cardiac function requires the measurement of stroke output over a range of end-diastolic pressures to enable the individual curve to be

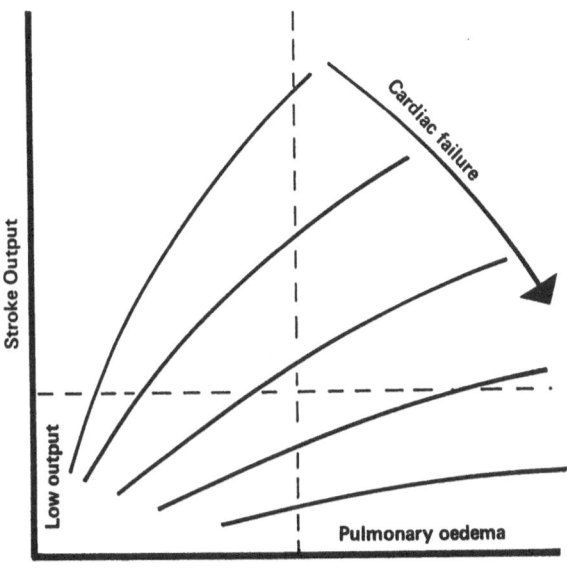

Fig. 4.1. Frank–Starling curves of left ventricular function. (Reproduced by courtesy of the Editor, *Scottish Medical Journal.*)

drawn. However, with lesser degrees of cardiac depression filling pressure remains normal and stroke output falls, and with more severe cardiac failure the rise in filling pressure that is found does not fully restore stroke output to normal. Therefore, although being only one dimension of a two dimensional situation, stroke output provides useful information about cardiac function. Provided the circulation is adequately filled, a low stroke output indicates reduced cardiac function, while an above normal stroke output is attributable to increased circulatory volume and filling pressure. A normal stroke output may result from the coincidence of depressed cardiac function and an overfilled circulation, but is much more likely to indicate normal cardiac function.

The Assessment of Cardiovascular Function

While measurement of stroke output gives an indication of left ventricular function, the measurement of cardiac output (stroke output × heart rate = minute output) reflects the function of the cardiovascular system as a whole, including the heart rate control system.

There would seem to be a case for the measurement of stroke and minute output in all clinical situations in which cardiovascular function is being considered. Unfortunately, measurement of volumetric cardiac output is extraordinarily difficult so most clinicians have never made the measurement or used it in making clinical decisions. Consequently, the view has developed that even if cardiac output could be measured simply and reliably the information would be of little clinical value.

The reader is asked to suspend judgement on the case for measuring stroke and minute output while the argument is developed that cardiac output is more usefully considered as a linear than as a volumetric measure.

In atrial fibrillation, patient management may be helped by a knowledge of minute output, but in addition there is a great deal of information of potential value encoded in the beat-to-beat variation of stroke output. In this chapter the available methods of measuring both volumetric and linear cardiac output will first be reviewed, and then the application of linear cardiac output measurement to atrial fibrillation will be discussed in chapters 5 and 6.

Measuring Volumetric Cardiac Output by the Fick Method

The Fick principle underlies the measurement of volumetric cardiac output by the direct Fick method and by dye-dilution and thermodilution. If a known quantity of a marker substance enters or leaves the circulation then from the difference in concentrations in the blood above and below the point of entry or exit the volume of blood flow past that point may be calculated. In the direct Fick method of measuring cardiac output, oxygen is the marker which enters the circulation through the lungs, oxygen saturation being measured in mixed venous blood from the pulmonary artery and from a systemic arterial sample. The method is therefore highly invasive, requiring right heart catheterisation and arterial sampling, and also a complete expired-gas collection for several minutes. The latter requires a high degree of patient cooperation unless the patient is intubated.

The direct Fick method, with its underlying assumption of steady state conditions, is the reference method for volumetric cardiac output measurement, but it is ill-suited to the purpose since the marker enters the circulation phasically with respiration and flow is pulsatile.

The Indirect Fick Technique

In the indirect Fick method carbon dioxide may be used as the marker, leaving the circulation through the lungs in a quantity which may be measured on-line with a spirometer and infrared absorption analyser (Russell et al. 1990). By this means also the end-expiratory carbon dioxide partial pressure is measured, and this is taken to be the same as the arterial partial pressure, from which arterial concentration of carbon dioxide is calculated. Mixed venous carbon dioxide concentration may be similarly calculated from the partial pressure in expired gas equilibrated by the rebreathing technique. Thus, using the Fick principle cardiac output may be measured non-invasively. In neither direct nor indirect methods is information obtained about beat-to-beat variation of output.

Thermodilution

Because of its convenience thermodilution has come to be the standard clinical method of measuring volumetric cardiac output, having superseded dye-dilution with which it was often compared on its introduction. A bolus of the indicator is injected into the circulation and the area under the concentration–time curve downstream is measured (Fig. 4.2).In dye- dilution the quantity of indicator is accurately known, and in thermodilution allowance is made for the amount lost from the circulation, but there remains the second problem inherent in methods based on the Fick principle: that of pulsatile flow. With continuous flow the fall in the concentration of the indicator is exponential, so that the total area under the curve may be calculated from the early part of the curve. This obviates the critical dependency on measurement of vanishingly small differences from the baseline concentration for calculating the area under the tail of the curve. However, pulsatile flow superimposes a stepwise fall on the exponential decline making estimation of the exponential rate of decline inaccurate (Fig. 4.2). The steps may be smoothed out by altering the time-constant of the thermistor circuitry, at the cost of altering the shape of and area under the curve.

A further, neglected consideration with pulsatile flow is that of sampling error. If the total area under the concentration– time curve is calculated from the first part of the curve, the eventual calculation of volumetric cardiac output per minute is based on a time sample of just a few seconds. This may introduce considerable error unless the time-sample is a multiple of the cardiac cycle length. Otherwise, if the time-sample commences with systole cardiac output will be overestimated and if it commences with diastole it will be underestimated. Clearly, the problem of the sampling error is accentuated in atrial fibrillation.

There are some theoretical doubts about the validity of applying the Fick principle to pulsatile flow, but even if it is theoretically correct, in practice, as a

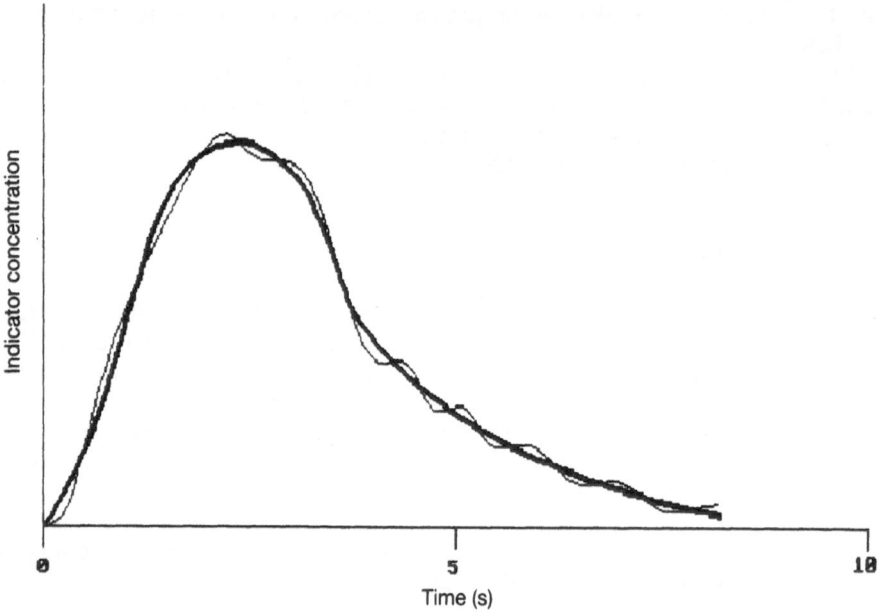

Fig. 4.2. Indicator-dilution curves for continuous flow (*thick line*), and pulsatile flow (*fine line*).

method of measuring volumetric cardiac output, thermodilution is compromised. Nevertheless, this is the method against which the Doppler ultrasound technique has most often been compared.

Reproducibility of Thermodilution

The performances of three different thermodilution cardiac output computers and a Doppler ultrasound method of measuring blood flow have been compared in an artificial circulation with continuous or pulsatile flow (Mackenzie et al. 1986; McLennan et al. 1986). As a reference, absolute flow rates were accurately measured by timed blood volume collection. The overall performance of each device was summarised as the percentage range of flow rates that could give rise to identical measurements, this statistic reflecting the ability of the device to detect a change in flow rate, embodying both the repeatability of the measurement and the slope of its relation to flow rate. The best thermodilution cardiac output computer gave a 30% range of flow rates which could give rise to an identical measurement (Fig. 4.3) ; the corresponding figures for the other computers were 40% and 62%. It was interesting to observe that in all three computers the slope of the regression line relating the measurement to the flow rate was greater than 1, flow rate being overestimated. This has the effect of improving the equipment's ability to detect a change in flow rate. These results in vitro provide a yardstick for comparison with linear cardiac output measured by Doppler ultrasound.

In vivo thermodilution has also been shown to overestimate flow compared with both dye dilution and the indirect Fick technique using carbon dioxide as the marker (Russell et al. 1990).

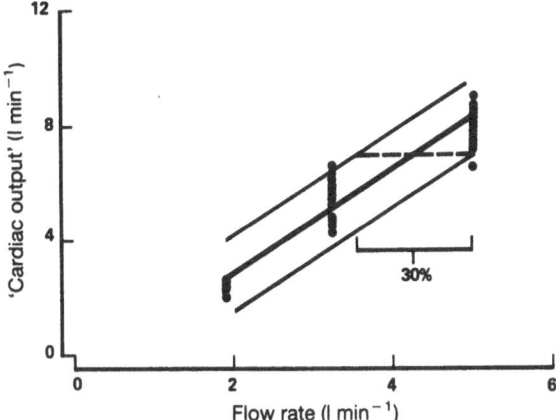

Fig. 4.3. Regression between blood flow rate and its measurement by thermodilution cardiac output computer($r = 0.96$, $p<0.001$). Regression line and 95% confidence limits of a single prediction are shown. Horizontal distance between confidence limits is expressed as percentage of standard flow rate of 5 l min^{-1} (Reproduced by permission of the Editor, *British Heart Journal.*)

Measurement of Cardiac Output by Echocardiography and Doppler Ultrasound

A number of groups have shown that it is possible to measure left ventricular stroke volume non-invasively using the combination of Doppler ultrasound and echocardiography (Magnin et al. 1981; Goldberg et al. 1982; Huntsman et al. 1983; Chandraratna et al. 1984; Rose et al. 1984; Ihlen et al. 1984; Mehta et al. 1985; Levy et al. 1985). The principle of the method is illustrated in Fig. 4.4. Imagine a single left ventricular stroke volume filling the aorta from the aortic valve orifice to a point around the aortic arch. Stroke volume is the product of this stroke distance and the aortic cross- sectional area. Stroke distance is not measured as such but can be calculated from the systolic velocity-integral at a point where the ultrasound beam is in line with the direction of flow. The diagram shows that ultrasound access to a wide area is possible from the suprasternal notch, and alternative directions of the transducer readily permit the measurement of stroke distance either in the aortic root or in the aortic arch, similar values being obtained in either site.

There are three important assumptions implicit in the method. Firstly, that stroke distance used in the calculation is representative of blood flow as a whole, as it would be if the flow profile were flat. That is a simplification since the flow profile is skewed as indicated by the line of arrowheads, or under some circumstances may be parabolic or even disturbed. The second assumption is that total stroke volume passes the point at which stroke distance is measured. This too is not the case since a proportion of stroke volume enters the coronary circulation from the aortic root and a larger proportion is lost to the head and neck before velocity is measured in the aortic arch. The third assumption is also unwarranted: it is that the aortic cross-section has a regular geometrical shape the area of which can be readily calculated and is constant throughout the cardiac cycle. In spite of the approximations inherent in the method, several groups have

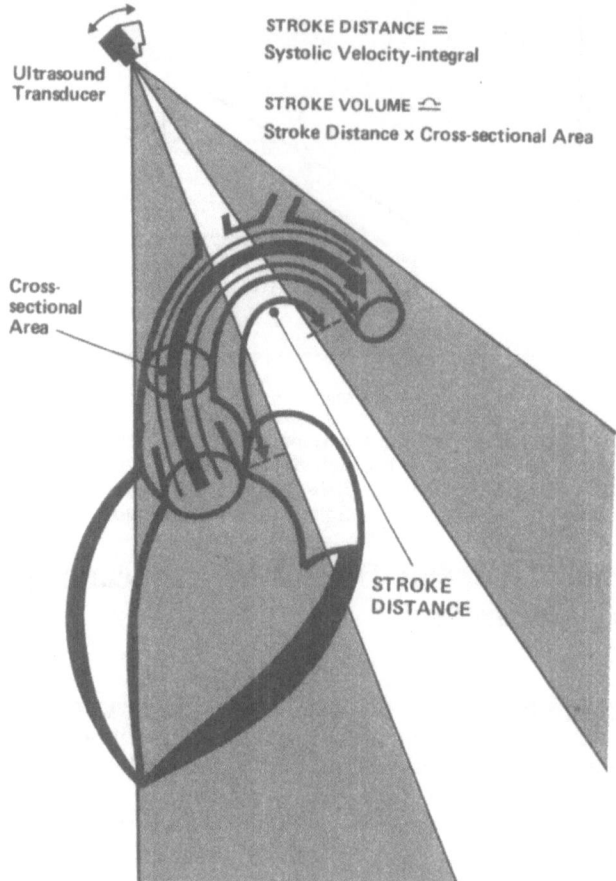

Fig. 4.4. Method of measuring cardiac output by ultrasound (Reproduced by permission of the Editor, *Scottish Medical Journal*.)

now achieved results that correlate closely with cardiac output measured invasively. Perfect correlation between cardiac output measured by ultrasound and by invasive methods cannot be obtained since the reference methods – Fick, dye-dilution or thermodilution – are themselves poorly reproducible with an error not less than about 0.5 litres, or 10% of a cardiac output of 5 litres per minute.

Measurement of Aortic Cross-Sectional Area

The accuracy of the ultrasonic measurement of cardiac output is limited by that of each of its two components: stroke distance and aortic cross-sectional area. The accuracy of measurement of aortic size is limited by axial resolution, which with modern transducers is about 1 mm, or 3% of a typical aortic diameter of 30 mm; the accuracy of measurement of cross-sectional area is therefore at the very best about 7%. But the measurement of aortic cross-sectional area is also greatly influenced by the alignment of the ultrasound beam, which if not at a right angle to the vessel may give rise to substantial error. In this respect cross-sectional echocardiography is superior to A- or M-mode. But which diameter should be used for the calculation of aortic cross-sectional area? All diameters have been used, that of the aortic valve orifice (Ihlen et al. 1984), of the sinus of Valsalva at the level of the maximum opening of the valve leaflets (Goldberg et al. 1982), of a point just above the valve leaflets (Chandraratna et al. 1984; Rose et al. 1984), and of the narrowest point of the sino-tubular junction (Huntsman et al. 1983). The use of such a variety of sites, between which there are substantial differences in cross-sectional area, leads to the suggestion that the aortic cross-sectional area that provides the best correlation with the reference measurement is determined to a large extent by the method of obtaining the systolic velocity- integral.

Studies of the reproducibility of measurements of aortic size are few, but when cross-sectional echocardiographic measurements were compared with direct measurement made at the time of surgery for aortic valve replacement, the correlation was poor with a coefficient of 0.68 (Mackay et al. 1985), and in a study using M-mode echocardiography the correlation coefficient was 0.7 (Francis et al. 1975). In a comparison of preoperative cross- sectional measurements and M-mode echocardiography of the aortic root and replacement aortic valve size, the correlation coefficients were 0.89 and 0.14 respectively, with standard errors of the mean of 0.68 and 1.49 mm (Cohen et al. 1984). In a study on 8 healthy subjects Robson et al. (1988) obtained coefficients of variation for measurement of aortic cross- sectional area and systolic velocity integrals of 8.4% and 6.4% respectively.

Measurement of Stroke Distance

In order to measure volume flow in the aorta an estimate of the mean spatial velocity is required. While mean spatial velocity is related to the systolic velocity-integral, the derivation of the former from the latter presents formidable technical difficulties which are usually circumvented. Of the measurements that may be made on aortic blood flow, the one that is least arbitrary, and least liable to artefactual distortion is that of maximum velocity obtained from the envelope of the spectral analysis of a continuous signal. Because the flow profile is nearly flat it is representative of the mean spatial velocity, and it has been used in its stead to give a good correlation with volumetric flow.

Maximum velocity of aortic blood flow is seen in midstream where, in the absence of outflow tract obstruction, flow is not usually disturbed. The systolic velocity-integral of maximum velocity is a distance: stroke distance. Although there is little direct evidence on this point, stroke distance would be expected to be identical at various sites in the aorta. If midstream blood moved further in the proximal than in the distal ascending aorta, disturbed lateral movement would be required distally to accommodate the excess. Since this is not seen, the systolic velocity-integral of maximum velocity may therefore be measured at the aortic root or in the arch, very similar values being seen at either site (Gisvold and Brubakk 1982; Fisher et al. 1983; Mayo and Rawles 1991).

Measurement of stroke distance is much more conveniently made in the aortic arch because ultrasound access is easier and causes less patient discomfort.

Another important reason for measuring stroke distance in the arch rather than in the root is that by using continuous wave Doppler without axial resolution correct alignment of the ultrasound beam with the direction of flow is assured at some point around the arch. Moreover, the reproducibility of the measurement is higher in the arch than in the ascending aorta (Mayo and Rawles 1991).

Studies of reproducibility of the measurement of stroke distance report values of 3.2%–3.8% (Gardin et al. 1984), 5.3% (McLennan et al. 1986), 6.0%–7.0% (Fraser et al. 1976), 8.4%–9.1% (Gisvold and Brubakk 1982), and 9.1% (Hatle and Angelsen 1982).

Mayo and Rawles (1991) show that reproducibility is greatly influenced by equipment design, being best (4.8%) with a broad beam of continuous wave ultrasound directed at the aortic arch, manual evaluation, and averaging of an indefinite number of systolic velocity integrals.

Linear Cardiac Output

The measurement of aortic cross-sectional area, besides limiting the accuracy of the measurement of volumetric cardiac output by ultrasound, imposes a need for highly trained personnel and expensive, sophisticated imaging equipment. However, the experience with the combination of Doppler ultrasound and echocardiography provides validation for the use of Doppler ultrasound by itself to measure what we have called "linear cardiac output". Linear cardiac output can be measured with excellent reproducibility by relatively unskilled personnel using non-imaging equipment.

Our initial experience was with the transcutaneous aortovelograph, a continuous wave Doppler apparatus which produces a broad beam of ultrasound directed at the aortic arch from the suprasternal notch. Fig. 4.5 shows the paper trace produced in real-time, showing aortic blood velocity against time. It is the result of spectral analysis by a group of band-pass filters, the amplitude in each frequency channel being recorded in a grey scale by a separate pen. The envelope of each systolic velocity complex represents the highest velocity within the ultrasound beam at any moment, and the area within the envelope is the systolic velocity-integral or stroke distance, pertaining to maximum and presumably midstream blood flow. The product of stroke distance and heart rate gives minute

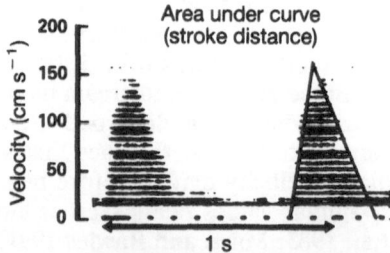

Fig. 4.5. Hard copy produced by transcutaneous aortovelograph, showing aortic blood velocity against time. (Reproduced by permission of the Editor, *Scottish Medical Journal*.)

distance; stroke and minute distance are linear measures of stroke volume and cardiac output respectively. As we have seen:

$$\text{Stroke volume} = \text{stroke distance} \times \text{aortic cross-sectional area} \qquad (4.1)$$

As an absolute measure, stroke volume is uninterpretable without knowledge of body size. By convention, stroke volume is adjusted for variation of body size by dividing it by body surface area, to give stroke index:

$$\text{Stroke volume/body surface area} = \text{stroke index} \qquad (4.2)$$

However, instead of body surface area, an area of part of the body that is representative of the whole could be used to characterise body size. Rearrangement of equation (4.1) gives

$$\text{Stroke volume/aortic cross-sectional area} = \text{stroke distance} \qquad (4.3)$$

Thus, comparing equations (4.2) and (4.3) we see that stroke distance and stroke index are very similar in form, being stroke volume divided by an area characterising body size. Since stroke distance is so closely related to stroke index it could be used in its stead as an absolute measure of cardiac output or for following serial changes. Having obtained stroke distance it seems unnecessary to determine its product with aortic cross-sectional area, adding to the technical complexity and degrading the reproducibility of the measurement in the process, only to then divide the result by body surface area to obtain stroke index.

For stroke distance to behave like stroke index there are two requirements: firstly, aortic cross-sectional area has to be related to body surface area and, secondly, aortic size has to be constant or else to change in such a way that a linear relationship between stroke distance and stroke volume is maintained, in particular with changes of blood pressure.

Relationship Between Body Surface Area and Aortic Cross-Sectional Area

That big people have big aortas is a common sense belief for which there is ample evidence. Body surface area and aortic diameter are highly correlated, though in a non-linear fashion and not always very closely, even in children during the period of growth (Feigenbaum 1976; Henry et al. 1978; Rogé et al. 1978). Some of the spread in the relationship between aortic diameter and body surface area is due to shortcomings of the latter as a means of assessing body size and metabolic rate. It is sufficient to note that a range of heights from 100 cm to 200 cm, and of weights from 50 kg to 150 kg, can all give rise to the same normal body surface area of 1.73 m^2 it is hard to believe that such a wide range of body habitus is metabolically equivalent. It is, perhaps, easier to believe that the cross-sectional area of the conduit which supplies the body's total metabolic requirements is more closely related to those requirements than is body surface area.

Effect of Age and Blood Pressure on Aortic Size and Stroke Distance

Another factor which will influence the spread of the relationship between body surface area and aortic cross-sectional area is the age structure of the population studied. In adults there is a progressive dilatation of the aorta with age over 40 which is sufficient to account for the observed fall in aortic blood velocity (Towfiq et al. 1986).

The effect of increased aortic size is to reduce aortic blood velocity for the same stroke volume, so if the relationship between stroke distance and stroke volume is plotted at different ages and constant blood pressure we find a series of straight lines radiating from the origin: the greater the age, the greater the aortic cross-sectional area and the lower the stroke distance for a given stroke volume (Fig. 4.6).

From a knowledge of the compliance of the aorta at various ages the effect of blood pressure on aortic size may be calculated. Fig. 4.7 shows that at each age there is a near-linear relationship between blood pressure and cross-sectional area, and at any constant blood pressure below 150 mmHg, cross-sectional area

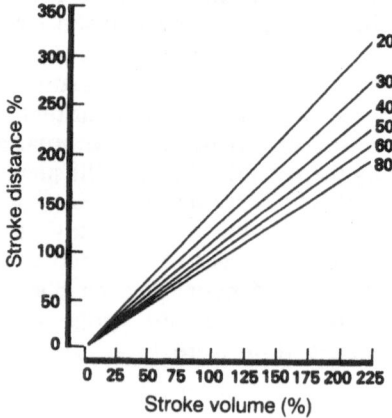

Fig. 4.6. Predicted relationship between stroke distance and stroke volume at various ages. Blood pressure constant.

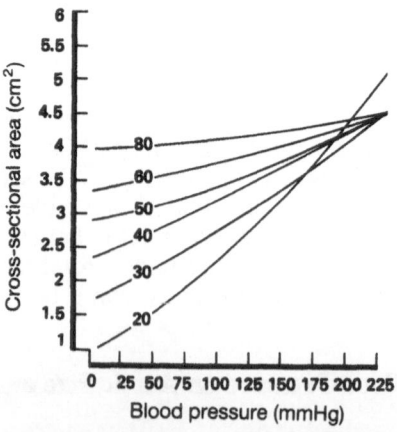

Fig. 4.7. Predicted effect of blood pressure on cross-sectional area of aorta at various ages. (Reproduced by permission of the Editor, *British Heart Journal*.)

increases with age. However, at very high pressures all aortas distend to a similar size, except for those of the youngest subjects (aged 20) which distend more than those of older subjects.

Fig. 4.8 shows the relationship between stroke distance and stroke volume not at constant blood pressure, but at constant peripheral resistance. Blood pressure now changes proportionately with stroke volume and the calculated effect on stroke distance is seen. At lower pressures the lines fan out from the origin, as they did when blood pressure was constant, the greater the age the less steep the slope. At higher pressures, as aortic sizes converge, so do the slopes, except for that of the 20-year-old which flattens below those of older age groups. In the youngest age the distensibility of the aorta results in a non-linear relationship between stroke distance and stroke volume over a range of blood pressures from 0 to 225 mmHg. The departure from exact linearity over a more limited range of blood pressures would have little practical significance even at this age, and this non-linearity disappears in older subjects until at age 80 the aorta behaves like a rigid pipe. More disadvantageous is the buffering effect of the highly elastic aorta at age 20, which reduces the slope of this relationship so that at a blood pressure of 100 mmHg only 34% of a change of stroke volume is reflected in a change in stroke distance, reducing the ability of the method to detect a change of flow rate (Fig. 4.9). The equivalent figures at age 50 and age 80 are 82% and 91% respectively. At every age the graph of the relationship between stroke distance and stroke volume in the region of a blood pressure of 100 mmHg has a slope of less than 1.

It may be concluded that with constant blood pressure there is a linear relationship between stroke distance and stroke volume at all ages, but that dilatation of the aorta with age makes age adjustment of velocity measurements necessary. The linear relationship between stroke distance and stroke volume is preserved if there is a moderate accompanying change in blood pressure, though the slope is substantially reduced in the youngest subjects.

Thus, the two conditions necessary for stroke distance to behave as stroke

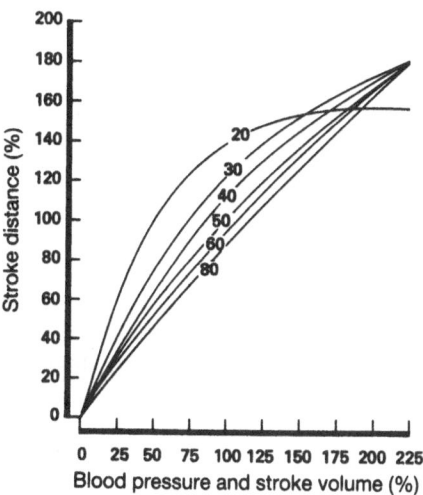

Fig. 4.8. Predicted relationship between stroke distance and stroke volume at various ages. Peripheral resistance constant. (Reproduced by permission of the Editor, *British Heart Journal.*)

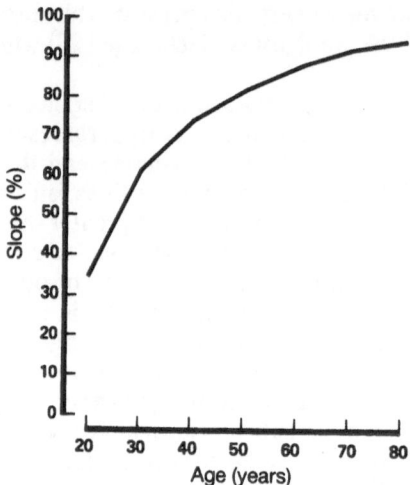

Fig. 4.9. The slope of the curves depicted in Fig. 4.8 at a blood pressure of 100 mmHg and a stroke volume of 100%. (Reproduced by permission of the Editor, *British Heart Journal*.)

index – aortic cross-sectional area being related to body surface area, and aortic size changing in such a way that a linear relationship between stroke distance and stroke volume is maintained – have been met.

Expectations of Linear Cardiac Output

If we use stroke distance by itself as a linear measure of cardiac output, three expectations of the measurement follow.

1. Within subjects stroke distance should be proportional to stroke volume, and minute distance to cardiac output.
2. Stroke distance should be independent of body size.
3. Stroke distance and its product with heart rate should provide absolute measures of cardiac output, analogous to stroke index and cardiac index.

Relationship Between Stroke Volume and Stroke Distance

The first of these three expectations, that minute distance should be proportional to cardiac output, was confirmed by incorporating a pig aorta into an artificial circulation (McLennan et al. 1986). A completely linear relationship was found between minute distance and absolute flow rate over a range of flows from 1 to 6 litres/minute (Fig. 4.10). The flow rate was altered by changing pulse volume, pulse rate being kept constant, so there is a similar linear relationship between stroke distance and stroke volume.

In studies in humans in which changes in stroke volume have been compared with changes in stroke distance measured by transcutaneous aortovelography,

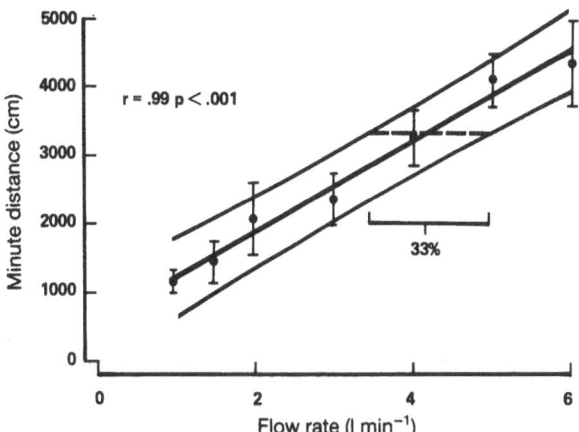

Fig. 4.10. Minute distance at various flow rates in a pig aorta included in an artificial circulation. Horizontal distance between confidence limits expressed as percentage of standard flow rate of 5 l min^{-1}. $r = 0.99$, $p<0.001$. Compare with Fig. 4.3. (Reproduced by permission of the Editor, *British Heart Journal.*)

stroke distance is closely proportional to stroke volume, the deviation from exact proportionality ranging from 9% to 13% (Light et al. 1986).

Stroke Distance and Body Size

Fig. 4.11 shows a complete absence of any relationship between body surface area and stroke distance in 140 normal adults. Neither was there any relationship between stroke distance and height or weight, nor between minute distance and these measures of body size.

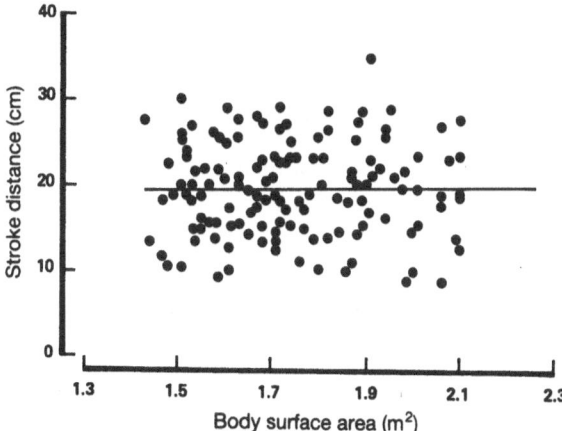

Fig.4.11. Stroke distance and body surface area in 140 healthy subjects. $r = 0.01$, p not significant. (Reproduced by permission of the Editor, *Scottish Medical Journal.*)

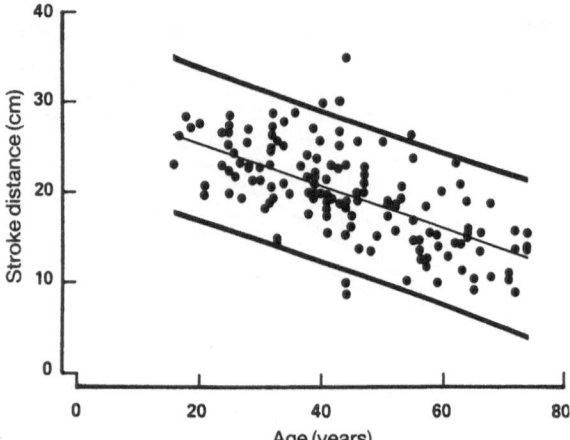

Fig.4.12. Regression line and 95% confidence limits for stroke distance and age in 140 healthy subjects. $y = 30.5 - 0.24x$; $r = 0.66$, $p < 0.001$. (Reproduced by permission of the Editor, *Scottish Medical Journal*.)

An Absolute Measure of Cardiac Output

The third expectation we had of stroke distance was that it should provide an absolute measure of cardiac output, enabling cardiac output in different individuals or groups to be compared meaningfully. By implication, there is a population norm for stroke and minute distance of which similar estimates should be obtained in different centres.

Fig. 4.12 shows stroke distance recorded from 140 healthy subjects. Very similar mean values have been recorded in other centres using different machines (Light et al. 1986). Due to the increase in aortic size with age there is a highly significant negative relationship with age for which the regression line and the 95% confidence limits for a single prediction are indicated. From this regression the normal stroke distance at any age may be predicted, for comparison with any individual or group, and this has been done for six patient groups of various mean ages (Fig. 4.13). The diagnoses of the patients in the groups were pure, with no overlap. Stroke distance is increased in pregnancy, normal in hypertension, reduced in cardiac failure, increased in anaemia, normal in convalescence from uncomplicated myocardial infarction, and reduced in atrial fibrillation. In each group stroke distance is altered in the same direction and to the same extent as would be expected of a measure of cardiac output.

Thus, all three expectations of the measurement have been fulfilled.

Comparison between Thermodilution and Doppler Ultrasound

From Fig. 4.10, the horizontal distance between the confidence limits is seen to be 33% of a standard flow rate of 5 litres per minute. This represents the ability of

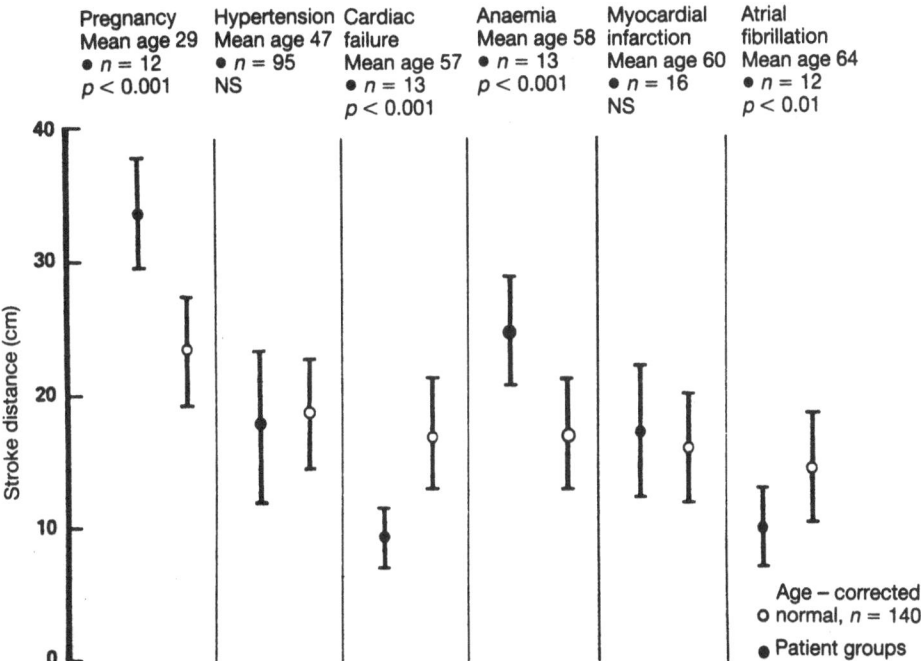

Fig.4.13. Mean (±1 SD) stroke distance in six patient groups compared with age-predicted normal values (Reproduced by permission of the Editor, *Scottish Medical Journal.*)

the Doppler method to detect a change in flow rate and may be directly compared with thermodilution (Fig. 4.3), where the corresponding figures were 30%, 40% and 62% for three different cardiac output computers. Although the repeatability of the Doppler measurement is as good as the best thermodilution results, the regression line has a slope of less than 1 with a positive intercept which slightly increases the horizontal distance. The pig aorta in this study behaved like a 50 year old aorta depicted in Figs. 4.8 and 4.9, the flattening of the slope being due to the change in aortic dimension with change in flow rate under conditions of constant peripheral resistance.

Relationship between Linear and Volumetric Cardiac Output

From the equations for stroke distance and stroke index (p. 85) it may be seen that the relationship between the linear and volumetric measures of cardiac output embodies the relationship between aortic cross-sectional area and body surface area, and also that between the systolic velocity-integral and mean spatial velocity. As we have seen, there is a good deal of uncertainty and variability in both these areas, so although we would expect linear and volumetric cardiac outputs to be highly correlated, we would not expect the correlation coefficient to be close enough to 1 to allow direct conversion from one to the other.

A correlation coefficient of 0.82 has been documented for the relation between stroke distance and stroke index (Northridge et al. 1990), though our own work suggests a value closer to 0.7. This relatively low value is not a reason for abandoning linear cardiac output measurement, since good theoretical and practical arguments have been advanced for its use. Rather, volumetric and linear measurements should be seen as equally valid ways of assessing cardiac output.

An analogy may be useful here. In any group of normal subjects there is a highly significant relationship between height and weight, though the correlation coefficient is much less than 1, so that direct conversion of one to the other is not possible. This closely mirrors the relationship between volumetric and linear cardiac output. If we wished to measure growth, and if the measurement of weight were invasive, technically difficult and poorly reproducible, then we could without detriment use height as an alternative.

Application of Linear Cardiac Output Measurement to Atrial Fibrillation

The main obstacles to the wider acceptance of linear cardiac output into clinical practice are conceptual conservatism and, currently, the lack of suitable up-to-date equipment that is dedicated to the single function of measuring stroke distance; multipurpose equipment is likely to have design compromises that adversely affect the performance of this function. There are many potential clinical and research applications for the measurement of linear cardiac output by Doppler ultrasound, some of which have been reviewed elsewhere (Rawles 1989). The potential range and volume of use for linear cardiac output measurement is sufficient justification for the development of equipment to replace the now obsolete transcutaneous aortovelograph.

In atrial fibrillation linear cardiac output measurement offers several advantages over any other method of assessing cardiac output. It is non-invasive, completely safe, very well tolerated by patients, and may be repeated as often as desired.

Average Stroke and Minute Distance in Atrial Fibrillation

Stroke and minute distances, as argued above, constitute an absolute measure of cardiac output. It is therefore legitimate to compare stroke and minute distance in patients with atrial fibrillation with age-predicted normal values for healthy subjects in sinus rhythm. This has been done for a group of 60 patients with atrial fibrillation of mixed aetiology (Rawles 1988). On average, stroke distance was 64% and minute distance 85% of the age-predicted normal values for sinus rhythm – a highly significant reduction in each case ($p<0.001$). Minute distance was reduced less than stroke distance because ventricular rate was above normal, averaging 103 beats per minute. As a rule, in atrial fibrillation *cardiac* function is more impaired than *cardiovascular* function, an increased ventricular rate partially compensating for reduced stroke output.

Beat-to-Beat Measurement of Stroke Output

The Doppler ultrasound measurement of stroke distance requires flow to be pulsatile for its determination; by contrast, pulsatility is an unwelcome complication in other methods of cardiac output measurement, contributing to the variability and inaccuracy of the results. With the Doppler technique the average output per minute (minute distance) is derived from the summation of individual stroke distances, whereas with other methods it is only possible to derive an average value for stroke output by division of cardiac output by heart rate.

We have seen how stroke distance, besides being an absolute measure of cardiac output, may readily be used for following serial changes, stroke distance being proportional to stroke volume within subjects. In atrial fibrillation, where the average age tends to be high, the detection of beat-to-beat changes in stroke volume is facilitated by low aortic compliance, which ensures that most of the changes in stroke volume will be translated into changes in stroke distance, with very little elastic buffering by the aorta (Fig. 4.9). The good reproducibility obtained with the transcutaneous aortovelograph also assists in the accurate quantification of beat-to-beat changes in cardiac output.

In the next two chapters the analysis of the beat-to-beat changes of stroke distance in atrial fibrillation will be described.

References

Chandraratna PA, Nanna M, McKay C et al. (1984) Determination of cardiac output by transcutaneous continuous-wave ultrasonic Doppler computer. Am Heart J 53:234–37

Cohen JL, Austin SM, Kim CS, Christakos ME, Hussain SM (1984) Two-dimensional echocardiographic preoperative prediction of prosthetic aortic valve size. Am Heart J 107:108–12

Feigenbaum H (1976) Echocardiography, 2nd edn. Lea and Febiger, Philadelphia

Fisher DC, Sahn DJ, Friedman MJ et al. (1983) The effect of variations on pulsed Doppler sampling site on calculation of cardiac output: an experimental study in open chest dogs. Circulation 67:370–6

Francis GS, Hagan AD, Oury J, O'Rourke RA (1975) Accuracy of echocardiography for assessing aortic root diameter. Br Heart J 37:376–8

Fraser CB, Light LH, Shinebourne EA, Buchtal A, Healy MJR, Beardshaw JA (1976) Transcutaneous aortovelography: reproducibility in adults and children. Eur J Cardiol 4/2:181–9

Gardin JM, Dabestani A, Matin K, Allfie A, Russell D, Henry WL (1984) Reproducibility of Doppler aortic blood flow measurements: studies on intraobserver, interobserver and day-to-day variability in normal subjects. Am J of Cardiology 54:1092–8

Gisvold SE, Brubakk AO (1982) Measurement of instantaneous blood- flow velocity in the human aorta using pulsed Doppler ultrasound. Cardiovasc Res 16:26–3

Goldberg SJ, Sahn DJ, Allen HD, Valdes-Cruz LM, Hoenecke H, Carnahan Y (1982) Evaluation of pulmonary and systemic blood flow by 2-dimensional Doppler echocardiography using fast Fourier transform spectral analysis. Am J Cardiol 50:1394–1400

Hatle L, Angelsen B (1982) Doppler ultrasound in cardiology. Physical principles and applications. Lea and Febiger, Philadelphia

Henry WL, Ware J, Gardin JM, Hepner SI, McKay J, Weiner M (1978) Echocardiographic measurements in normal subjects. Growth-related changes that occur between infancy and early childhood. Circulation 57:278–85

Huntsman LL, Stewart DK, Barnes SR, Franklin SB, Colocousis JS, Hessel EA (1983) Noninvasive Doppler determination of cardiac output in man. Clinical validation. Circulation 67:593–602

Ihlen H, Amlie JP, Dale J et al. (1984) Determination of cardiac output by Doppler echocardiography. Br Heart J 51:54–60

Levy BI, Payen DM, Xhaard M, McIlroy MB (1985) Non-invasive ultrasonic cardiac output measurement in intensive care unit. Ultrasound Med Biol 11:841–9

Light LH, Cross G, Rawles JM, Haites NE (1986) Convenient monitoring of cardiac output and global left ventricular function by transcutaneous aortovelography – an update. In: Spencer MP (ed) Cardiac Doppler diagnosis, vol. II. Martinus Nijhoff, Dordrecht pp 73–91

Mackay A, Been M, Rodrigues E, Murchison J de Bono DP (1985) Preoperative prediction of prosthesis size using cross-sectional echocardiography in patients requiring aortic valve replacement. Br Heart J 53:507–9

Mackenzie JD, Haites NE, Rawles JM (1986) Method of assessing the reproducibility of blood flow measurement: factors influencing the performance of thermodilution cardiac output computers. Br Heart J 55:14–24

Magnin PA, Stewart JA, Myers S, von Rammm O, Kisslo JA (1981) Combined Doppler and phased array echocardiographic estimation of cardiac output. Circulation 63:388–92

Mayo A, Rawles JM (1991) Comparison of four different Doppler instruments used to measure linear and volumetric cardiac output: a study of reproducibility and agreement. Ultrasound Med Biol (in press)

McLennan FM, Haites NE, Mackenzie JD, Daniel MK, Rawles JM (1986) Reproducibility of linear cardiac output measurement by Doppler ultrasound alone. Br Heart J 55:25–31

Mehta N, Iyawe VI, Cummin ARC, Bayley S, Saunders KB, Bennett ED (1985) Validation of a Doppler technique for beat-to-beat measurement of cardiac output. Clinical Science 69:377–82

Northridge DB, Findlay IN, Wilson J, Henderson E, Dargie HJ (1990) Non-invasive determination of cardiac output by Doppler echocardiography and electrical bioimpedance. Br Heart J 63:93–7

Rawles JM (1988) A mathematical model of left ventricular function in atrial fibrillation. Int J Biomed Comput 23:57–68

Rawles JM (1989) Stroke distance – an improved measure of cardiovascular function. Scot Med J 34:394–8

Robson SC, Murray A, Peart I, Heads A, Hunter S (1988) Reproducibility of cardiac output measurement by cross-sectional and Doppler echocardiography. Br Heart J 59:680–4

Roge CLL, Silverman NH, Hart PA, Ray RM (1978) Cardiac structure growth pattern determined by echocardiography. Circulation 57:285–90

Rose JS, Nanna M, Rahimtoola SH, Elkayam U, McKay C, Chandraratna AN (1984) Accuracy of determination of changes in cardiac output by transcutaneous continuous-wave Doppler computer. Am J of Cardiol 54:1099–1101

Russell AE, Smith SA, West MJ et al. (1990) Automated non-invasive measurement of cardiac output by the carbon dioxide rebreathing method: comparisons with dye dilution and thermodilution. Br Heart J 63:195–9

Towfiq B, Weir J, Rawles JM (1986) Effect of age and blood pressure on aortic size and stroke distance. Br Heart J 55:560–8

The Haemodynamics of Atrial Fibrillation: The Development of a Model of Left Ventricular Function

The Relationship between Time and Volume of the Pulse

Every medical student is taught that the pulse in atrial fibrillation is irregularly irregular in time and volume. We have seen how that statement needs to be qualified with regard to the timing of the pulse, but what about its volume? Is pulse volume totally irregular or are there patterns which we can recognise? In sinus rhythm pulse volume varies very little from beat to beat but in atrial fibrillation there is marked variability which is characteristic of the arrhythmia. Pulse volume reflects stroke output, which, as seen in the last chapter, may be measured as stroke distance.

Fig. 5.1 shows a sequence of stroke distances, together with the electrocardiogram, recorded from a patient with atrial fibrillation. There is striking variation in stroke distance which clearly results from variation in pulse rate, or more precisely, in the R-R intervals between beats. What is the relationship between stroke distance and the duration of preceding R-R intervals?

Every stroke distance – or left ventricular stroke volume – is determined by the preload, afterload and contractility at the time of each beat. These determinants of stroke output are in turn affected by the beat-to-beat variation in ventricular

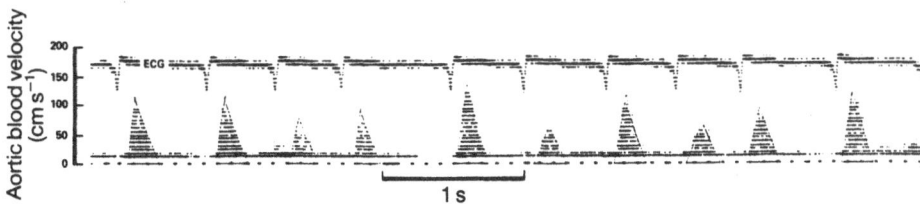

Fig. 5.1. Doppler ultrasound recording of a sequence of stroke distances, together with the electrocardiogram, in a patient with atrial fibrillation. (Reproduced by permission of the Editor, *British Heart Journal*.)

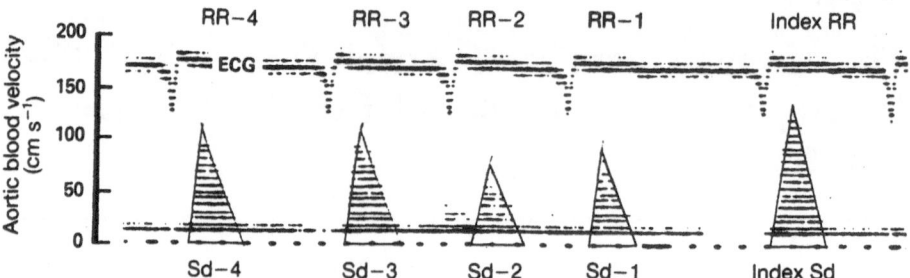

Fig. 5.2. A sequence of stroke distances showing the nomenclature for the analysis of the relation between stroke distance and preceding R-R intervals. (Reproduced by permission of the Editor, *International Journal of Biomedical Computing*.)

rate. If stroke distance could be described mathematically in terms of preceding R-R intervals, then this relationship would be a description of the physiological rules governing the operation of the ventricle – that is, left ventricular function – albeit a description in unfamiliar terms. The relationship between irregularity of timing and irregularity of volume of the pulse therefore has the potential to indicate the functional state of the left ventricle. In addition, if changes in preload, afterload and contractility are dissociated from each other, perhaps the separate contributions of these elements to left ventricular function may be determined.

Fig. 5.2 shows the nomenclature that will be used throughout this chapter. Each stroke distance in a continuous sequence of about 300 beats recorded over 3 minutes is considered in turn as the index beat in relation to the preceding four R-R intervals, referred to as RR−1 to RR−4, respectively.

Is Pulse Volume Irregularly Irregular?

Fig. 5.4 shows autocorrelograms of R-R intervals and stroke distances from a patient with atrial fibrillation. It will be recalled (from Chapter 2) that in autocorrelation each R-R interval is correlated with its successors. If the sequence of beats is truly random then none of the autocorrelation coefficients will be statistically significant. Correlation coefficients, which may range from −1 to +1, are plotted against the order of autocorrelation, 1 to 10. In a similar way each stroke distance may be correlated with its successors, 1 to 10, to give the autocorrelogram for stroke distances.

Fig. 5.3 shows how the cross-correlogram is derived; stroke distances are correlated with the R-R intervals containing them (Index R-R), preceding them (RR−1), or pre- preceding them (RR−2) and so on, and also with those that follow (RR+1 to RR+10).

In Fig. 5.4 there is a random sequence of R-R intervals, none of the autocorrelation coefficients being significantly different from 0. The sequence of stroke distances is, however, non-random, the autocorrelogram showing a highly significant negative first order coefficient ($r = -0.28$ $p<0.001$). This indicates a tendency for stroke distances to be alternately big and small – a statistical pulsus alternans.

In the cross-correlogram each stroke distance is correlated with the R-R

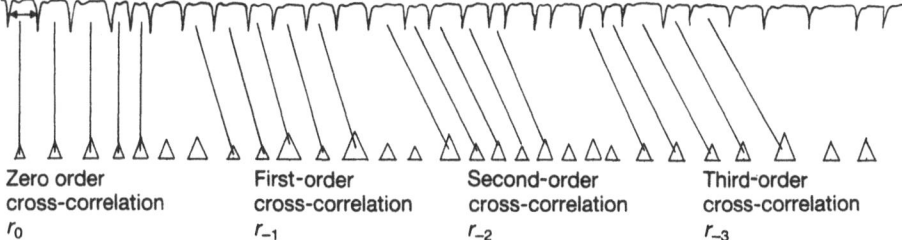

Fig. 5.3. The derivation of the cross-correlogram from the relationships between R-R intervals of the electrocardiogram (*top*) and stroke distances (*below*).

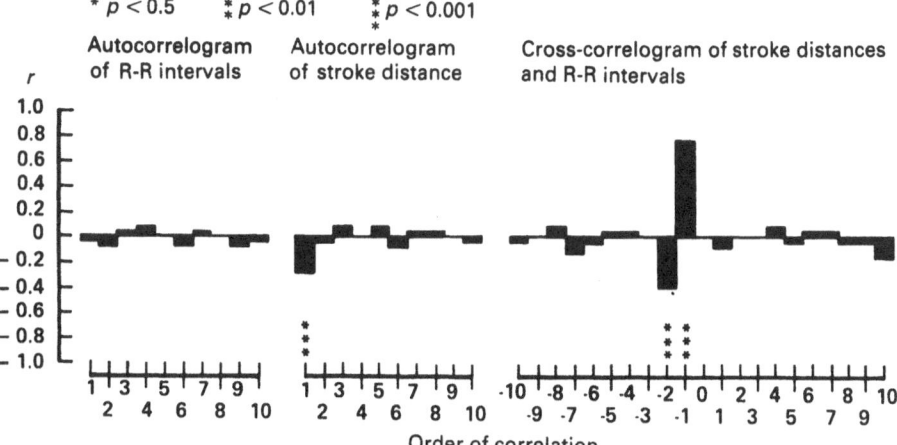

Fig. 5.4. Autocorrelograms for R-R intervals and stroke distances, and cross-correlograms between them for a patient with random rhythm and a non-random sequence of stroke distances (alternans) (Reproduced by permission of the Editor, *British Heart Journal*.)

intervals that precede it (-1 to -10) or follow it ($+1$ to $+10$). The zero-order correlation coefficient is not significant, neither is there any correlation between stroke distances and following R-R intervals. There is, however, a highly significant positive correlation ($r = 0.78, p<0.001$) between stroke distances and immediately preceding R-R intervals (RR-1 in our nomenclature), and a negative correlation ($r = -0.38, p<0.001$) between stroke distances and the pre-preceding R-R intervals (RR-2). This patient, then, has a random rhythm but a non- random sequence of stroke distances.

The results from a patient with a non-random rhythm and non- random pulse volume sequence are shown in Fig. 5.5. The non-random rhythm is indicated by a significant positive first order autocorrelation coefficient ($r = 0.24, p<0.05$). A positive first-order coefficient means that consecutive intervals are more like each other than chance alone would dictate. This patient also demonstrates pulsus alternans ($r = -0.2, p<0.05$) and, as in the previous patient, significant relationships between stroke distances and the two preceding R-R intervals are seen, positive to RR-1 ($r = 0.60, p<0.001$) and negative to RR-2 ($r = -0.28, p<0.05$).

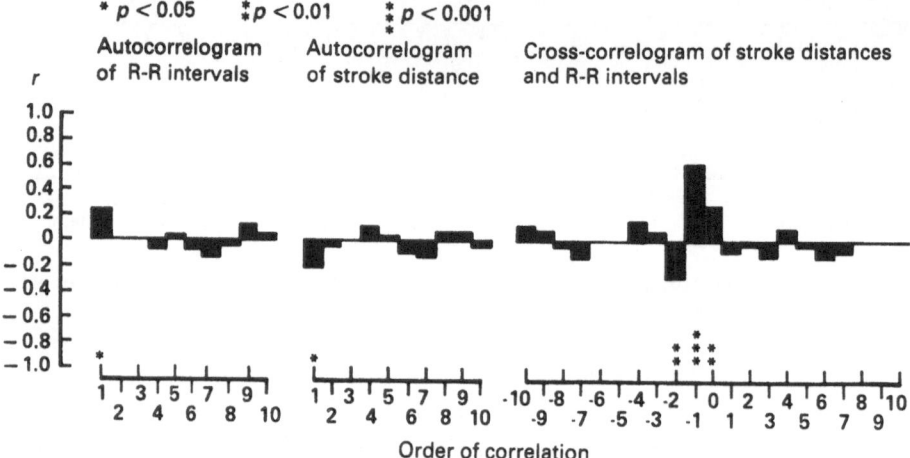

Fig. 5.5. As figure 5.4 for a patient with non-random rythm (chaining) and non-random sequence of stroke distances (alternans) (Reproduced by permission of the Editor, *British Heart Journal*.)

Fig. 5.6 shows the auto- and cross- correlograms from a patient with a non-random rhythm. The first- order autocorrelation coefficient for R-R intervals is significant and negative ($r = -0.42$, $p<0.001$), indicating that cardiac cycle lengths tend to alternate. Pulsus alternans is also present, indicated by a negative first order coefficient on the autocorrelogram of stroke distance ($r = -0.51$, $p<0.001$). It is also apparent that there is a significant positive second-order coefficient ($r = 0.22$, $p<0.01$) indicating that stroke distances from every second beat are positively related.

The cross-correlogram shows a very complex pattern of inter- relationships, but as before the positive correlation between stroke distances and RR-1 ($r =$

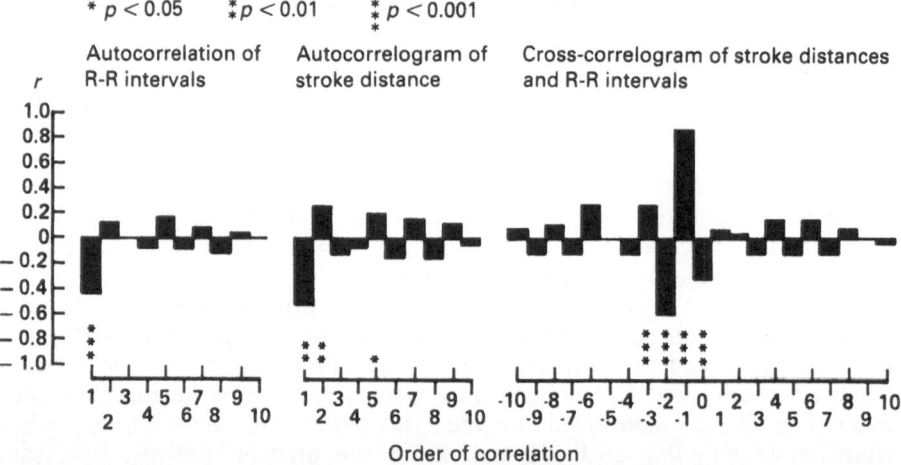

Fig. 5.6. As Fig. 5.4, for a patient with non-random rhythm (alternating cycle length) and non-random sequence of stroke distances (alternans). (Reproduced by permission of the Editor, *British Heart Journal*.)

0.88, $p<0.001$), and the negative correlation between stroke distances and RR−2 ($r = -0.59$, $p<0.001$) may be identified.

In a series of 74 patients with atrial fibrillation, 30 (41%) had statistically significant first order autocorrelation coefficients for stroke distances, 21 negative and 9 positive (Rawles and Rowland 1986). Thus, the sequence of stroke outputs is frequently non-random even in the presence of a random sequence of R-R intervals, the commonest pattern of non- randomness being alternation of stroke distances, a form of pulsus alternans.

Derivation of a Mathematical Model of Left Ventricular Function

In each of the cross-correlograms in Figs. 5.4–5.6 there is a positive correlation between stroke distances and the preceding R-R intervals (RR−1), and a negative correlation between stroke distances and the pre-preceding R-R intervals (RR−2). A highly significant positive relation between stroke distance and RR−1 is always found, and a significant negative univariate relation to RR−2 is present in about 75% of cases. Weakly positive or negative correlations between stroke distance and third and fourth previous R-R intervals are occasionally seen.

Cross-correlograms therefore provide the preliminary evidence that it is the two preceding R-R intervals that largely determine the variation in stroke distance.

Fig. 5.7 shows a scatter plot of stroke distances against the previous R-R intervals (RR−1). The relationship is not linear and is best described as quadratic, the equation having the form

$$Sd = a' + b(RR-1) + c(RR-1)^2 \tag{5.1}$$

A scatter plot of stroke distances and pre-preceding R-R intervals (RR−2) is similarly shown in Fig. 5.8. This relationship is also quadratic, with the form

$$Sd = a'' + d(RR-2) + e(RR-2)^2 \tag{5.2}$$

The two equations may be combined into the multiple regression equation

$$Sd = a + b(RR-1) + c(RR-1)^2 + d(RR-2) + e(RR-2)^2 \tag{5.3}$$

which is the mathematical model finally adopted (Rawles 1988). The fit is closest with the quadratic terms included in the equation, but is not substantially improved by the inclusion of additional predictor variables such as RR−3 or RR−4. Many other mathematical models have been tried, but none substantially improves on this quadratic polynomial equation that includes RR−1 and RR−2.

On average, 68% of the variance of stroke distance is explained in terms of RR−1 and RR−2 by this equation; 58% is contributed by RR−1, and 10% by RR−2. The third and fourth previous R-R intervals, RR−3 and RR−4, between them contributed less than 1% to the variance of stroke distance, and were therefore dropped from the model.

The mean multiple correlation coefficient was 0.82 (range 0.68–0.94).

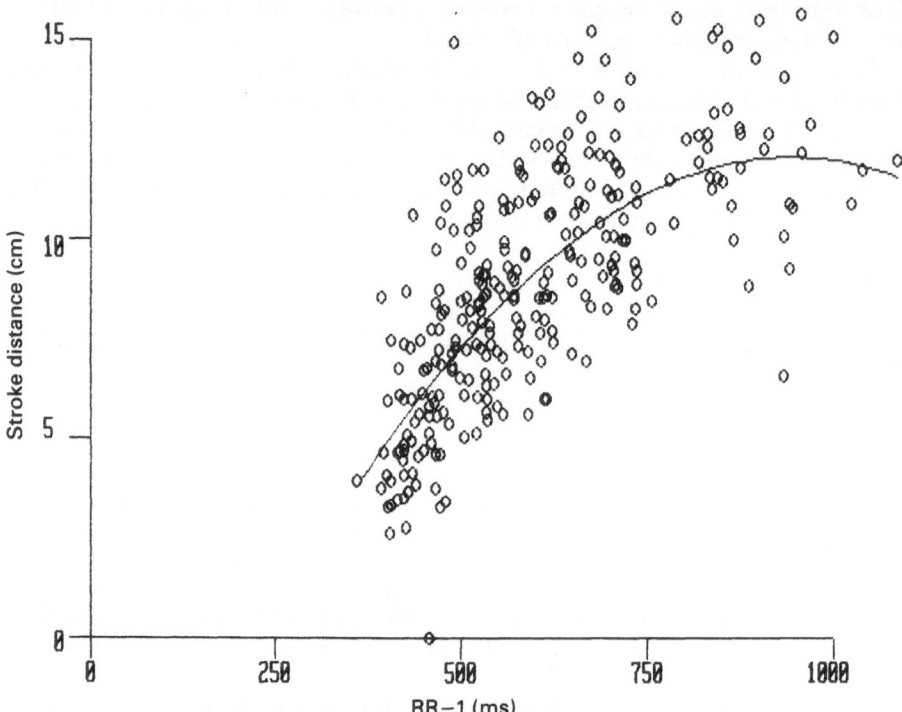

Fig. 5.7. Scatter plot of stroke distances against RR−1. The best fit quadratic regression line is also shown.

The Physiological Basis for the Mathematical Model of Left Ventricular Function

The value of a mathematical model of a biological phenomenon is greatly increased if the form of the model corresponds to known physiological mechanisms. In the case of atrial fibrillation, inequality of the pulse volume has aroused much interest ever since its recognition (Einthoven and Korteweg 1915; Lewis 1925). A good deal is therefore known about the haemodynamic mechanisms that operate in atrial fibrillation; these mechanisms correspond to the components of the model, thereby adding to its validity.

Effect of RR−1 on Stroke Distance: Preload

Einthoven and Korteweg (1915) described a positive curvilinear relationship between the size of the pulse recorded on a polygraph in atrial fibrillation, and the duration of the previous cardiac cycle, an observation that has been amply confirmed. Many different methods of assessing stroke output have been employed, but however it is manifest, stroke performance in atrial fibrillation is always correlated to the preceding R-R interval. Non-invasive measurements

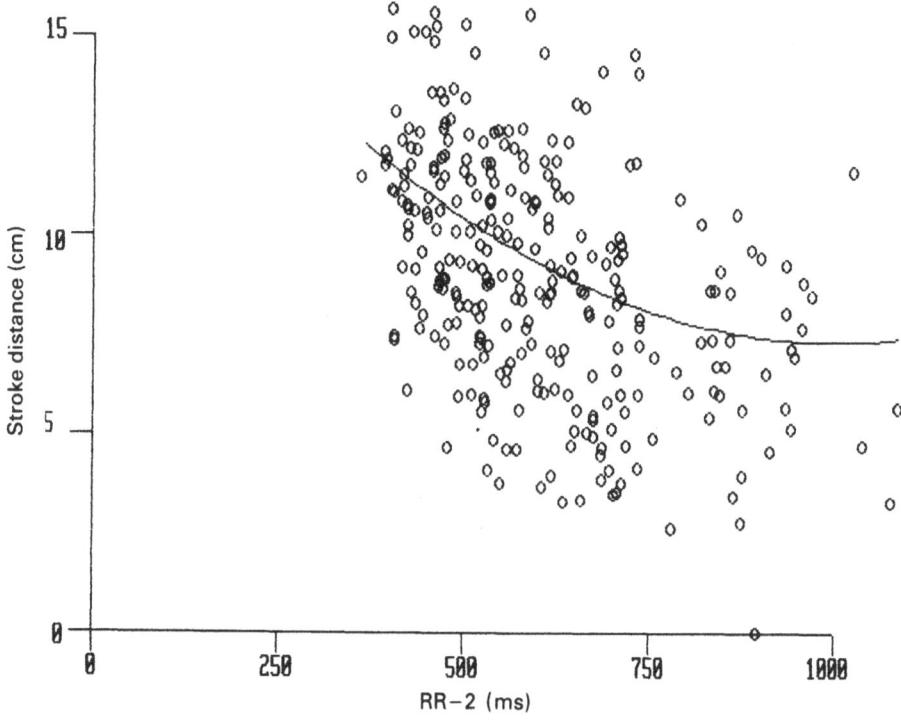

Fig. 5.8. Same patient as Fig. 5.7. Scatter plot of stroke distances against RR−2 with the best fit quadratic regression line.

used to indicate stroke performance have included left ventricular ejection time (Katz and Feil 1923; Tavel et al. 1972; Chirife et al. 1974; Kligfield 1974), systolic time intervals in patients with prosthetic heart valves (Gibson et al. 1971; Johansen et al. 1971; Meijler et al. 1971), the velocity of circumferential fibre shortening or left ventricular ejection fraction determined echocardiographically (Gibson 1972; Karliner et al. 1974; Belenkie 1979; Fujii et al. 1981; Iwase et al. 1988–9), or using radionuclides (Benjelloun et al. 1983; Schneider et al. 1983). Invasive methods that have been used to assess stroke performance include direct measurement of blood pressure (Braunwald et al. 1960; Edmands et al. 1970; Greenfield et al. 1968), velocity of aortic blood flow (Benchimol et al. 1969; Gabe et al. 1969), and myocardial tension (Meijler et al. 1968; Rogel and Mahler 1969).

There are three possible mechanisms working alone or in combination that could explain the positive relationship between stroke performance and the duration of the preceding R-R interval: alteration of preload, contractility, or afterload. There is much evidence that the influence of RR−1 on stroke volume is largely mediated by preload through the Frank–Starling effect. This comes about since RR−1 determines the duration of diastolic filling of the left ventricle, and hence end- diastolic volume and pressure (Dodge et al. 1957; Braunwald et al. 1960). As Einthoven and Korteweg (1915) showed, the relationship between RR−1 and pulse volume is biphasic, increasing linearly, then remaining at a plateau value.

A similar plateau is seen in the relationship between RR−1 and left ventricular end-diastolic diameter determined either angiographically (Greenfield et al. 1968) or echocardiographically (van Dam et al. 1980); the preceding rise in left ventricular diameter corresponds to rapid and complete ventricular filling in early diastole (Noble et al. 1969; DeMaria et al. 1976).

Iwase et al. (1988–9) reported a positive correlation between left ventricular stroke and filling volumes and preceding R-R intervals if these were shorter than 600 milliseconds; stroke and filling volumes were independent of R-R interval duration when this exceeded 600 milliseconds. In another echocardiographic study left ventricular emptying was proportional to the duration of the preceding R-R interval up to a duration of 800 milliseconds, but systolic and diastolic dimensions were unchanged when preceding R-R intervals were in the range 800–1200 milliseconds (Suzuki et al. 1981).

In mitral stenosis, when ventricular filling takes longer, the plateau section of the curve has a later onset (Cieslinski et al. 1984) or is not seen, stroke volume continuing to rise up to the longest RR−1 intervals (Tavel et al. 1972; Kligfield 1974). The end-diastolic diameter of the left ventricle correlated better than previous cycle length with circumferential fibre shortening velocity and ejection fraction in most patients with atrial fibrillation studied echocardiographically (Belenkie 1979), again suggesting that the influence of RR−1 on systolic function is indirect through its effect on preload.

In many of our patients with atrial fibrillation, the graph of the stroke distance/RR−1 relationship is biphasic, reaching a maximum value and then declining. A declining value of ejection fraction with the longest RR−1 intervals is depicted by Caird and Williams (1981) using a quadratic polynomial equation in a multiple regression analysis. This shape of curve, with a fall from the plateau values with very long intervals, is unlikely to be a statistical artefact from the use of a quadratic polynomial model, since it was also seen by us with a very different exponential model. A decline of the curve, rather than it remaining at a plateau value, may result from the development of mitral regurgitation with overdistension of the ventricle in the longest diastolic intervals (Sanada et al. 1985).

Effect of RR−1 on Stroke Distance: Contractility

In sinus rhythm, a premature extrasystole is associated with altered contractility itself and the beat or beats that follow: the Bowditch or force–frequency effect (Hoffman et al. 1956; Kruta and Braveny 1960). The consequence of this is illustrated in Fig. 5.9. The duration of the coupling interval separating the extrasystole from the preceding normal beat is positively related to the contractility of the extrasystole and its stroke output, shown as the area within the velocity–time complex, the stroke distance. It is also negatively related to the contractility of the post-extrasystolic beat, and its stroke distance. Thus, the premature beat with its short coupling interval is weakened, and the post-extrasystolic beat is potentiated (Benchimol et al. 1974), even in the absence of a compensatory pause (Kuijer et al. 1990). This force- frequency effect is a fundamental property of cardiac muscle (Blinks and Koch-Weser 1961) and plays an important role in determining beat-to-beat changes in stroke volume in atrial fibrillation (Meijler et al. 1968; Rogel and Mahler 1969). In atrial fibrillation each index beat in turn may be considered as an extrasystole, the contractility of which

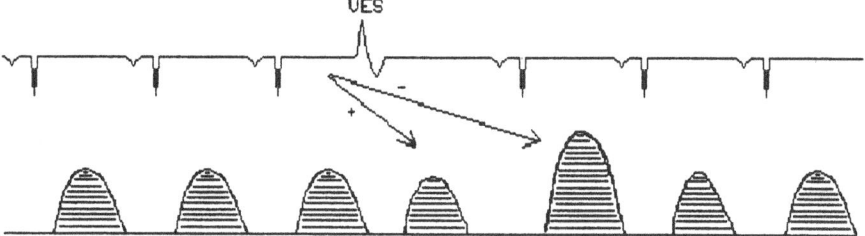

Fig. 5.9. The electrocardiogram and stroke distances in sinus rhythm. The effect of a ventricular extrasystole (VES) on two succeeding stroke distances is shown.

is therefore determined by RR−1, the interval separating it from the previous beat. In Fig. 5.10a the stroke distance of the index beat is smaller than average because RR−1 is short, and the converse is illustrated in Fig. 5.10b.

In humans, beat-to-beat changes in contractility of the left ventricle in atrial fibrillation have been evaluated by Gibson et al. (1971) in patients with aortic Starr–Edwards prostheses. The interval between ventricular activation and the opening sound of the prosthesis (pre-ejection time) shortened with increasing length of the preceding R-R interval. This reduction of pre- ejection time in proportion to RR−1 was thought to result from increasing contractility or diminished afterload, rather than increased stroke volume, since the measurement was little affected by changes in ventricular filling.

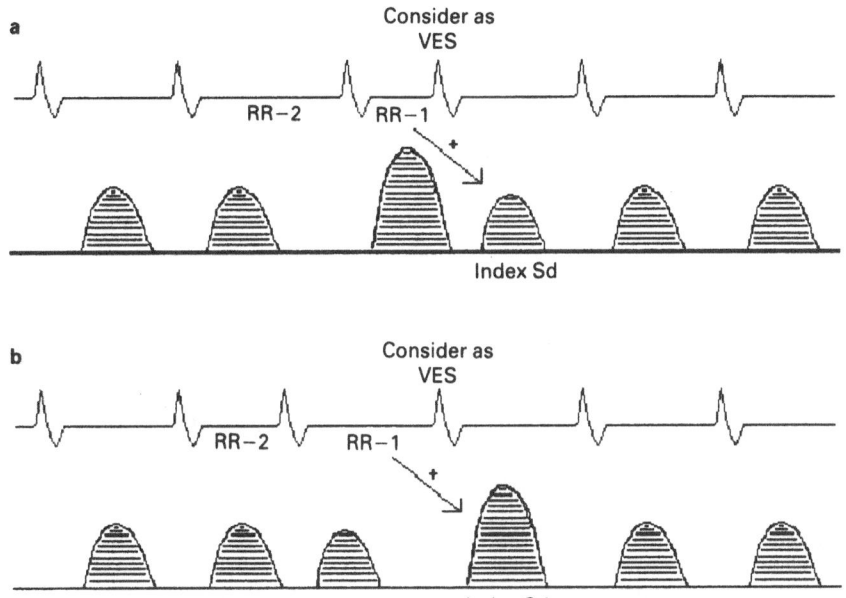

Fig. 5.10. A heartbeat in atrial fibrillation considered as a ventricular extrasystole (VES) with a short (a) or a long (b) coupling interval RR−1. Stroke distance is positively related to RR−1.

The Effect of RR−1 on Stroke Distance: Afterload

The third possible explanation for the positive relationship between RR−1 and stroke distance (Sd) is diminishing aortic diastolic pressure and afterload with increasing time since the previous R-wave. This is unlikely to be an important factor in the majority of our patients, in whom it has been shown that there is no significant independent relationship between Sd and Sd−1 (Rawles and Haites 1984). If the size of the previous stroke has no influence on stroke distance it is unlikely that variation in its timing will have any influence either.

In atrial fibrillation, stroke output is therefore positively related to the previous R-R interval by increased preload, augmented contractility and diminished afterload. Although these are inseparably associated in the Sd/RR−1 relationship, preload may be the most important influence in humans (Karliner et al. 1974) though possibly not in the dog (Edmands et al. 1970).

The relationship between the filling time of the left ventricle and preload is complex. Filling of the left ventricle does not occur at a uniform rate during diastole, but is maximal when the mitral valve first opens. Neither is left ventricular diastolic pressure linearly related to left ventricular volume. The pressure/volume relationship, or compliance, varies widely between subjects and is not described by any simple mathematical expression (Yettram et al. 1990). The relationship between stroke distance and the duration of the preceding R-R interval is therefore a very indirect representation of the Frank–Starling curve of left ventricular function.

Effect of RR−2 on Stroke Distance

As well as each beat in atrial fibrillation acting as an extrasystole, it may also be considered as the post extrasystolic beat in relation to the beat before (Fig. 5.11a). The second previous R-R interval (RR−2) now becomes the coupling interval that has a negative relationship with contractility and output of the index beat (Index Sd). The explanation for the negative relationship between stroke distance and RR−2 is therefore potentiation of contractility with shortening of RR−2, or reduction of contractility with lengthening of RR−2 (Fig. 5.11b).

The existence of a negative relationship between the strength of ventricular contraction and the second previous R-R interval in atrial fibrillation has been described by many authors (Greenfield et al. 1968; Meijler et al. 1968; Rogel and Mahler 1969; Edmands et al. 1970; Gibson et al. 1971; Johansen et al. 1971; Karliner et al. 1974). Gibson et al. (1971) showed that the opening time of aortic valve prostheses shortened in proportion to RR−2 in 14 of 19 patients with atrial fibrillation. This was not due to alterations in ventricular filling or to changes in end-diastolic gradient across the prosthesis, confirming that this effect was mediated by a change in contractility.

In a study of myocardial relaxation in atrial fibrillation, Nakamura et al. (1986) showed that a rising left ventricular dP/dt correlated with the length of the preceding R-R interval ($r = 0.48$), and more strongly with the ratio of the preceding to the pre-preceding interval ($r = 0.82$). However, the relaxation time associated with a falling dP/dt was fairly constant in each patient and was not correlated with previous R-R intervals, indicating that relaxation is independent of contractility (van Dam et al. 1980).

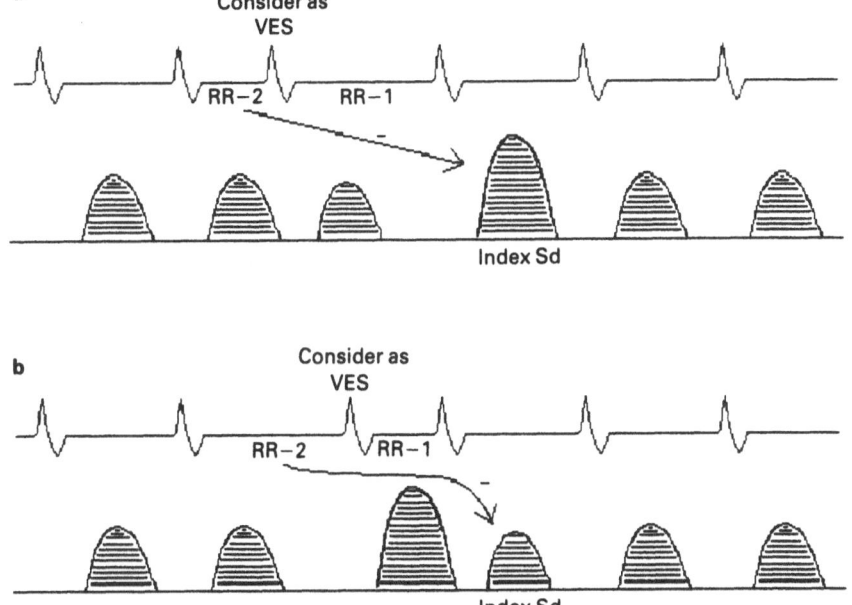

Fig. 5.11. A heartbeat in atrial fibrillation considered as a post-extrasystolic beat, the ventricular extrasystole (VES) having a short (*a*) or a long (**b**) coupling interval RR−2. Stroke distance is negatively related to RR−2.

Balance of Effects of RR−1 and RR−2

In sinus rhythm the output of a premature beat is reduced in proportion to its prematurity, while the output of the post- extrasystolic beat is increased. The sum of the outputs of the two beats is usually less than twice that of the regular beats before the extrasystole, the increased contractility of the post- extrasystolic beat not fully compensating for the reduced contractility and filling of the premature beat (Cohn and Kryda 1981). Translated to atrial fibrillation, this observation implies that the effect on stroke distance of RR−1 is more pronounced than the effect of RR−2. However, although RR−1 always predominates, the balance of influence between RR−1 and RR−2 varies widely from patient to patient, partly in relation to overall left ventricular function.

In this non-invasive study, mean age-adjusted stroke distance provides the best estimate of overall left ventricular function. Between patients there are positive correlations between mean age-adjusted stroke distance and the slope of the Sd/RR−1 curve, and with the ratio between the absolute slopes of Sd/RR−1 and Sd/RR−2 curves.

The slope of the Sd/RR−1 curve, as we have seen, reflects the relationship between the duration of diastolic filling of the left ventricle and its output, and is therefore related to the Frank–Starling curve of left ventricular function. As left ventricular function declines stroke distance falls, and the contribution of RR−1 to the variance of stroke distance becomes less in proportion to that of RR−2.

Fujii et al. (1981) plotted stroke volumes, determined echocardiographically, against preceding R-R intervals in 14 patients with atrial fibrillation. In patients with left ventricular failure these left ventricular function curves, closely related to our graphs of Sd/RR−1, were below and to the right of the curves of patients without failure, and moved upwards and to the left after treatment with digitalis and diuretics.

It has been observed that correlation between the pre-preceding interval and left ventricular performance – equivalent to the slope of the Sd/RR−2 curve – may be reduced in patients with left ventricular failure (Gibson et al. 1971), raising the possibility that the force generated by the left ventricle is at a maximum and no further increase can occur as a result of shortening this interval. In our study there was no evidence that this was so, the absolute value of the Sd/RR−2 slope being unrelated to mean stroke distance, expressed as a percentage of the age-predicted normal value. Indeed, on the related phenomenon of post-extrasystolic potentiation, Kuijer et al. (1990) reported that potentiation is especially evident in patients with left ventricular damage.

Thus, in patients with a high mean stroke distance and good left ventricular function, the preceding R-R interval (RR−1) makes the largest contribution to the beat-to-beat variation of stroke distance. When mean stroke distance is low, and left ventricular function is poor, the contribution of RR−1 falls, while that of the pre-preceding interval (RR−2) increases relative to that of RR−1. In 4 patients RR−2 did not make any significant contribution to the variance of stroke distance, and at the other extreme the contribution from RR−2 was 77% of that from RR−1; in every case the contribution of RR−1 exceeded that of RR−2.

Pulsus Alternans

Both Einthoven and Korteweg (1915) and Lewis (1925, p. 441) described pulsus alternans in atrial fibrillation, and there have been other case reports (Gould 1969), notably of alternation of blood pressure in the pulmonary artery but not the left ventricle (Gould et al. 1973). We found alternation of stroke distance in 34 of 74 patients studied (Rawles and Rowland 1986). It occurs because of the opposing effects of a given R-R interval on the two subsequent stroke volumes (Fig. 5.12); it does not signify impaired left ventricular function as it does in sinus rhythm. In the dog it is seen particularly with rapid ventricular rates (Mahler and Rogel 1970).

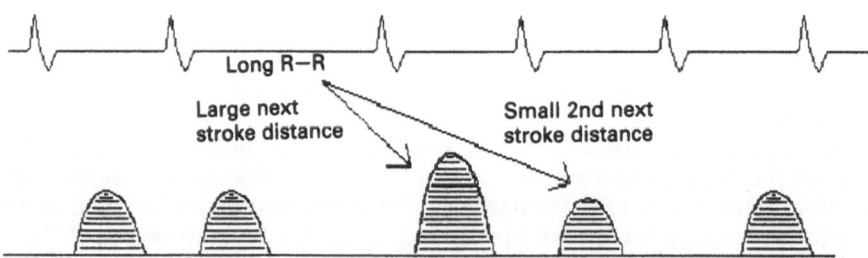

Fig. 5.12 The mechanism of pulsus alternans in atrial fibrillation.

Thus, to summarise, the positive relationship between stroke distance and $RR-1$ indicates predominantly the effect of preload and the Frank-Starling effect, and the negative relationship between stroke distance and $RR-2$ shows the extent to which contractility may be increased by the Bowditch effect. These physiological mechanisms underlie the mathematical model and explain most of the beat-to-beat variation in stroke distance.

Application of the Mathematical Model of Left Ventricular Function

From a 3-minute recording of a sequence of stroke distances and R-R intervals made at the patient's bedside, much information of potential clinical value may be obtained.

Firstly, from mean stroke distance and mean ventricular rate linear cardiac output or minute distance can be calculated. Mean stroke and minute distances may be compared with the age- predicted normal values which have been established in a series of 140 healthy subjects (Mowat et al. 1983). Mean stroke distance in atrial fibrillation is usually lower than the predicted normal value for sinus rhythm, but ventricular rate is usually higher than in sinus rhythm so minute distance is not depressed by as much as stroke distance (Haites et al. 1985). As discussed in chapter 4, mean stroke distance gives a good indication of overall cardiac function.

Graphical Presentation of Multiple Regression Equation

The regression equation (5.3, p. 99), which, as we have seen, is based on well recognised physiological mechanisms, may be calculated for a patient with atrial fibrillation within a few minutes of making the recording. For each patient the five coefficients of the multiple regression equation are stored in a microcomputer together with mean, minimum and maximum values of stroke distances and R-R intervals. A program has been developed that shows graphs of the relationship between stroke distance and $RR-1$ and $RR-2$, $RR-2$ being assigned a mean value while $RR-1$ is altered, and vice versa. From each graph the slope at the intercept of mean R-R interval and mean stroke distance can be determined; this is referred to as "the slope".

Fig. 5.13a shows a composite graph of the multiple regression equation for a 91-year-old patient with atrial fibrillation whose mean stroke distance at 7.4 cm was 89% of the predicted normal value in sinus rhythm. By multiple regression analysis there were only two significant predictor variables – $RR-1$ and its square – but these together accounted for 67% of the variance of stroke distance, giving a multiple correlation coefficient (R) of 0.82. Inclusion of $RR-2$ in the regression did not materially increase the multiple correlation coefficient, and the Sd/$RR-2$ graph is essentially flat. The dashed line indicates the mean stroke distance, and its point of intersection with the Sd/$RR-1$ curve corresponds to the mean R-R interval of 1040 milliseconds.

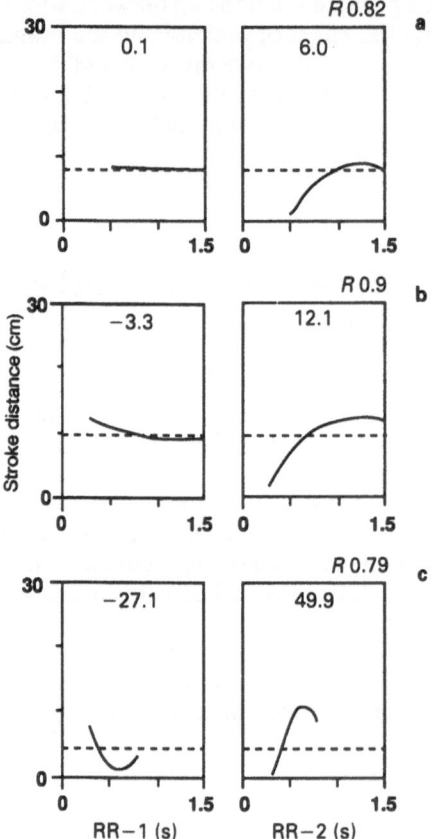

Fig. 5.13. Composite graphs from 3 patients with atrial fibrillation showing regressions of stroke distance against RR−1 and RR−2. *Dashed lines* show mean stroke distances. *R*, multiple correlation coefficients. (Reproduced by permission of the Editor, *International Journal of Biomedical Computing.*)

Fig. 5.13b shows similar graphs from a patient whose mean stroke distance was 65% of the predicted normal value. There is a significant contribution from RR−2 towards the variation of stroke distance, the slopes of the Sd/RR−2 and Sd/RR−1 curves being −3.3 and 12.1 centimetres per second respectively.

The graphs from a patient whose mean stroke distance was only 41% of the predicted normal value are shown in Fig. 5.13c. Compared with the previous patient the slope of the Sd/RR−2 curve is greater relative to that of Sd/RR−1.

These three graphical representations of multiple regression equations are from patients with progressively bad left ventricular function as judged by mean stroke distance. They illustrate the increasing influence of RR−2 on stroke distance as left ventricular function declines.

In each patient both graphs are curved but the overall gradient of the Sd/RR−1 graph is positive, and that of the Sd/RR−2 graph is negative, corresponding to the positive and negative cross-correlations previously noted between stroke distances and the two preceding R-R intervals. The slope of the graph, regardless of sign, is always greater for Sd/RR−1 than for Sd/RR−2.

In summary, we have seen that computer analysis of a short recording of a sequence of stroke distances taken non-invasively at the bedside yields a fairly complete description of left ventricular function in atrial fibrillation. The mathematical model employed describes stroke distance in terms of the two preceding R-R intervals which, though not pure, are related to preload and contractility respectively; these relationships may be depicted graphically, the graphs giving insight into left ventricular functioning. There is a wide individual variation in the relative importance of the preceding and pre-preceding R-R intervals in causing changes in stroke distance.

Additional information that may be obtained by manipulation of each patient's model of left ventricular function will be discussed in the next chapter.

Haemodynamic Consequences of Atrial Fibrillation

Venous Return

The return of blood to the heart is down a pressure gradient: from the periphery, via the great veins, to the right atrium. The pressure in the right atrium varies during the cardiac cycle, so flow in the great veins is pulsatile. In the jugular vein, waves of systolic and diastolic flow are recognised. In sinus rhythm the systolic flow wave results from a fall in right atrial pressure due to relaxation of the atrium, at the same time as the tricuspid valve descends with contraction of the right ventricle. In atrial fibrillation relaxation of the right atrium does not occur so the systolic flow wave is solely related to right ventricular contraction, and the systolic wave is small, distorted and sometimes reversed compared with that in sinus rhythm (Kalmanson et al. 1971).

The atrial myocardium is in a tetanic-like state of contraction in atrial fibrillation, and although contraction is not coordinated wall tension is high, mean atrial pressures are raised, and atrial compliance is greatly reduced (White et al. 1982). Pulmonary hypertension may be a consequence of increased left atrial pressure, particularly where atrial fibrillation is associated with rheumatic heart disease, when symptoms of breathlessness are more closely related to pulmonary hypertension than to any reduction of cardiac output (Ferrer et al. 1952); pulmonary congestion during exercise is a particularly disabling consequence of reduced atrial compliance. The symptomatic benefit of restoring sinus rhythm may be greater from reduction of left atrial filling pressure than from increasing cardiac output (Shapiro and Klein 1968).

Ventricular Filling

In sinus rhythm ventricular filling occurs in two waves, in early and late diastole (Noble et al. 1969). The proportion of ventricular filling in late diastole associated with atrial contraction varies from 20%–50% depending on underlying ventricular function (Channer and Jones 1988), being greater when left ventricular function is impaired, as in hypertrophic obstructive cardiomyopathy or after myocardial infarction (Rahimtoola et al. 1975; DeMaria et al. 1976). Experimental induction of atrial fibrillation in dogs results in reduction of blood pressure, increased right atrial pressure, and reduced left ventricular stroke volume and cardiac output (Oldham et al. 1967; Hilwig 1972).

In humans, the significance of the loss of the atrial contribution to ventricular filling (Abelson 1968) and stroke volume that occurs in atrial fibrillation may be gauged by comparing the results of atrial and ventricular pacing at identical pacing rates. Cardiac output, aortic and ventricular pressures, stroke power and dp/dt of arterial pressure are 5%–15% lower during ventricular than during atrial pacing, and loss of the atrial contribution is greater still if there is heart disease or a pacing rate above 120 beats per minute (Benchimol 1969). A further 7% reduction in cardiac output may result from irregular ventricular cycles in atrial fibrillation (Naito et al. 1983).

Nicod et al. (1986) remarked on the importance of the "atrial kick" in increasing the effective mitral valve area in mitral stenosis. So to the other haemodynamic disadvantages must be added that of a smaller effective mitral valve area when atrial fibrillation supervenes in mitral stenosis.

During very rapid ventricular rates the filling time of the left ventricle is short. There are two mechanisms for prolonging filling time – premature opening and delayed closure of the mitral valve – both of which are seen. Filling may continue into the pre-ejection period of the following cycle, but after a beat with reduced filling aortic closure and mitral opening occur early (Cieslinski et al. 1984).

Mitral Regurgitation

Mitral regurgitation is commonly detected in atrial fibrillation, even in the absence of mitral valve disease (Daley et al. 1955). It may occur during diastole at the end of particularly long R-R intervals, when it results from overdistension of the left ventricle (Sanada et al. 1985). More commonly, mitral reflux may be detected in early systole (Skinner et al. 1964). The atrial contraction in sinus rhythm normally results in a brief reversal of the atrioventricular pressure gradient at the end of diastole. As a consequence the mitral valve is already closed at the onset of ventricular systole, but this is not the case in atrial fibrillation where coordinated atrial contractions are absent. A mitral valve that is still open at the beginning of systole is not the whole explanation, however, since with regular ventricular pacing in atrial fibrillation mitral reflux disappears (Naito et al. 1983). Systolic mitral regurgitation in atrial fibrillation occurs in beats with a short preceding R-R interval, is accentuated by a long pre-preceding interval, and is attributed to asynchrony due to localised wall motion abnormalities in the apical area of the left ventricle (Mawatari et al. 1987, 1988).

Coronary Blood Flow

Within 5 seconds of the onset of atrial fibrillation blood flow to the atria increases dramatically (White et al. 1982), reaching flow rates per gram of tissue comparable to those of the left ventricle. This augmented flow is regulated by the metabolic needs of atrial tissue and does not represent maximal vasodilation

(McHale et al. 1983). Elsewhere, coronary flow may be diminished, especially to subendocardial ventricular myocardium, because of increased coronary vascular resistance; overall, coronary blood flow is reduced. The increased coronary vascular resistance is specific to atrial fibrillation, and is not seen in response to atrial or ventricular pacing at comparable rates (Saito et al. 1978; Ertl et al. 1987).

During atrial fibrillation the velocity integral of coronary blood flow varies beat-to-beat with the duration of the diastolic interval in which the flow occurs (Benchimol and Matsuo 1971).

Regional Blood Flow

In humans, induction of atrial fibrillation leads to a drop in systolic blood pressure and cardiac output (Lau et al. 1990). In the dog the fall in cardiac output at the onset of atrial fibrillation is associated with reduced splanchnic and renal cortical flow; brain blood flow also decreases (Friedman et al. 1987). Petersen et al. (1989) reported on cerebral blood flow after cardioversion in 9 patients who had had atrial fibrillation for less than 3 months. After correction for changes in end-tidal CO_2 tension, cerebral blood flow was significantly increased with restoration of sinus rhythm.

Systolic Time Intervals

Induction of atrial fibrillation in the dog leads to a prolonged pre-ejection period (PEP) and isovolumic contraction time; the left ventricular ejection time (LVET) is reduced (Knudson et al. 1978). In humans, the pre-ejection period lengthens with increasing heart rate, while the LVET is less affected. The PEP/LVET ratio therefore increases as heart rate rises above 75 beats per minute (Boudoulas et al. 1978).

Endocrine Consequences of Atrial Fibrillation

Atrial natriuretic factor concentration decreases after cardioversion of chronic atrial fibrillation, and increases after the induction of supraventricular tachycardia during electrophysiological testing (Roy et al. 1987). The reduced level of atrial natriuretic peptide is maintained for a period of 30 days after cardioversion (Petersen et al. 1988). The raised plasma concentrations of atrial natriuretic peptide found in atrial fibrillation are independent of left atrial dimensions, and are associated with increased concentrations of aldosterone (Berglund et al. 1990). These endocrine changes may underlie the increase in blood haematocrit seen during paroxysms of atrial fibrillation (Imataka et al. 1987). Electron microscopic granule counts of the cells that produce natriuretic peptide reflect turnover of the hormone. The count is raised in mitral stenosis, and is correlated with left atrial dimension; there is a tendency towards higher counts in atrial fibrillation (Doubell et al. 1990).

References

Abelson D (1968) Ultrasonic Doppler auscultation of the heart, with observations on atrial flutter and fibrillation. JAMA 204:438–43

Belenkie I (1979) Beat-to-beat variability of echocardiographic measurements of left ventricular end diastolic diameter and performance. J Clin Ultrasound 7:263–8

Benchimol A (1969) Significance of the contribution of atrial systole to cardiac function in man. Am J Cardiol 23:568–71

Benchimol A, Matsuo S (1971) Continuous measurements of phasic aortic and coronary flow velocity during atrial fibrillation in man. Am J Med 51:466–73

Benchimol A, Stegall HF, Maroko PR, Gartlan JL, Brener L (1969) Aortic flow velocity in man during cardiac arrhythmias measured with the Doppler catheter–flowmeter system. Am Heart J 78:649– 59EP

Benchimol A, Desser KB, Wang TF, Mori K (1974) Left ventricular blood flow velocity during atrial arrhythmias in man. Am J Cardiol 34:271–5

Benjelloun H, Itti R, Philippe L, Lorgeron JM, Brochier M (1983) Beat-to-beat assessment of left ventricular ejection in atrial fibrillation. Eur J Nucl Med 8:206–10

Berglund H, Boukter S, Theodorsson E, Vallin H, Edhag O (1990) Raised plasma concentrations of atrial natriuretic peptide are independent of left atrial dimensions in patients with chronic atrial fibrillation. Br Heart J 64:9–13

Blinks JR, Koch-Weser J (1961) Analysis of the effects of changes in rate and rhythm upon myocardial contractility. J Pharmacol Exp Ther 134:373–89

Boudoulas H, Lewis RP, Sherman JA, Bush CA, Dalamangas G, Forester WF (1978) Systolic time intervals in atrial fibrillation. Chest 74:629–34

Braunwald E, Frye RL, Aygen MM et al. (1960) Studies on Starling's law of the heart. III. Observations in patients with mitral stenosis and atrial fibrillation on the relationships between left ventricular end-diastolic segment length, filling pressure, and the characteristics of ventricular contraction. J Clin Invest 39:1874–84

Caird FI, Williams BO (1981) Left ventricular performance in atrial fibrillation in the elderly. Age Ageing 10:231–6

Channer KS, Jones JV (1988) Atrial systole: its role in normal and diseased hearts. Clin Sci 75:1–4

Chirife R, Foerster JM, Bing OH (1974) Left ventricular ejection time by densitometry in patients at rest and during exercise, atrial pacing and atrial fibrillation. Comparison with central aortic pressure measurements. Circulation 50:1200–4

Cieslinski A, Hui WK, Oldershaw PJ, Gregoratos G, Gibson D (1984) Interaction between systolic and diastolic time intervals in atrial fibrillation. Br Heart J 51:431–7

Cohn K, Kryda W (1981) The influence of ectopic beats and tachyarrhythmias on stroke volume and cardiac output. J Electrocardiol 14:207–18

Daley R, McMillan IKR, Gorlin R (1955) Mitral incompetence in experimental auricular fibrillation. Lancet ii:18–20

DeMaria AN, Miller RR, Amsterdam EA, Markson W, Mason DT (1976) Mitral valve early diastolic closing velocity: relation to sequential diastolic flow and ventricular compliance. Am J Cardiol 37:693–700

Dodge HT, Kirkham FT, King CV (1957) Ventricular dynamics in atrial fibrillation. Circulation 15:335–47

Doubell AF, Greeff MP, Rossouw DJ, Weich HF (1990) Electron microscopic analysis of the specific granule content of human atria. An investigation of the role of atrial pressure and atrial rhythm in the release of atrial natriuretic peptide. S Afr Med J 78:207–11

Edmands RE, Greenspan K, Fisch C (1970) The role of inotropic variation in ventricular function during atrial fibrillation. J Clin Invest 49:738–46

Einthoven W, Korteweg AJ (1915) On the variability of the size of the pulse in cases of atrial fibrillation. Heart 6:107–20

Ertl G, Meesman M, Krumpiegel K, Kochsiek K (1987) The effects of atrial fibrillation on coronary blood flow and performance of ischaemic myocardium in dogs with coronary artery stenosis. Clin Sci 73:437–44

Ferrer MI, Harvey RM, Cathcart RT, Cournand A, Richards DW (1952) Haemodynamic studies in rheumatic heart disease. Circulation 6:688–710

Friedman HS, O'Connor J, Kottmeier S, Shaughnessy E, McGuinn R (1987) The effects of atrial fibrillation on regional blood flow in the awake dog. Can J Cardiol 3:240–5

Fujii J, Watanabe H, Kuboki M, Kato K (1981) Left ventricular function curve determined by echocardiography in patients with atrial fibrillation. Jpn Heart J 22:561–73

Gabe IT, Gault JH, Ross J et al. (1969) Measurement of instantaneous blood flow velocity and pressure in conscious man with a catheter-tip velocity probe. Circulation 40:603–14

Gibson DG (1972) Beat-to-beat analysis of left ventricular pressure–volume relations in atrial fibrillation in man. Br Heart J 34:204–5

Gibson DG, Broder G, Sowton E (1971) Effect of varying pulse interval in atrial fibrillation on left ventricular function in man. Br Heart J 33:388–93

Gould L (1969) Pulmonary artery alteration with atrial fibrillation. JAMA 207:1515–6

Gould L, Reddy R, Gomprecht RF (1973) Brachial artery alternation with atrial fibrillation. Angiology 24:365–7

Greenfield JC, Harley A, Thompson HK, Wallace AG (1968) Pressure- flow studies in man during atrial fibrillation. J Clin Invest 47:2411–21

Haites NE, McLennan FM, Mowat DHR, Rawles JM (1985) Assessment of cardiac output by the Doppler ultrasound technique alone. Br Heart J 53:123–9

Hilwig RW (1972) Hemodynamic relationships in dogs with sinus arrhythmia and atrial fibrillation. Am J Vet Res 33:475–83

Hoffman BF, Bindler E, Suckling EE (1956) Post extrasystolic potentiation of contraction in cardiac muscle. Am J Physiol 185:95–102

Imataka K, Nakaoka H, Kitahara Y, Fujii J, Ishibashi M, Yamaji T (1987) Blood hematocrit changes during paroxysmal atrial fibrillation. Am J Cardiol 59:172–3

Iwase M, Aoki T, Maeda M, Yokota M, Hayashi H (1988–9) Relationship between beat-to-beat interval and left ventricular function in patients with atrial fibrillation. Int J Card Imaging 3:217–26

Johansen B, Hoelen AJ, Schneider H, Meijler FL (1971) Clinical implications of frequency–force relation. Cardiovasc Res 5[Suppl 1]:112–19

Kalmanson D, Veyrat C, Chiche P (1971) Atrial versus ventricular contribution in determining systolic venous return. A new approach to an old riddle. Cardiovasc Res 5:293–302

Karliner JS, Gault JH, Bouchard RJ, Holzer J (1974) Factors influencing the ejection fraction and the mean rate of circumferential fibre shortening during atrial fibrillation in man. Cardiovasc Res 8:18–25

Katz LN, Feil HS (1923) Clinical observations on the dynamics of ventricular systole. I. Auricular fibrillation. Arch Intern Med 32:672–92

Kligfield P (1974) Systolic time intervals in atrial fibrillation and mitral stenosis. Br Heart J 36:798–805

Knudson MB, Amend JF, Stone EC (1978) Systolic time intervals in induced atrial fibrillation in the dog: Effects of ectopic ventricular activation. Cardiology 63:65–72

Kruta V, Braveny P (1960) Potentiation of contractility in the heart muscle of the rat and some other mammals. Nature 187:327–8

Kuijer PJP, van der Werf T, Meijler FL (1990) Post-extrasystolic potentiation without a compensatory pause in normal and diseased hearts. Br Heart J 63:284–6

Lau CP, Leung WH, Wong CK, Cheng CH (1990) Haemodynamics of induced atrial fibrillation: a comparative assessment with sinus rhythm, atrial and ventricular pacing. Eur Heart J 11:219–24

Lewis T (1925) The mechanism and graphic registration of the heart beat. Shaw & Sons Ltd, London

Mahler Y, Rogel S (1970) Computer analysis of myocardial tension and pressure variations in atrial fibrillation. J Appl Physiol 29:77–81

Mawatari K, Kuroiwa N, Sanada J et al. (1987) Assessment of abnormal left ventricular systolic blood flow in atrial fibrillation. J Cardiol 17:625–33

Mawatari K, Sanada JI, Tanaka Y, Kuroiwa N, Nakamura K, Hashimoto S (1988) Left ventricular systolic blood flow dynamics and left ventricular wall motion abnormalities in atrial fibrillation. J Cardiol 18:803–11

McHale PA, Rembert JC, Greenfield JC (1983) Effect of atrial fibrillation on atrial blood flow in conscious dogs. Am J Cardiol 51:1722–7

Meijler FL, Strackee J, van Capelle FJL, du Perron JC (1968) Computer analysis of the R-R interval–contractility relationship during random stimulation of the isolated heart. Circ Res 22:695–702

Meijler FL, Strackee J, Johansen B, Schneider H (1971) Control of cardiac function by RR intervals in patients with atrial fibrillation. Br Heart J 33:148

Mowat DHR, Haites NE, Rawles JM (1983) Aortic blood velocity measurement in healthy adults using a simple ultrasound technique. Cardiovasc Res 17:75–80

Naito M, David D, Michelson EL, Schaffenburg M, Dreifus LS (1983) The hemodynamic consequences of cardiac arrhythmias: evaluation of the relative roles of abnormal atrioventricular sequencing, irregularity of ventricular rhythm and atrial fibrillation in. Am Heart J 106:284–91

Nakamura Y, Konishi T, Nonogi H, Sakurai T, Sasayama S, Kawai C (1986) Myocardial relaxation in atrial fibrillation. J Am Coll Cardiol 7:68–73

Nicod P, Hillis LD, Winniford MD, Firth BG (1986) Importance of the "atrial kick" in determining the effective mitral valve orifice area in mitral stenosis. Am J Cardiol 57:403–7

Noble MIM, Milne ENC, Goerke RJ et al. (1969) Left ventricular filling and diastolic pressure–volume relations in the conscious dog. Circ Res 24:269–83

Oldham HN, Vasko JS, Brawley RK, Henney RP, Morrow AG (1967) The hemodynamic effects of atrial fibrillation. J Surg Res 7:587–90

Petersen P, Kastrup J, Vilhelms R, Schutten HJ (1988) Atrial natriuretic peptide in atrial fibrillation before and after electrical cardioversion therapy. Eur Heart J 9:639–41

Petersen P, Kastrup J, Videbaek R, Boysen G (1989) Cerebral blood flow before and after cardioversion of atrial fibrillation. J Cereb Blood Flow Metab 9:422–5

Rahimtoola SH, Ehsani A, Sinno MZ, Loeb HS, Rosen KM, Gunnar RM (1975) Left atrial transport function in myocardial infarction. Am J Med 59:686–94

Rawles JM (1988) A mathematical model of left ventricular function in atrial fibrillation. Int J Biomed Comput 23:57– 68

Rawles JM, Haites NE (1984) Computer analysis of aortic blood velocity measurements determined by Doppler ultrasound in atrial fibrillation. Proceedings of the sixth Nordic meeting on medical and biological engineering. Biological Engineering Society: London

Rawles JM, Rowland E (1986) Is the pulse in atrial fibrillation irregularly irregular? Br Heart J 56:4–11

Rogel S, Mahler Y (1969) Myocardial tension in atrial fibrillation. J Appl Physiol 27:822–5

Roy D, Paillard F, Cassidy D et al. (1987) Atrial natriuretic factor during atrial fibrillation and supraventricular tachycardia. J Am Coll Cardiol 9:509–14.

Saito D, Haraoka S, Ueda M, Fujimoto T, Yoshida H, Ogino Y (1978) Effect of atrial fibrillation on coronary circulation and blood flow distribution across the left ventricular wall in anesthetized open-chest dogs. Jpn Circ J 42:417–23

Sanada J, Kawahira M, Kubo H, Kuroiwa N, Nakamura K, Hashimoto S (1985) Late diastolic mitral regurgitation: pulsed Doppler echocardiographic study. J Cardiogr 15:817–28

Schneider J, Berger HJ, Sands MJ, Lachman AB, Zaret BL (1983) Beat-to-beat left ventricular performance in atrial fibrillation: radionuclide assessment with the computerized nuclear probe. Am J Cardiol 51:1189–95

Shapiro W, Klein G (1968) Alterations in cardiac function immediately following electrical conversion of atrial fibrillation to normal sinus rhythm. Circulation 38:1074–84

Skinner NS, Mitchell JH, Wallace AG, Sarnoff SJ (1964) Hemodynamic consequences of atrial fibrillation at constant ventricular rates. Am J Med 36:342–50

Suzuki H, Yaginuma T, Kondo K, Tsuchiya M, Shiina A, Hosoda S (1981) Echocardiographic evaluation of the left ventricular contraction in cases with atrial fibrillation. J Cardiogr 11:509–6

Tavel ME, Baugh DO, Feigenbaum H, Nasser WK (1972) Left ventricular ejection time in atrial fibrillation. Circulation 46:744–2

Van Dam I, van Zwieten G, Vogel JA, Meijler FL (1980) Left ventricular (diastolic) dimensions and relaxation in patients with atrial fibrillation. Eur Heart J 1 [Suppl A]:149–6

White CW, Kerber RE, Weiss HR, Marcus ML (1982) The effects of atrial fibrillation on atrial pressure–volume and flow relationships. Circ Res 51:205–15

Yettram AL, Grewal BS, Gibson DG, Dawson JR (1990) Relation between intraventricular pressure and volume in diastole. Br Heart J 64:304–8

The Haemodynamics of Atrial Fibrillation: The Effect of Ventricular Rate on Cardiac Output

Further Information from Multiple Regression Analysis

We have seen how in atrial fibrillation stroke distance (Sd) may be very largely explained in terms of the two preceding R-R intervals by means of the following equation:

$$Sd = a + b(RR-1) + c(RR-1)^2 + d(RR-2) + e(RR-2)^2$$

Each patient's regression equation is unique by virtue of the combination of the five coefficients a to e, and may be manipulated to give information of potential clinical value. By holding $RR-2$ constant and altering $RR-1$ the independent effect of $RR-1$ on stroke distance may be displayed graphically, this relation being explained, at least in part, by the Frank–Starling effect (Fig. 6.1, bottom right).

Similarly, by holding $RR-1$ constant and altering $RR-2$ the independent effect of $RR-2$ on stroke distance is seen, this relation being due to the force-frequency, or Bowditch, effect (Fig. 6.1, bottom middle).

By assigning $RR-1$ and $RR-2$ identical values within the regression equation the effect on stroke distance of a change in mean R-R interval may be calculated and displayed (Fig. 6.1, bottom left). Since the contribution of $RR-1$ is always greater than that of $RR-2$ the graph of Sd/Mean R-R bears a much closer resemblance to the Sd/$RR-1$ graph than to that of Sd/$RR-2$. It invariably has a positive slope.

The relationship between stroke distance and mean R-R interval (Sd/Mean R-R) may be readily transformed to show that between minute distance and mean ventricular rate (Md/Mean VR) (Fig. 6.1 top). The way in which beat-to-beat behaviour affects cardiac output, particularly at high ventricular rates, will be discussed later in this chapter. For the moment we should note that this graph, showing predicted linear cardiac output for any ventricular rate, is highly relevant for the clinician managing a patient with atrial fibrillation. The whole of the analysis depicted in Fig. 6.1 may be called the "haemodynamic profile".

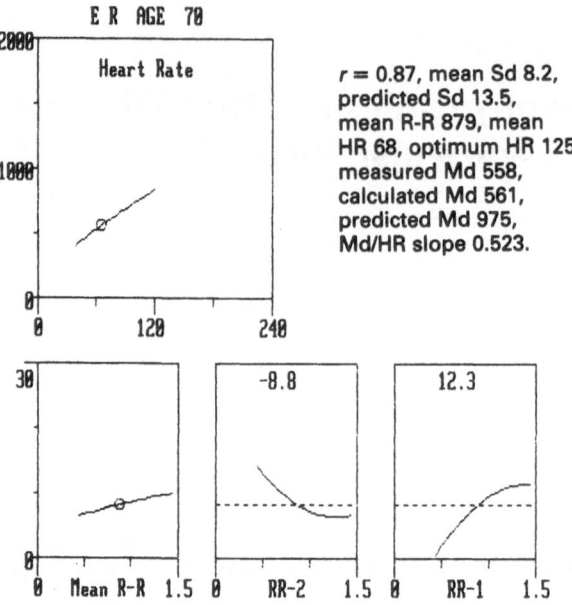

Fig. 6.1. Haemodynamic profile of a 70-year-old man.

Three Haemodynamic Profiles

Fig. 6.1 shows the complete analysis for a 70-year-old man. The multiple correlation coefficient, R, is 0.87 indicating that 76% of the variance of stroke distances is explained in terms of RR−1 and RR−2 in the multiple regression equation. Mean stroke distance at 8.2 cm is 61% of the age-predicted value of 13.5 cm; mean R-R interval of 879 milliseconds corresponds to a mean ventricular rate of 68. The two curvilinear graphs, of Sd/RR−1 and Sd/RR−2, combine to give almost a straight-line graph for Sd/Mean R-R. This transforms into a virtual straight-line graph for Md/Mean VR, the slope of which at the point of mean minute distance and ventricular rate (indicated by a circle) is 0.523% per beat per minute. Slope is calculated as minute distance, expressed as a percentage of the age-predicted normal value, per beat per minute change in ventricular rate. In this case an increase in mean ventricular rate of 10 beats per minute would result in an increase of 5.23% of the age-predicted minute distance of 975 cm, or 51 cm, some 9% of the measured minute distance. The ventricular rate at which minute distance is predicted to be maximal (Optimum HR) is 125. Minute distance measured over 3 minutes was 558 cm, compared with the predicted normal value of 975 cm. Minute distance calculated by substitution of mean R-R interval into the regression equation is very slightly higher than measured minute distance (561 vs. 558 cm).

The haemodynamic profile of a 91-year-old man is shown in Fig. 6.2, in exactly the same format. The main points to note are that mean stroke distance is 89% of the predicted normal value, and the ventricular rate is only 58, indications of good

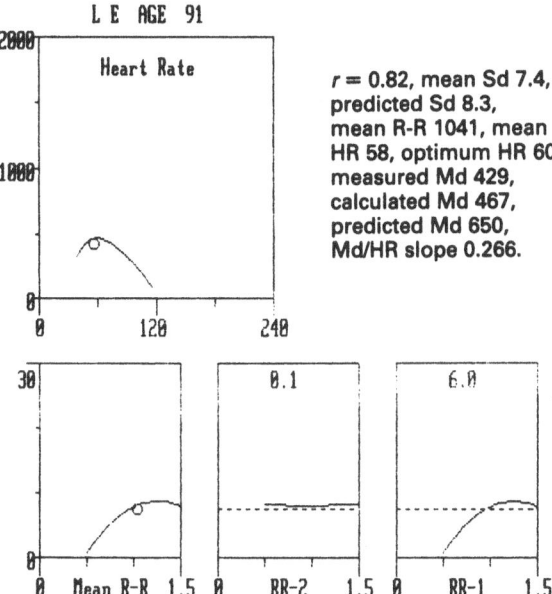

r = 0.82, mean Sd 7.4,
predicted Sd 8.3,
mean R-R 1041, mean
HR 58, optimum HR 60,
measured Md 429,
calculated Md 467,
predicted Md 650,
Md/HR slope 0.266.

Fig. 6.2. Haemodynamic profile of a 91-year-old man.

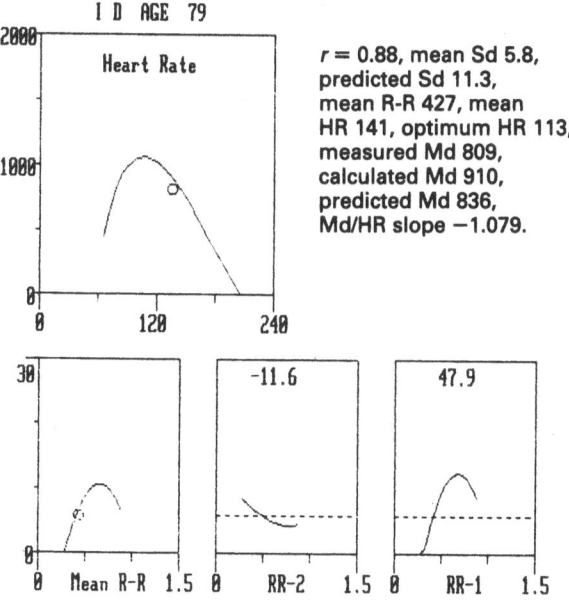

r = 0.88, mean Sd 5.8,
predicted Sd 11.3,
mean R-R 427, mean
HR 141, optimum HR 113,
measured Md 809,
calculated Md 910,
predicted Md 836,
Md/HR slope −1.079.

Fig. 6.3. Haemodynamic profile of a 79-year-old man.

left ventricular function. Unusually, the Sd/RR−2 curve is almost flat, nearly all of the variance of stroke distance being due to changes in RR−1. The Sd/Mean R-R graph is therefore almost identical with that of Sd/RR−1. Mean ventricular rate at 58 is slightly below the value at which minute distance is predicted to be maximal (60), the Md/Mean VR slope being positive.

A profile from a 79-year-old with poor left ventricular function (mean stroke distance 51% of predicted) is seen in Fig. 6.3. Mean ventricular rate at 141 is well above the rate at which minute distance is predicted to be maximal (113), the slope of the Md/Mean VR curve being steeply negative.

These three examples show patients in whom mean ventricular rate is the same as, less than, and more than that predicted for maximal cardiac output. Apart from any insights into ventricular functioning, the haemodynamic profile indicates whether or not ventricular rate is appropriate, and whether a change in ventricular rate will bring about an increase or a decrease in cardiac output. Thus, to increase output an increase in heart rate would be required in the first case (Fig. 6.1), and a decrease in the third case (Fig. 6.3). Any change beyond a couple of beats per minute would result in a fall in output in the second case (Fig. 6.2).

Relationship between Stroke and Minute Output

In Chapter 4 the idea that cardiac output may be considered as a distance, rather than as a volume, was developed. The basic measurement is the systolic velocity-integral of aortic blood flow, or stroke distance; the product of stroke distance and ventricular rate is minute distance. Stroke and minute distance are linear analogues of stroke volume and cardiac output respectively. Linear and volumetric measurements may be used interchangeably in the following argument.

To understand the relationship between minute distance and heart rate it is helpful to consider in detail how the Md/Mean VR graph derives from that of Sd/Mean R-R. In Fig. 6.4 (top left) the latter relationship is represented by a straight line with the equation

$$\text{Stroke distance} = a + b(\text{Mean R- R}) \tag{6.1}$$

where a and b are the intercept and slope of the line respectively, and Mean R-R is in minutes. It follows that

$$\text{Minute distance} = \text{Ventricular rate}[a + b(\text{Mean R-R})]$$

But Mean R-R = 1/Ventricular rate, therefore

$$\text{Minute distance} = a(\text{Ventricular rate}) + b \tag{6.2}$$

This, too, is the equation for a straight line, the intercept, a, of equation (6.1) becoming the slope of equation (6.2) and vice versa (Fig. 6.4, top right). Thus, for the slope of the Md/Mean VR curve to be negative, the Sd/Mean R-R graph has to have a negative intercept.

The relationship between mean R-R interval and stroke volume or distance is commonly biphasic, with rapid ventricular filling in early diastole followed by a plateau in late diastole, as shown diagrammatically in Fig. 6.4 (middle left). The plateau section, which has a positive intercept of 15 cm gives rise to the initial

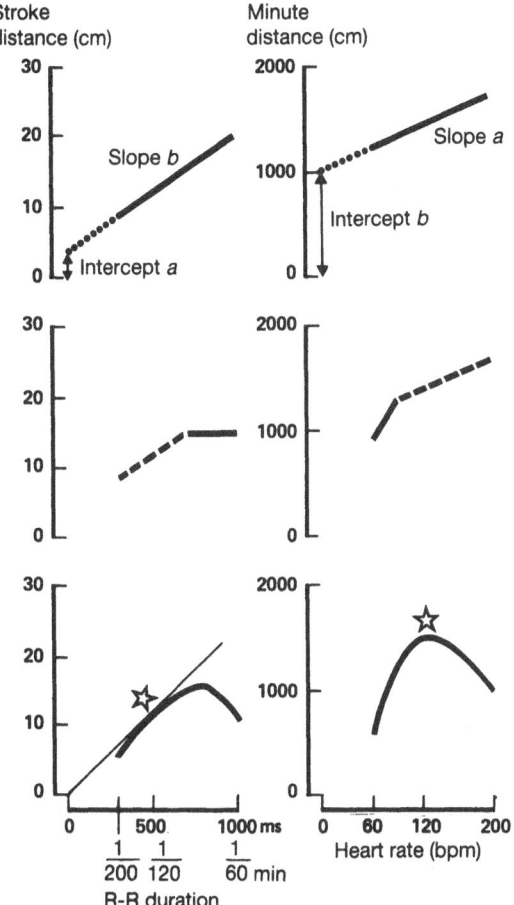

Fig. 6.4. The derivation of minute distance/mean ventricular rate graphs (*right*) from stroke distance/mean R-R interval graphs (*left*).

portion of the Md/Mean VR curve in Fig. 6.4 (middle right), which has a steeply positive slope of 15. The initial part of the Sd/Mean R-R graph has a positive intercept of 5 cm and gives rise to the right-hand portion of the Minute distance/ Heart rate curve which has a less positive slope of 5.

In reality, the relationship between stroke distance and mean R-R interval is usually curved as shown in Fig. 6.4 (bottom left), the corresponding curved Md/ Mean VR graph being shown in Fig. 6.4 (bottom right). The heart rate at which minute distance is maximal, when the slope of the Md/Mean VR curve is 0, corresponds to the R-R interval at which the slope of the Sd/Mean R-R curve passes through the origin and has an intercept value of 0; these points are marked by an asterisk.

Reduction of ventricular rate will only lead to an increase in minute distance if the slope of the Md/Mean VR curve is negative, which in turn depends on the slope of the Sd/Mean R-R curve having a negative intercept at the corresponding point.

The Contribution of the Force–Frequency Effect to Maintenance of Cardiac Output at High Ventricular Rates

In the mathematical model that we have been using, the graph showing the relationship between stroke distance and mean R-R interval is derived by summation of the independent effects of the two previous R-R intervals on stroke distance.

The intercept of the first part of the Sd/RR−1 relationship is always negative, so if the duration of the previous R-R interval were the sole determinant of stroke distance, the Md/Mean VR curve would always have a negative slope at high heart rates. However, because of operation of the force–frequency effect, stroke distance is usually negatively related to the pre- preceding R-R interval, RR−2. The intercept of the first part of the Sd/RR−2 curve is therefore usually positive, as in Figs. 6.1 and 6.3. When the independent effects of 2 previous R-R intervals on stroke distance are summated the intercept of the first part of the Sd/Mean R-R curve may be either negative, as in Figs. 6.2 and 6.3, or positive, as in Fig. 6.1. The slope of the resulting Md/Mean VR curve at high heart rates is therefore either negative (Figs. 6.2 and 6.3) or positive (Fig. 6.1). The operation of the force–frequency effect and the increase in contractility at high heart rates thus plays a major part in determining the ventricular rate for maximum cardiac output.

The Predicted Effect of Heart Rate on Cardiac Output in Mitral Stenosis

We have seen that it is the intercept of the Sd/Mean R-R curve that determines the slope of the Md/Mean VR graph at the corresponding point, a positive intercept giving rise to a positive slope, and a negative intercept giving rise to a negative slope. Further, it is the left-hand part of the Sd/Mean R-R curve that determines the cardiac output at high ventricular rates, and, conversely, the right-hand part of the Sd/Mean R-R graph, when R-R intervals are longest, that determines cardiac output at low ventricular rates. We have also seen how the Sd/RR−1 graph is commonly biphasic with stroke distance reaching a plateau with the longest R-R intervals, this behaviour also being seen in the Sd/Mean R-R graph; Fig. 6.2 is a good example. If the intercept of that region of the graph marked by long R-R intervals is positive, then so too is the slope of the Md/Mean VR graph in the region of low ventricular rates.

In mitral stenosis the plateau section of the Sd/RR−1 graph is delayed or absent since the ventricle is still filling at the end of diastole. If the Sd/RR−1 gradient is still steep when R-R intervals are long then the intercept will be negative and, as a consequence, the slope of the Md/Mean VR curve will be negative at low ventricular rates. An example from a woman with mitral stenosis is given in Fig. 6.5.

When the stenosed mitral valve first opens, early in diastole, ventricular filling occurs rapidly, and the force–frequency effect is operating normally. The form of the first part of the Sd/Mean R-R curve is therefore normal, with a positive intercept in this example. The resultant Md/Mean VR graph is U-shaped (Fig. 6.5). From the graph we may predict that minute distance would be increased by reduction of mean ventricular rate from 94 to 53; surprisingly, an increase in

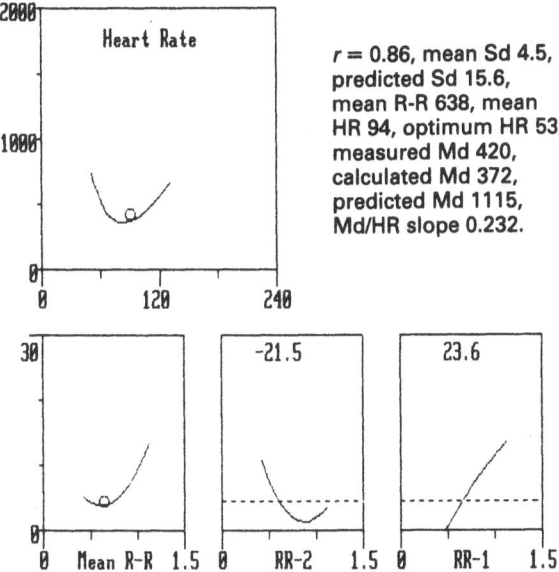

r = 0.86, mean Sd 4.5,
predicted Sd 15.6,
mean R-R 638, mean
HR 94, optimum HR 53,
measured Md 420,
calculated Md 372,
predicted Md 1115,
Md/HR slope 0.232.

Fig. 6.5. Haemodynamic profile of a 61-year-old woman with mitral stenosis.

linear cardiac output of similar magnitude is also predicted with an *increase* in ventricular rate to about 130 beats per minute.

Effect of Irregularity *Per Se* on Cardiac Output

By substituting the mean R-R interval in the multiple regression equation for each patient it is possible to calculate the value of minute distance when the mean ventricular rate is identical with that measured but the rhythm is regular, all R-R intervals being of equal duration. The difference between minute distance calculated in this way and measured directly is an indication of the effect on minute distance of irregularity of rate *per se*. On average, measured minute distance is 4% less than calculated minute distance, suggesting that irregularity of heart rate by itself is not an important cause of reduced cardiac output in atrial fibrillation.

The effect of irregularity on cardiac output is determined by the symmetry and shape of the Md/Mean VR graph and on the frequency distribution of beats with stroke distance above and below average. This is illustrated by the Md/Mean VR graph in Fig. 6.2. If the instantaneous rate of all beats were to be the same as the mean rate of 58, then the measured minute distance would be the same as the calculated minute distance at 467 cm. However, the mean ventricular rate comprises rates above and below the mean, rates which are associated with minute distances lower than that indicated by the Md/Mean VR curve at its highest point. Calculated minute distance is therefore some 8% higher than measured minute distance.

In Fig. 6.5 the opposite situation is depicted. Because the patient's mean

ventricular rate is at the trough of a U-shaped Md/Mean VR curve, the mean
ventricular rate comprises beats with instantaneous rates above and below the
mean that have outputs higher than predicted at the trough of the curve. The
calculated minute distance is 11% less than measured minute distance.

The Ventricular Rate for Maximum Cardiac Output

The Predicted Ventricular Rate for Maximal Cardiac Output

In a group of 60 patients with atrial fibrillation whose haemodynamic profiles
were computed the mean ventricular rate was 103 (Rawles 1990). The average
predicted ventricular rate for maximal minute distance derived from the regression equations was 122 beats per minute. A plot of actual mean ventricular
rates against predicted ventricular rates for maximal minute distance is shown in
Fig. 6.6. In 43 cases (72%) the predicted ventricular rate for maximal minute
distance is more than the measured ventricular rate (above the line of identity in
Fig. 6.6), but in 17 cases (28%) it was lower (below the line of identity in Fig. 6.6).
All 17 patients with a ventricular rate above that predicted for maximal minute
distance had a ventricular rate over 90 beats per minute. In none of the 20 patients
with a ventricular rate less than 90 did the ventricular rate exceed the predicted
ventricular rate for maximal cardiac output.

There is no relationship between the predicted ventricular rate for maximum
cardiac output and age, or diagnosis. In particular, patients with mitral stenosis
do not differ significantly in this respect from those with other diagnoses.

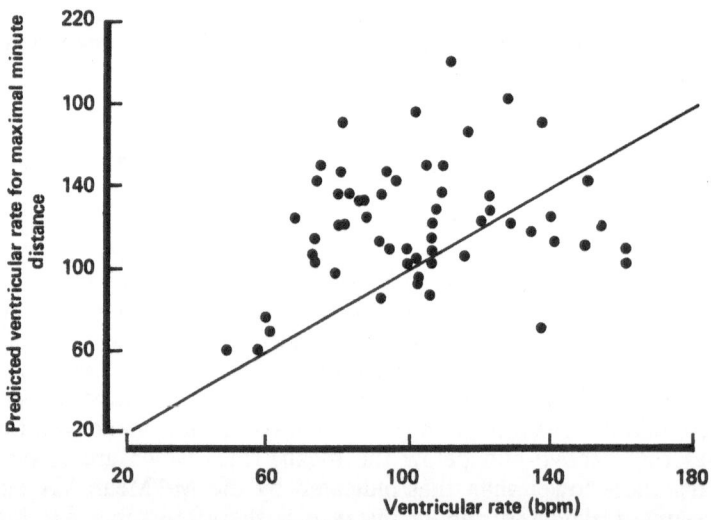

Fig. 6.6. Plot of predicted ventricular rates for maximal minute distance against mean ventricular
rates. The line of identity is shown. (Reproduced by permission of the Editor, *British Heart Journal*.)

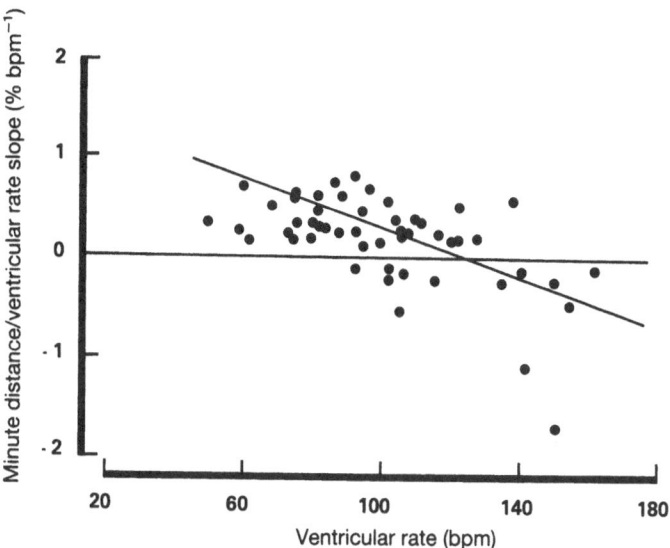

Fig. 6.7. The slope of the Md/Mean VR graph plotted against mean ventricular rate. The regression line is shown. (Reproduced by permission of the Editor, *British Heart Journal.*)

Relationship Between Slope of the Md/Mean VR Graph and Ventricular Rate

Fig. 6.7 shows the relationship between the slope of the Minute distance/Mean VR graph and ventricular rate. For lower heart rates the slope is positive, and for higher rates it is negative; the regression line intercepts the line of slope 0 at a ventricular rate of 122. Assuming an inverted U-shaped Md/Mean VR graph, a gradient of 0 is found at the summit of the graph, where minute distance is maximal. On average then, the rate for maximal cardiac output is 122; reduction of ventricular rate below this will reduce minute distance, and above this rate minute distance will be increased, but there is wide individual variation.

Group Relationship Between Mean R-R Interval and Stroke Distance

Mean stroke distances and mean R-R intervals from a group of 70 patients with atrial fibrillation are plotted in Fig. 6.8. The relationship between them is best described by a quadratic equation ($R = 0.54$ $p<0.001$). From this is derived Fig. 6.9, which shows the relationship between heart rate and minute distance for the group as a whole. The mean ventricular rate for maximal cardiac output is 117 beats per minute, close to the value of 122 beats per minute derived from analyses of the beat-to-beat variation of stroke distance within subjects. This result, obtained independently, provides a corroboration of the computer model.

In atrial fibrillation, therefore, the ventricular rate for maximal cardiac output at rest is about 120 beats per minute on average, with a wide individual variation. The ventricular rate for maximal cardiac output is not significantly different in patients with mitral valve disease from those with other diagnoses.

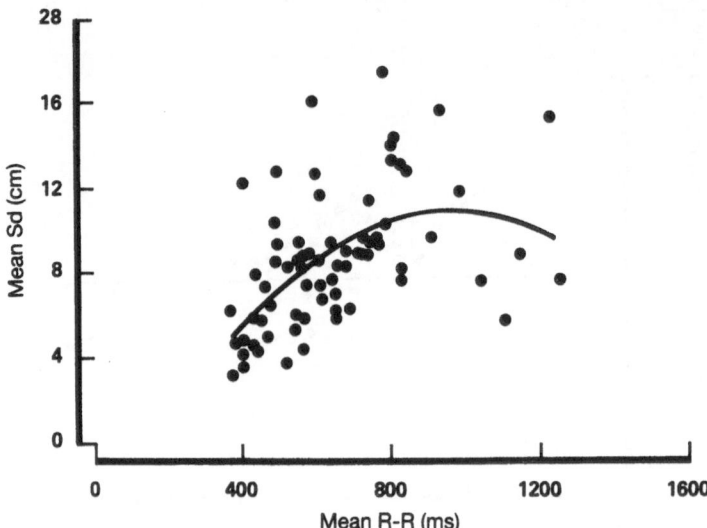

Fig. 6.8. Mean stroke distances and mean R-R intervals in 70 patients with atrial fibrillation. The best fit quadratic regression line is shown. $r = 0.54$, $p<0.001$.

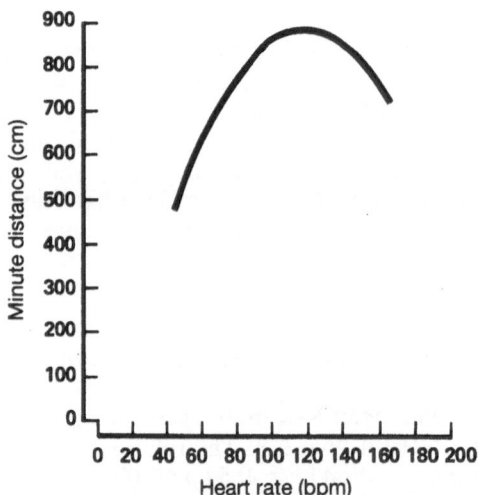

Fig. 6.9. Relationship between minute distance and mean ventricular rate in atrial fibrillation, derived from the regression line in Fig. 6.8.

The Magnitude of the Effect of Ventricular Rate on Cardiac Output

The average Md/Mean VR slope for cases where the slope was negative was −1.1% per beat per minute. This indicates that a reduction in mean ventricular rate of 1 beat per minute would lead to an increase in minute distance of 1.1% of the age- predicted normal value. This is equivalent to an increase in minute distance of 1.5% since in all cases the measured minute distance was less than predicted. On average, reduction of ventricular rate by 10 beats per minute in these patients would lead to a 15% increase in cardiac output.

The Ventricular Rate for Maximum Cardiac Output: Sinus Rhythm

In sinus rhythm, if other factors remain constant, increasing the heart rate results in a rise in cardiac output until a maximum value is achieved at a rate of about 120–140 beats per minute (Guyton et al. 1973). As in atrial fibrillation, there is a good deal of individual variation in the heart rate at which maximal cardiac output is achieved (Ross et al. 1965; Sugimoto et al. 1966) which may be explained by the variable effect of the duration of previous R-R intervals on contractility and on ventricular filling, the latter being the more important factor. This accounts for the greater influence of heart rate on cardiac output in the presence of augmented venous return (Sugimoto et al. 1966).

In mitral stenosis the pattern of ventricular filling is abnormal so the cardiac output/heart rate graph might be expected to differ from normal too. However, as discussed earlier, the main abnormality is continued filling in late diastole, which influences cardiac output at low heart rates, whereas the pattern of ventricular fillng in early diastole determines the heart rate at which cardiac output is maximal. In 11 patients with mitral stenosis paced at mean ventricular rates of 95, 124 and 146 beats per minute, mean cardiac indices were 2.29, 2.35 and 2.08 litres per minute respectively (Arani and Carleton 1967), indicating that in patients with mitral stenosis the ventricular rate for maximal cardiac output is in the same range as in other subjects.

Thus, in sinus rhythm, the average ventricular rate for maximal cardiac output is similar to that in atrial fibrillation, and the values are similar whether or not mitral stenosis is present.

Comparison of Atrial Fibrillation and Sinus Rhythm

Atrial fibrillation results in reduction of stroke volume due to loss of atrial contraction, but an increased ventricular rate partially compensates for this and the reduction of cardiac output due to the arrhythmia is estimated to be 20%–30% (Ferrer et al. 1952; Wade et al. 1952). In the present series, stroke and minute distance were reduced by 36% and 15% respectively, and the mean ventricular rate at 103 beats per minute was increased by 48% compared with age-predicted normal values for sinus rhythm (Rawles 1988). The tachycardia of atrial fibrillation is therefore usually a beneficial phenomenon partially counteracting an even greater fall in cardiac output that would otherwise occur, except in those patients where ventricular rate exceeds that for maximal output.

In our patients with atrial fibrillation, who had a mean ventricular rate of 103, mean minute distance was 85% of the age- predicted normal value for patients in sinus rhythm, but if ventricular rate had been reduced to the normal resting value of 70 it may be estimated that minute distance would have been only about 70% of normal.

What is Meant by "Control" of Ventricular Rate?

Applied to the ventricular rate in atrial fibrillation the term "controlled" is generally used loosely to mean "not excessive". Few authors define precisely what is meant by a controlled rate but Bigger (1980 p731) specifies 50–90, while Storstein (1984) suggests an even narrower range of 60–80 beats per minute, both authors referring to patients at rest. The mid point of both these ranges is 70, the accepted normal rate in sinus rhythm. However, in sinus rhythm the contribution made to stroke volume by atrial contraction is estimated to be 20–50%, depending on left ventricular compliance and other aspects of left ventricular function (Rahimtoola et al. 1975; DeMaria et al. 1976). In atrial fibrillation, therefore, the ventricular rate needs to be 20–50% higher than in sinus rhythm to maintain a normal cardiac output. The recommendation of a target ventricular rate of 70 in atrial fibrillation appears to disregard completely the haemodynamic consequences of the dysrhythmia.

The prediction of the effect of ventricular rate on cardiac output described earlier in this chapter is derived from analysis of recordings made from patients with atrial fibrillation at rest. The results should not be extrapolated to exercise, in which, as well as a change of heart rate, there may be alterations of posture, venous return, blood pressure, cardiac contractility, and autonomic tone. The heart rate for maximal cardiac output during exercise therefore differs from that at rest. In sinus rhythm rates of up to 180 are achieved during exercise, and it seems likely that rates at least as great as this may be required in atrial fibrillation to maximise cardiac output.

Some writers use the term "control" to mean avoidance of excessive ventricular rates during exercise. Thus David et al. (1979) reported a mean ventricular rate of 139 during exercise in patients with therapeutic serum levels of digoxin; they said "digitalis alone is ineffective in controlling the heart rate during exercise in many patients with chronic atrial fibrillation". Beasley et al. (1985) compared ventricular rates in patients with atrial fibrillation and therapeutic serum levels of digoxin with control subjects in sinus rhythm. At rest and at three levels of exercise ventricular rates were respectively 72, 105, 118 and 130 in sinus rhythm and 79, 135, 162 and 184 in atrial fibrillation. They considered that in atrial fibrillation the heart rate during exercise was inappropriately high and pointed out that it is not predictable from a knowledge of the resting value. Ventricular rate could therefore be "controlled" at rest but "uncontrolled" during exercise.

Channer et al. (1987) conducted a randomised cross-over trial of additional digoxin or verapamil in 14 patients with atrial fibrillation already taking digoxin. Although the additional therapy resulted in lower maximal heart rates, 6 minute walking distances were unchanged. Lewis et al. (1988) administered single doses of digoxin, verapamil and diltiazem, alone and in combination, to 6 patients with chronic atrial fibrillation. Immediately after exercise lower heart rates were seen with the combination of digoxin and calcium antagonists than with digoxin alone or with placebo. Reduction of the ventricular rate during exercise was associated with a small increase in stroke volume but the benefits of this were offset by a rate-related reduction in cardiac output. The highest mean cardiac output was after digoxin alone, when the mean ventricular rate was 159. Mean walking distances were not significantly different after any of the treatments.

These data indicate that exercise tolerance is not increased by reduction of ventricular rate. During exercise, as is the case at rest, a higher ventricular rate is required in atrial fibrillation than in sinus rhythm to maintain cardiac output in the face of the loss of the atrial contribution to stroke output.

A New Concept of Control of the Ventricular Rate in Atrial Fibrillation

In atrial fibrillation it may be necessary to reduce a rapid ventricular rate for the greater comfort of the patient, or, particularly if mitral stenosis is present, for the reduction of left atrial pressure and relief of pulmonary congestion. However, in the management of atrial fibrillation the effect that a reduction in ventricular rate would have on cardiac output should be considered; ventricular rate should not be reduced arbitrarily regardless of its effect on cardiac output or exercise tolerance.

In this chapter a method has been described that, for patients at rest, provides an answer to the question "If the ventricular rate in atrial fibrillation is reduced will cardiac output rise or fall?" In principle the method could be applied to patients during exercise, but with existing equipment the quality of the recordings would be inadequate for accurate interpretation. Even at rest continuous records of sufficient quality could not be obtained in some subjects.

To summarise, a "controlled" ventricular rate in atrial fibrillation may be considered as a rate below that predicted for maximal cardiac output, when the slope of the cardiac output/ventricular rate curve is positive; reduction in rate would lead to a reduction in output. When ventricular rate is "controlled" the circulatory reflexes may adjust the rate according to circulatory needs; their role in atrial fibrillation is considered in Chapter 8. Conversely, an "uncontrolled" ventricular rate is above that predicted for maximal cardiac output and a reduction in rate would lead to increased output; reflex activity may exacerbate the tachycardia. As so defined, ventricular rate was always controlled when less than 90, always uncontrolled when more than 140, and uncontrolled in 27% of patients with ventricular rates between 90 and 140.

In atrial fibrillation, a target ventricular rate of 90 beats per minute at rest would be within the range where it may be adjusted by the cardioregulatory reflexes, and cardiac output would be least compromised. During exercise the heart rate that results in maximal cardiac output is likely to be greater than the target rate of 90–115 that has been recommended (Bigger 1980 p. 731).

References

Arani DT, Carleton RA (1967) The deleterious role of tachycardia in mitral stenosis. Circulation 36:511–6

Beasley R, Smith DA, McHaffie DJ (1985) Exercise heart rate at different serum digoxin concentrations in patients with atrial fibrillation. Br Med J 290:9–11

Bigger JT (1980) Management of arrhythmias. In: Braunwald E, ed. Heart disease. WB Saunders, Philadelphia

Channer KS, Papouchado M, James MA, Pitcher DW, Rees JR (1987) Towards improved control of atrial fibrillation. Eur Heart J 8:141–7

David D, Di Segni E, Klein HO, Kaplinsky E (1979) Inefficacy of digitalis in the control of heart rate in patients with chronic atrial fibrillation: beneficial effects of an added beta adrenergic blocking agent. Am J Cardiol 44:1378–82

DeMaria AN, Miller RR, Amsterdam EA, Markson W, Mason DT (1976) Mitral valve early diastolic closing velocity: relation to sequential diastolic flow and ventricular compliance. Am J Cardiol 37:693–700

Ferrer MI, Harvey RM, Cathcart RT, Cournand A, Richards DW (1952) Haemodynamic studies in rheumatic heart disease. Circulation 6:688–710

Guyton AC, Jones CE, Coleman TG (1973) Circulatory physiology: Cardiac output and its regulation. WB Saunders, Philadelphia, p 307

Lewis RV, Irvine N, McDevitt DG (1988) Relationships between heart rate, exercise tolerance and cardiac output in atrial fibrillation: the effects of treatment with digoxin, verapamil and diltiazem. Eur Heart J 9:777–81

Rahimtoola SH, Ehsani A, Sinno MZ, Loeb HS, Rosen KM, Gunnar RM (1975) Left atrial transport function in myocardial infarction. Am J Med 59:686–94

Rawles JM (1988) A mathematical model of left ventricular function in atrial fibrillation. Int J Biomed Comput 23:57–68

Rawles JM (1990) What is meant by a "controlled" ventricular rate in atrial fibrillation? Br Heart J 63:157–61

Ross J, Linhart JW, Braunwald E (1965) Effects of changing heart rate in man by electrical stimulation of the right atrium: studies at rest, during exercise, and with isoproterenol. Circulation 32:549–58

Storstein L (1984) Role of digitalis in ventricular rate control in atrial fibrillation. In: Kulburtus HE, Olsson SB, Schlepper M, (eds) Atrial fibrillation. AB Hassle, Molndal p 288

Sugimoto T, Sagawa K, Guyton AC (1966) Effect of tachycardia on cardiac output during normal and increased venous return. Am J Physiol 211:288–292

Wade G, Werko L, Eliasch H, Gidlund A, Lagerlof H (1952) The haemodynamic basis of the symptoms and signs in mitral valvular disease. Q J Med 21:361–83

The Use of the Haemodynamic Model of Atrial Fibrillation for Evaluating Drug Action

More Haemodynamic Profiles

We have seen how the beat-to-beat changes in stroke distance in atrial fibrillation may be explained in terms of the two previous R-R intervals, the regression equation being, in effect, a description of left ventricular function. We have seen also how the regression equation may be manipulated to enable prediction of minute distance (a linear analogue of cardiac output) at any ventricular rate.

Fig. 7.1 shows the haemodynamic profile of a 64 year old man with controlled atrial fibrillation, mean ventricular rate 81. It may be recalled that in controlled atrial fibrillation the slope of the Minute distance/Mean ventricular rate (Md/Mean VR) graph is positive, in this case 0.317% per beat per minute. Reduction of mean ventricular rate by 10 beats per minute would be expected to reduce minute distance by 3.17% of the predicted minute distance of 1068 cm, equivalent to 34 cm or 7.3% of the measured minute distance.

Fig. 7.2 shows the haemodynamic profile after digoxin, which has reduced the mean ventricular rate to 65. The profiles have very similar appearances and the new minute distance is close to that predicted by the Md/Mean VR graph in Fig. 7.1, where the new mean value is indicated by the smaller circle.

Fig. 7.3 shows the Md/Mean VR graph for a 79-year-old man with uncontrolled atrial fibrillation, his mean ventricular rate being 154. Here the Md/Mean VR slope is −0.484% per beat per minute. The larger circle indicates the mean ventricular rate and minute distance for the recording on which the analysis was based. The smaller circle shows the mean ventricular rate and minute distance after the patient had been given intravenous verapamil. Ventricular rate has decreased slightly, but minute distance has increased considerably more than was predicted from the Md/Mean VR graph. This suggests that besides reducing ventricular rate, verapamil has brought about a change in the loading conditions resulting in an increase in output disproportionate to the reduction in ventricular rate.

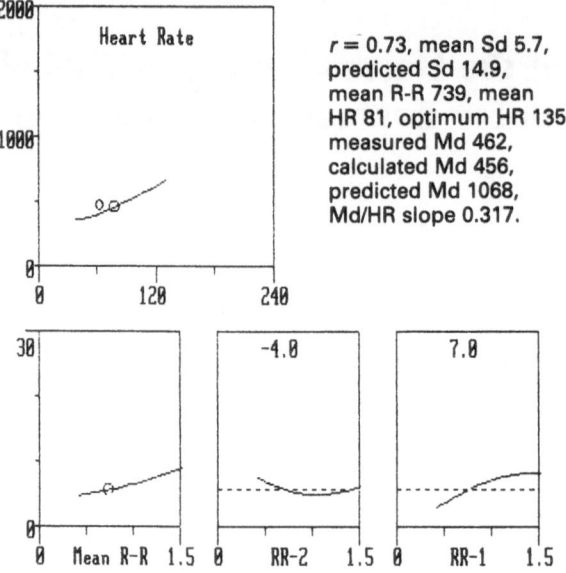

Fig. 7.1. Haemodynamic profile of a 64-year-old man.

Fig. 7.4 shows the Md/Mean VR graph of a patient with atrial fibrillation with a mean ventricular rate of 128, indicated by the larger circle. This just exceeds the predicted rate of 122 at which minute distance would be maximal, the slope being slightly negative −0.029% per beat per minute. The smaller circle to the left and just below the line shows the minute distance and mean ventricular rate after digoxin. In spite of the tachycardia the rate was then controlled by our definition, the slope at that time being positive, the minute distance being close to the predicted value. This suggests that digoxin merely reduced the ventricular rate; there is no evidence in this case of digoxin having an inotropic action, otherwise, minute distance would have increased more than indicated by the Md/Mean VR graph. The lower small circle in Fig. 7.4 indicates the minute distance and mean ventricular rate after the administration of sotalol. This drug brought about a substantial fall in minute distance with no reduction in ventricular rate. This might be explained by a brisk sympathetic response to the reduction of cardiac output over-riding the chronotropic but not the negative inotropic effect of the beta- blocker. On a later occasion the patient was observed to be in sinus rhythm, the minute distance and ventricular rate then being indicated by the small circle above and to the left. In sinus rhythm a much higher minute distance, and a lower ventricular rate, were recorded than were present in atrial fibrillation.

The Md/Mean VR graph of a subject with controlled atrial fibrillation and a mean ventricular rate of 75 is indicated in Fig. 7.5 (larger circle). After atropine,

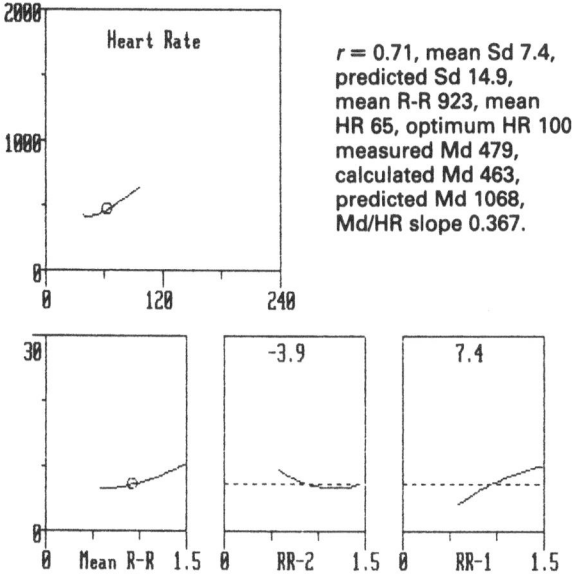

$r = 0.71$, mean Sd 7.4, predicted Sd 14.9, mean R-R 923, mean HR 65, optimum HR 100, measured Md 479, calculated Md 463, predicted Md 1068, Md/HR slope 0.367.

Fig. 7.2. Haemodynamic profile of same subject as in Fig. 7.1., after digoxin.

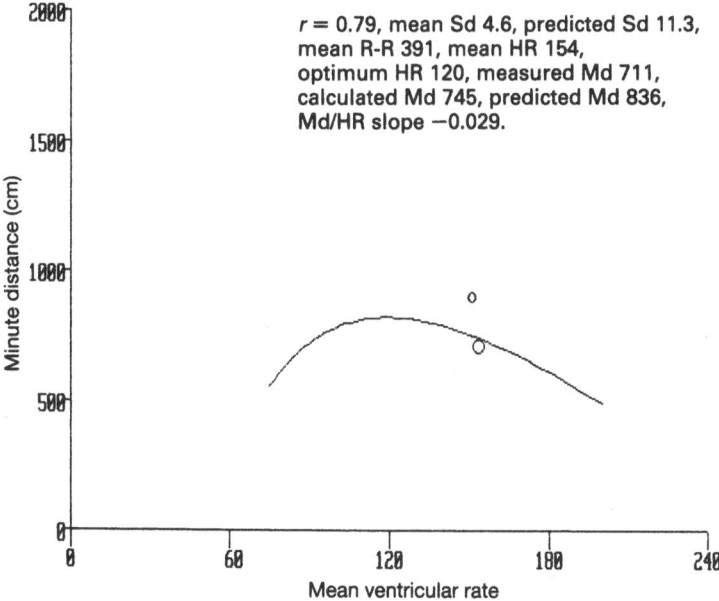

$r = 0.79$, mean Sd 4.6, predicted Sd 11.3, mean R-R 391, mean HR 154, optimum HR 120, measured Md 711, calculated Md 745, predicted Md 836, Md/HR slope −0.029.

Fig. 7.3. Md/Mean VR graph of a 79-year-old man with uncontrolled atrial fibrillation. The *smaller circle* shows the minute distance and mean VR after verapamil.

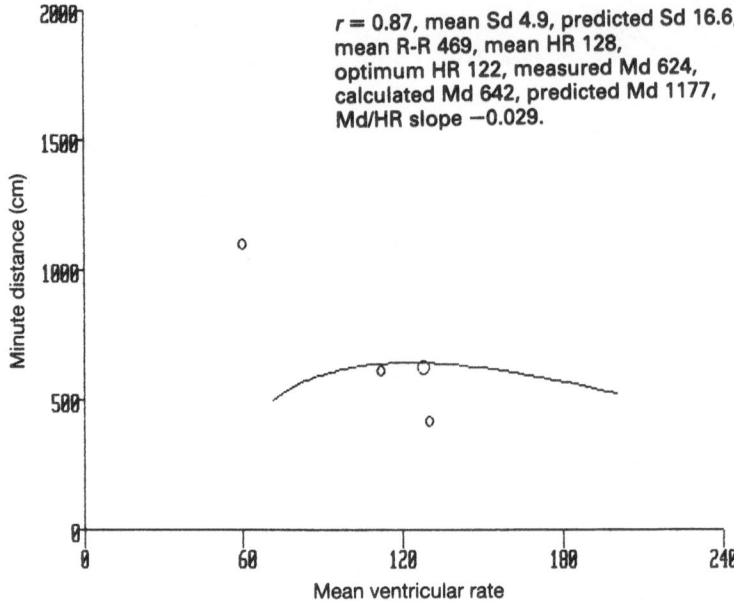

r = 0.87, mean Sd 4.9, predicted Sd 16.6,
mean R-R 469, mean HR 128,
optimum HR 122, measured Md 624,
calculated Md 642, predicted Md 1177,
Md/HR slope −0.029.

Fig. 7.4. Md/Mean VR graph of a 57-year-old with atrial fibrillation. The *lower smaller circle* below the line indicates minute distance and mean VR after sotalop, and above the line, when in sinus rhythm.

ventricular rate increased to 100 and minute distance rose to close to its predicted value (smaller circle).

These examples, besides indicating the clinical value of haemodynamic profiles in patient management, suggest another application for them. The administration of atropine and digoxin both brought about changes in ventricular rate associated with changes in minute distance which were close to what was predicted. On the other hand, the changes in minute distance resulting from the administration of sotalol and verapamil, or with reversion to sinus rhythm, were greater than predicted from the Md/Mean VR graph in atrial fibrillation. Comparison of minute distance after administration of a cardio-active drug can be made with that predicted from the Md/Mean VR graph recorded in a control period. A significant disparity might suggest that the drug has some action over and above its effect on ventricular rate. A patient with atrial fibrillation might therefore be used as his or her own control in a single-subject study with the object of detecting and quantifying inotropic and vasodilator actions. There is one further requirement before that can be done: confidence intervals for the prediction of minute distance for a given ventricular rate have to be calculated.

In Fig. 7.6 are shown the Md/Mean VR graphs for 9 patients with atrial fibrillation. The 95% confidence intervals for the regression lines are also shown. In some patients the confidence intervals are narrow, but in a few they are unacceptably wide with the result that predictions are too imprecise to be of value in detecting a significant departure from what is expected.

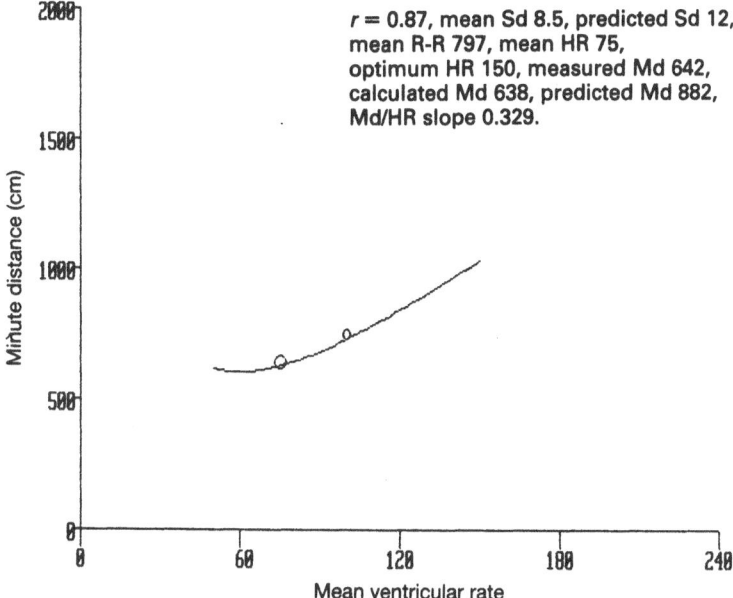

Fig. 7.5. Md/Mean VR graph of a 76-year-old man with uncontrolled atrial fibrillation. The *smaller circle* indicates minute distance and mean VR after atropine was administered.

The width of confidence intervals depends on several factors to do with the recording and analysis of the sequence of stroke distances and R-R intervals. Most important is the technical quality of the recording, which should give clearly defined velocity–time complexes of ample size. Unfortunately many patients with atrial fibrillation have poor cardiac function and low aortic blood velocity, close to the noise level. Obstructive airway disease with hyperinflation of the lungs, which is not uncommon in elderly patients with atrial fibrillation, impairs the transmission of ultrasound and the quality of the recordings. To some extent the analysis of the signal with a digitising pad off-line, which is the system we have used, compensates for deficiencies in the recording, particularly the presence of spurious noise. This is because an experienced interpreter can read the Doppler signal through the noise in a way that is quite impossible for a computer algorithm in an automatic evaluation system. However, the key word is experienced – an inexperienced operator may badly misinterpret poor quality Doppler signals. Moreover, the careful digitising of 3-minute Doppler recordings is tedious and time-consuming, and errors are inevitably introduced by inattention.

Finally, shortcomings in the model used to describe changes in stroke distance will contribute to the error of the predictions made on the basis of the model. Examples of possible deficiencies in some individuals might be omission of the third and fourth previous R-R intervals when these are making a substantial contribution to the variation in stroke distance, or use of the quadratic equation when a cubic or exponential equation would provide a better fit to the data.

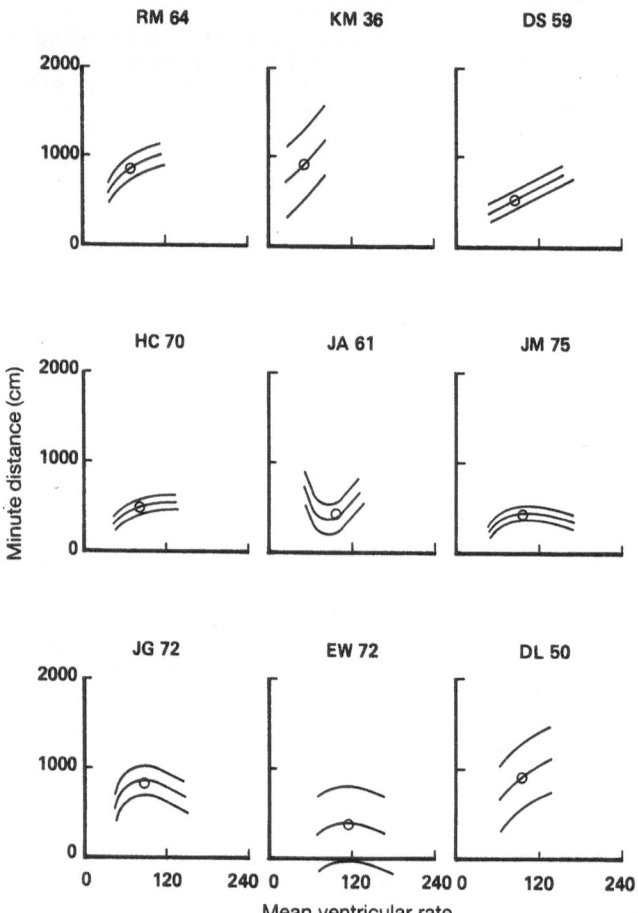

Fig. 7.6. Md/Mean VR graph with 95% confidence intervals for 9 patients participating in a study of enoximone.

Additional Haemodynamic Data Derived from Stroke Distance

The measurement of linear cardiac output, described at length in Chapter 4, permits the calculation of other haemodynamic variables that are helpful in understanding the circulation at any moment. Peripheral vascular resistance is calculated as:

Peripheral vascular resistance = Mean blood pressure/Cardiac output

The Doppler equivalent is:

Linear resistance = Mean blood pressure/Minute distance

where minute distance is measured in metres, and mean blood pressure in millimetres of mercury. Linear resistance, like stroke and minute distance, is related to age, so it is often more convenient to express all three variables as percentages of the age-predicted normal values (Haites et al. 1985).

Stroke work may be calculated as the product of stroke distance and mean blood pressure:

Stroke work = Stroke distance × Mean blood pressure

Hybrid units of centimetres and millimetres of mercury are used for convenience.

A Pilot Study of the Effect of Enoximone in Atrial Fibrillation

We have conducted a single blind pilot study of enoximone in 9 patients with atrial fibrillation, evaluating the action of the drug at each stage non-invasively with Doppler ultrasound (Rawles et al. 1987). The results were interpreted in relation to the model of left ventricular function described in Chapter 5. As well as evaluating the drug clinically and by linear cardiac output measurement, this study provided a means of assessing the utility of the model.

Enoximone (Perfan; Merrell Dow Pharmaceuticals) is a phosphodiesterase inhibitor with inotropic and vasodilator properties which may be given intravenously or by mouth (Dage et al. 1987).

The patients had all been in established atrial fibrillation for longer than 3 months and all had congestive cardiac failure, NYHA class 2–3. The age range was from 36–75 years 7 female, 2 male) and the diagnoses were ischaemic or hypertensive heart disease (6) and mitral valve disease (3). During this study all patients continued their usual medication, including digoxin.

After a 3-minute baseline Doppler ultrasound recording, 50 mg enoximone was administered intravenously. Ultrasound recordings were repeated 30 minutes and 4 hours later. Thereafter, all patients were given oral therapy for the next 3 weeks, the patients being blind to the treatment given. Enoximone was given in a dosage of 50 mg tds, followed by enoximone 100 mg tds, and then placebo, each treatment period being 1 week in duration. At the end of each treatment week the patient returned for clinical assessment and an ultrasound recording.

Of the 9 patients, 5 (3 NYHA class 2, 2 NYHA class 3) had symptomatic improvement during the low-dose and high-dose oral treatment phases. Two patients moved from NYHA class 3 to NYHA class 2 during treatment and in both the improvement was maintained while on placebo therapy. The other 3 patients all had a return of their symptoms to the pre-treatment level while on placebo.

A summary of the results is given in Table 7.1. Stroke distance was increased following intravenous but not oral enoximone, but ventricular rate and minute distance were increased by both routes of administration. Mean blood pressure fell after intravenous enoximone and after oral enoximone in the higher dosage. Linear resistance was reduced after either intravenous or oral administration. Stroke work was only significantly increased 30 minutes after intravenous injection of enoximone, and not after oral administration. None of the measure-

Table 7.1. Mean (SD) haemodynamic measurements after intravenous, oral and placebo enoximone in 9 patients with atrial fibrillation. Probability values are given for comparisons with base-line

	Base-line	Intravenous		Oral		Placebo
		+30 minutes	+4 hours	50 mg tds	100 mg tds	
Ventricular	80 (17)	89 (18)	83 (16)	96 (15)	96 (5)	85 (20)
rate (bmp)	–	<0.001	NS	<0.01	<0.01	NS
Mean blood	106 (15)	100 (14)	98 (13)	104 (13)	100 (14)	102 (15)
pressure (mmHg)	–	<0.05	<0.05	NS	<0.05	NS
Linear	273 (92)	180 (70)	189 (64)	209 (20)	206 (109)	276 (139)
resistance (%)	–	<0.01	<0.001	<0.05	<0.01	NS
Stroke	37 (16)	49 (21)	47 (19)	37 (11)	41 (15)	37 (16)
distance (%)	–	<0.01	<0.05	NS	NS	NS
Minute	39 (11)	57 (18)	53 (18)	49 (13)	54 (21)	42 (19)
distance (%)	–	<0.01	<0.01	<0.01	<0.01	NS
Stroke	615 (326)	751 (415)	718 (318)	600 (218)	631 (262)	593 (291)
work (cm mmHg)	–	<0.05	NS	NS	NS	NS

ments differed significantly at the end of the final week on placebo compared with the run-in values.

From these results there is clear evidence of vasodilatation, with the increased stroke work 30 minutes after intravenous administration being indicative of positive inotropic action at that time. The effect of enoximone in increasing mean ventricular rate in atrial fibrillation is particularly striking after oral administration.

The Effect of Enoximone on the Haemodynamic Model

The Md/Mean VR graphs for all 9 patients at baseline are shown in Fig. 7.6. The most marked haemodynamic changes took place 30 minutes after intravenous injection of enoximone. Visual inspection of the haemodynamic profiles recorded at that time did not show any obvious qualitative differences from those recorded in the run-in period. The slopes of both the Sd/RR−1 and Sd/RR−2 graphs were increased significantly then, and even more so after oral enoximone. However, the ratio of the slopes of these two graphs did not change significantly at any time in the study. The expectation that an inotropic drug might preferentially affect one or other of the two main constituents of the haemodynamic profile was therefore not borne out. Whether this is because enoximone is not a pure inotrope, or whether it is because the slopes of either Sd/RR−1 or Sd/RR−2 are not pure representations of these elements of left ventricular function is not clear.

The mean minute distance and ventricular rates for all 9 patients at each stage of the study are seen in Fig. 7.7. At 30 minutes after intravenous enoximone minute distance was increased by 46% compared with the run-in period, and was accompanied by a 12% increase in mean ventricular rate. In six out of 9 individuals minute distance exceeded the predicted minute distance for the ventricular rate at that time, the Md/Mean VR point being outside the confidence intervals established during the run-in period. Thus, in two-thirds of subjects the

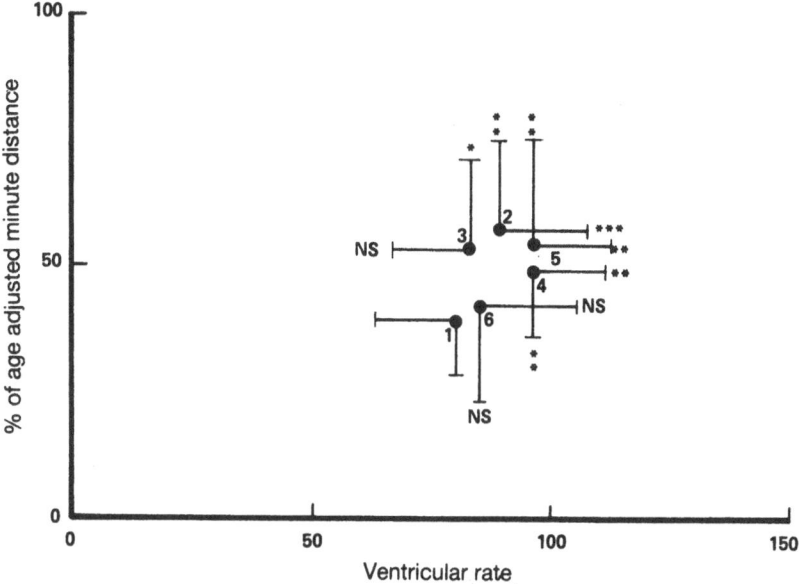

Fig. 7.7. Plot of average minute distances (SEM) and ventricular rates (SEM) for 9 patients at each stage of enoximone study. 1, baseline; 2, 30 min after enoximone IV; 3, 4h after enoximone IV; 4, enoximone 50 mg tds; 5, enoximone 100 mg tds; 6, placebo.

model was sufficiently sensitive, and the predictions sufficiently precise, for the detection of an improvement in left ventricular function beyond that attributable just to a change in ventricular rate.

Control of Ventricular Rate While on Enoximone

The average ventricular rate was 80 during the run-in period, and it was controlled in all cases, the slopes of the Md/Mean VR graphs being positive. Thirty minutes after intravenous enoximone the average ventricular rate was 89 but it was still controlled in 8 cases out of 9. The highest average ventricular rates were recorded while enoximone was taken orally. On the higher dose the average rate was 96 and this was uncontrolled in 3 subjects who had mean rates of 96, 108 and 122. As in the study reported in Chapter 6, uncontrolled ventricular rates were not seen where the mean rate was less than 90, and ventricular rate was uncontrolled in only 8 out of 27 occasions when the mean rate exceeded 90 beats per minute. In 7 of these 8 cases of uncontrolled atrial fibrillation, the mean ventricular rate was closer than 10 beats per minute to the rate at which minute distance was predicted to be maximal.

The lesson to be learnt from this study is that in atrial fibrillation reduction of a rapid ventricular rate is not necessarily associated with increased cardiac output, but cardiac output may be increased with an inotropic/vasodilator agent which may increase ventricular rate. In most cases the rate will be close to that at which cardiac output is predicted to be maximal.

A Placebo Controlled Trial of Enoximone and Digoxin in Atrial Fibrillation

We have carried out a randomised double blind trial of enoximone, digoxin, enoximone and digoxin, and placebo in 8 patients with chronic atrial fibrillation. Five patients were female and 3 were male, age range 52–75 years. During a run-in period the dose of digoxin needed to obtain therapeutic serum levels was established. There were four treatment periods each of 2 weeks, in random order, in which were administered digoxin (in the previously determined dosage) plus placebo enoximone, enoximone 100 mg tds and placebo digoxin, digoxin and enoximone, and double placebo. Patients were assessed on four occasions at the end of each treatment period: after a clinical and Doppler ultrasound examination each patient walked as far as he or she could during 6 minutes, and the distance covered was recorded.

Results are given in Table 7.2. Neither symptom scores nor 6-minute walking distances were significantly different in any of the treatment periods; neither stroke nor minute distances during administration of digoxin, enoximone or the combination were significantly different from measurements made during the double placebo period. However, by two-way analysis of variance administration of digoxin, but not enoximone, was associated with increased minute distance ($p < 0.05$). The average ventricular rate was lower on digoxin and higher on enoximone than with double placebo, and systolic blood pressure was higher on digoxin. Comparing digoxin with enoximone, ventricular rate was lower, and stroke distance and stroke work were significantly higher with digoxin.

In Fig. 7.8 average ventricular rates and minute distances are plotted for each of the treatment periods. It is seen that compared with placebo, both digoxin and

Table 7.2. Mean (SD) haemodynamic measurements after placebo, digoxin, enoximone, and digoxin with enoximone in 8 patients with atrial fibrillation. Probability values are given for comparisons with placebo, and in italics for the comparisons between digoxin and enoximone

	Placebo	Digoxin		Enoximone	Digoxin + enoximone
Ventricular rate (bpm)	95 (24) –	74 (13) <0.05	*<0.001*	106 (24) <0.05	84 (13) NS
Systolic blood pressure (mmHg)	126 (19) –	137 (17) <0.01	*NS*	131 (19) NS	137 (18) NS
Diastolic blood pressure (mmHg)	84 (12) –	79 (10) NS	*NS*	82 (15) NS	82 (15) NS
Stroke distance (cm)	6.7 (1.5) –	10.8 (6.5) NS	*<0.05*	7.4 (3.4) NS	9.1 (4.8) NS
Minute distance (cm)	622 (193) –	799 (529) NS	*NS*	798 (478) NS	764 (442) NS
Linear resistance (mmHg/m)	16.9 (4.9) –	16.7 (9.1) NS	*NS*	16.4 (9.6) NS	16.6 (7.3) NS
Stroke work (cm mmHg)	656 (117) –	1059 (648) NS	*<0.05*	709 (304) NS	901 (445) NS
Walking distance (m)	465 (132) –	456 (90) NS	*NS*	470 (104) NS	483 (111) NS

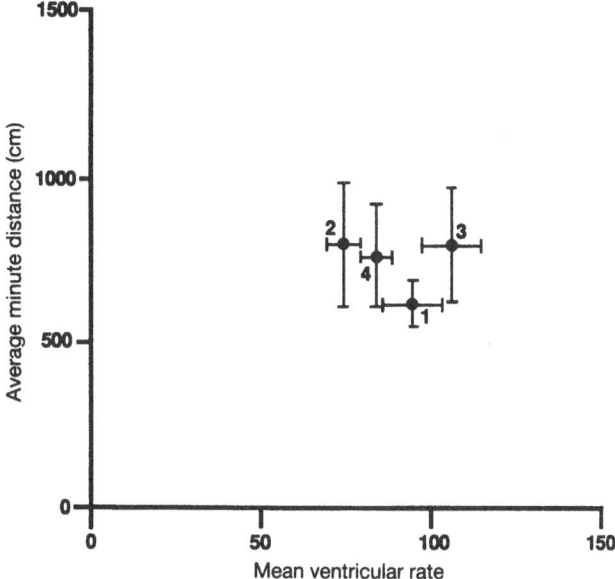

Fig. 7.8. Plot of average minute distances (SEM) and ventricular rates (SEM) for 8 patients at each stage of digoxin–enoximone trial. 1, placebo; 2, digoxin; 3, enoximone; 4, digoxin plus enoximone.

enoximone gave higher average minute distances, the former with a significantly lower ventricular rate than the latter. These results may be interpreted in the light of the population curve of minute distance and mean ventricular rates seen in Fig. 6.9. Both digoxin and enoximone appear to give greater cardiac outputs than would be predicted from the measurements made during the placebo period, but treatment with enoximone was associated with a higher ventricular rate.

In this study, for reasons discussed earlier, the confidence intervals for the prediction of minute distance for individual patients were generally too wide for the predictions to be of any value. Nevertheless, the insight gained from the use of the model aids the interpretation of the results in Fig. 7.8.

Out of 32 recordings, 6 showed uncontrolled atrial fibrillation: 3 during placebo, 2 while on enoximone and 1 on the combination of enoximone and digoxin. The lowest mean ventricular rate that was uncontrolled in this series was 103 beats per minute.

Conclusions

In this and the previous three chapters we have seen many examples of measurement of linear cardiac output in patients with atrial fibrillation. The fundamental measurement is of aortic blood velocity and its integral, stroke distance. The physical principles of measurement by Doppler ultrasound are well understood; stroke distance is conceptually simple, and physiologically meaningful. On that foundation may be built a whole system of haemodynamic measurement which includes Doppler equivalents of stroke work and peripheral vascular resistance.

While the average measurements of linear cardiac output, mean blood pressure, and linear resistance tell us about the overall functioning of the circulatory system, the beat-to-beat changes of stroke distance carry additional information about the functioning of the left ventricle of the heart. It may well be that the relationship between stroke distance and the two preceding R-R intervals does not provide such a clear-cut distinction between the effect of contractility and preload as was hoped. Nevertheless, the approach gives insight into left ventricular functioning and allows the relationship between ventricular rate and cardiac output to be explored. Here we find that the tachycardia so common in atrial fibrillation is often close to the rate at which cardiac output is maximal, leading us to redefine what is meant by "controlled" or "uncontrolled" atrial fibrillation. These observations lead also to a consideration of the reflex neurological control of ventricular rate in atrial fibrillation, the subject of the next chapter.

The model of left ventricular function described in Chapter 5 makes it possible to predict the effect of a change of heart rate on cardiac output, assessed as minute distance. The predictions are generally borne out unless there are concomitant changes, especially of inotropic state. The model therefore provides a means of investigating such changes of state, non-invasively, and with few subjects. The use of the model in this way is limited by technical factors, but the outlook for improved automated methods of evaluation of Doppler recordings in atrial fibrillation is promising.

Doppler ultrasound has opened a new window on the circulation: the view has only been glimpsed.

References

Dage RC, Kariya T, Hsieh CP et al. (1987) Pharmacology of enoximone. Am J Cardiol 60:10C–14C

Haites NE, McLennan FM, Mowat DHR, Rawles JM (1985) Assessment of cardiac output by the Doppler ultrasound technique alone. Br Heart J 53:123–9

Rawles JM, Pai GR, Copland SA (1987) Doppler haemodynamic assessment of enoximone in atrial fibrillation. Heart Vessels 3 (Suppl 3):45 (abstract)

Chapter 8

The Autonomic Control of Ventricular Rate in Atrial Fibrillation

Regulation of the heart rate is just one of many control mechanisms that operate to ensure that haemodynamic needs are met. In sinus rhythm, heart rate is controlled by the action of the autonomic nerves on the sinoatrial node, by the prevailing level of catecholamines, and by other factors such as body temperature. Of the mechanisms regulating heart rate, the baroceptor–heart rate reflex has been particularly well studied, mainly perhaps because of its accessibility.

The heart rate also varies with respiration, and while this respiratory sinus arrhythmia may not serve any vital function, it reflects cardiac vagal tone (Eckberg 1983). Sinus arrhythmia is therefore of interest as a means of observing and quantifying autonomic action.

One of the problems of studying biological control mechanisms is that there are usually several layers of control, so that the function of one mechanism is not necessarily revealed when it is inactivated, since a mechanism at another level may take its place.

Atrial fibrillation itself presents just such an example of a fall-back control mechanism – regulation of ventricular rate is exercised through the medium of the atrioventricular node when the sinoatrial node is inactivated in atrial fibrillation.

There is abundant evidence that the autonomic nervous system, acting on the atrioventricular node, does alter ventricular rate in atrial fibrillation. Ventricular rate usually increases after atropine (Horan and Kistler 1961), indicating that ventricular rate is held down by tonic vagal action on the atrioventricular node. Ventricular rate increases with exercise (Aberg et al. 1972; Beasley et al. 1985), though to what extent this is due to autonomic adjustment or to circulation of catecholamines, is not known. Ventricular rate is slower by night than by day in atrial fibrillation, as in sinus rhythm, reflecting diurnal variation in autonomic tone (Channer et al. 1987). Bleecker and Engel (1973) describe how 6 patients with atrial fibrillation learned to increase or decrease their ventricular rates at will; their ability to do this was blocked by atropine.

Evidence of anatomical integrity of the baroreflex arc in atrial fibrillation is given in a case report of a hypertensive patient with an implanted carotid sinus nerve stimulator. Bilateral stimulation of the carotid sinus nerves resulted in reduction of ventricular rate after a delay of 1.5 s (Borst and Meijler 1984). In 1890 Branham described transient slowing of the heart rate during manual compression of an arteriovenous fistula. Bassan (1980) reported a patient with atrial fibrillation and Branham's sign evoked by compression of the femoral artery supplying an arteriovenous communication in the thigh. The patient's ventricular rate fell from 120 to 60 almost immediately, demonstrating that the heart rate regulating system was intact and apparently functioning normally even in atrial fibrillation.

Both sinoatrial and atrioventricular nodes are richly innervated, but the sinoatrial node is predominantly supplied by the right vagus nerve, whereas the atrioventricular node is supplied more from the left. This asymmetrical distribution of the vagus nerves may be reflected in the different influence of the right and left vagi on atrial repolarisation and fibrillation (Balsano et al. 1967; James et al. 1973). The parasympathetic nervous system has a greater influence than the sympathetic on the sinoatrial node, but the sympathetic has a greater influence than the vagus over the atrioventricular node. Besides these differences in autonomic innervation of the two nodes, there are apparent dissimilarities in the mechanism of heart rate control exercised by the sinoatrial node in sinus rhythm, and by the atrioventricular node in atrial fibrillation. Differences in the operation of the heart rate control system between atrial fibrillation and sinus rhythm are therefore to be expected. An example is the heart rate response to carotid sinus stimulation. Usually there is little change of ventricular rate in atrial fibrillation, while bradycardia occurs in sinus rhythm; exceptionally carotid sinus massage may cause bradycardia or asystole in atrial fibrillation while causing little change in sinus rhythm (Soloff and Zatuchni 1954; Hellestrand et al. 1982).

The operation of the heart rate control mechanisms in sinus rhythm and atrial fibrillation, and the differences between them, are the subject matter of this chapter.

Respiratory Sinus Arrhythmia

Respiratory sinus arrhythmia is commonly defined loosely as variation of heart rate in phase with respiration (Bellet 1971, p. 140), or, more rigorously but arbitrarily, as a difference of 0.16 s between the longest and shortest heart rate interval in a respiratory cycle (Myerburg 1972, p. 73). This is equivalent to a variation of 10% above and below a mean heart rate of 75 beats per minute, i.e. an *amplitude* of 10%. However, respiratory variation of heart rate is not an all-or-nothing phenomenon, its amplitude being continuously variable from none to an upper limit which in normal subjects depends on age and other factors; the latter definition is thus inappropriately rigid.

The generally accepted view is that respiratory sinus arrhythmia is reflexly mediated, cardiac vagal fibres constituting the efferent arm; sinus arrhythmia is abolished by atropine and is absent in cases of autonomic neuropathy (Ewing and Clarke 1982).

On the afferent origin of the reflex there have been diverse views (Melcher 1976). Hering (1871) showed that moderate inflation of the lungs was associated with tachycardia and concluded that sinus arrhythmia was a reflex of pulmonary origin. Several authors proposed a central nervous origin, based on observations in animals that sinus arrhythmia may persist in the absence of respiratory movements due to paralysis or opening of the chest. Bainbridge (1920) proposed that respiratory sinus arrhythmia resulted from stimulation of stretch receptors that he had previously identified on the filling side of the heart (Bainbridge 1915). Others have suggested that the oscillation of systemic arterial pressure associated with breathing might induce respiratory sinus arrhythmia through the baroreflex. The demonstration that local stretch of the sinoatrial node causes alteration of the depolarisation rate (Brooks et al. 1966) raises the possibility that respiratory sinus arrhythmia has a mechanical cause.

Experiments by Melcher (1976), involving positive and negative pressure ventilation in human volunteers, led him to the conclusion that respiratory sinus arrhythmia results in part from stimulation of cardiac volume receptors by the inspiratory rise in filling pressure. Also, during inspiration the baroreflex is inhibited. Reflexes from the lungs or chest wall may contribute but are of secondary importance. Respiratory sinus arrhythmia is then the net effect of a central integration of afferent inflow from the systemic arterial baroreceptors, the low pressure receptors in the heart, and possibly the stretch receptors in the lungs.

The reflex serves to increase cardiac output by raising heart rate at the time of increased venous return during inspiration.

Quantifying Respiratory Sinus Arrhythmia

If instantaneous heart rate during quiet, spontaneous respiration is plotted against time, it is seen to fluctuate in a sinusoidal manner in phase with respiration, though at times the pattern disappears, obscured by unrelated changes in heart rate (Fig. 8.1). In order to distinguish a recurring pattern of change from the background variation, a signal averaging approach is employed, whereby changes in heart rate over multiple respiratory cycles are averaged (Rawles et al. 1989a). Analyses may be carried out on a 90-second ECG recording in sinus rhythm, or a 3-minute recording in atrial fibrillation; measurement and statistical analysis is then rapidly performed with a digitising pad and

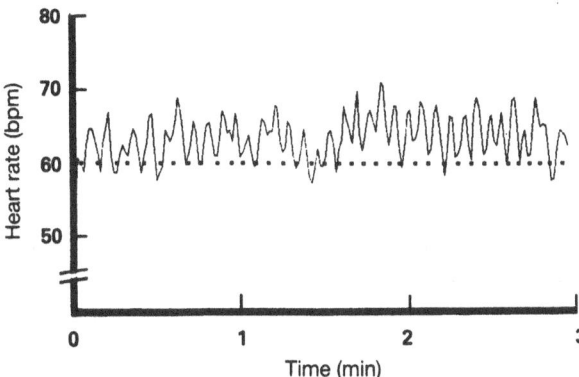

Fig. 8.1. Respiratory sinus arrhythmia. The *small dots* represent end-expiration. (Reproduced by permission of the Editor, *Clinical Science*.)

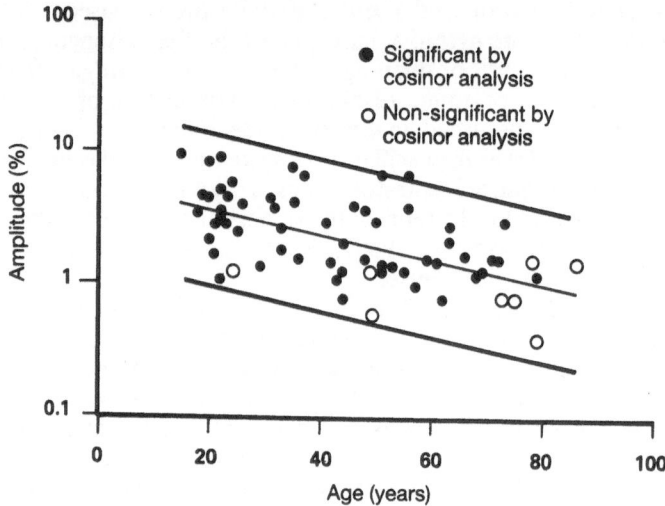

Fig. 8.2. Amplitude (%) of respiratory sinus arrhythmia in 70 healthy subjects. $r = -0.55$, $p < 0.001$.(Reproduced by permission of the Editor, *Clinical Science*.)

microcomputer. The best fit of a cosine function curve to the heart rate data is determined by cosinor analysis, a form of non-linear regression (Halberg et al. 1972). This constitutes a powerful statistical test for the presence of respiratory variation of ECG intervals utilising every interval in the recording, in contrast to the clinical definition which only uses the longest and shortest intervals. The method presupposes that the changes approximate to the form of a cosine function curve. Besides confirming the presence of sinus arrhythmia, cosinor analysis enables its amplitude and phase to be quantified. Analyses that give statistically significant results may not be qualitatively different from those that do not, but merely fail to reach the threshold for statistical significance, which for P-P intervals in sinus rhythm is an amplitude of about 1%.

By means of cosinor analysis the very frequent presence at all ages of respiratory variation of P-P intervals has been demonstrated, achieving conventional statistical significance in 84% of 70 subjects aged 15–86 (Rawles et al. 1989a). The amplitude of respiratory sinus arrhythmia declines progressively with age (Fig. 8.2), as others have reported (Hellman and Stacy 1976; Smith 1982; Hrushesky et al. 1984; O'Brien et al. 1986). The shortest cardiac intervals, or quickest heart rates, generally occurred at about the time of end- inspiration, confirming the well-known observation that heart rate increases in inspiration.

Cheyne–Stokes Respiration

Periodic, or Cheyne-Stokes respiration, is characterised by alternating periods of apnoea and hyperventilation, and may be seen in patients with raised intracranial pressure or congestive cardiac failure. In the latter it arises where circulation time is prolonged, so that there are abnormally long time-delays between detection of hypercapnoea by chemoreceptors in the brainstem, its over-correction by hyperventilation, the detection of hypocapnoea, and the subsequent apnoeic response (Bellet 1971, p. 920). The system for regulating carbon dioxide excretion through the lungs becomes unstable and oscillates between apnoea and hyperventilation (Mackay and Glass 1977; Glass and Mackay 1979). During hyperventilation bradycardia, and even atrioventricular block, may occur (Eyster 1906; Resnik and Lathrop 1925; Mears 1956); these rate and rhythm changes, though not the Cheyne–Stokes breathing, may be blocked by atropine (Matthews and Wood 1940).

Respiratory Variation of P-R Intervals

The parasympathetic supply to the sinoatrial and atrioventricular nodes is predominantly from the right and left vagus nerves respectively, which are thought to act independently but in parallel. Since the vagus nerve mediates changes in sinoatrial node function, manifest as respiratory sinus arrhythmia, comparable changes in atrioventricular node function during respiration might also be expected. However, the vagus has opposing effects on the atrioventricular node: by direct (dromotropic) action atrioventricular delay is increased, but indirectly, as a consequence of the reduction of heart rate, atrioventricular delay may be shortened (chronotropic action). The overall effect of respiratory variation of vagal tone on the P-R interval is therefore somewhat unpredictable.

Cosinor analysis of P-R intervals, with adjustment for variation of heart rate, demonstrated a respiratory effect in 39% of 70 cases (Rawles et al. 1989a). The distribution of phase angles for respiratory variation of P-R intervals, whether or not statistical significance was achieved, was not significantly different from that for P-P intervals. Thus, during respiration both P-P and P-R intervals frequently tend to be longer at about the time of end-expiration, indicating parallel changes in sinoatrial and atrioventricular node function.

Modelling the Effect of the Vagus on the Atrioventricular Node

In chapter 3 is described a model of the special conducting system of the heart that explains many puzzling aspects of its behaviour, particularly that of the atrio-ventricular node. Both sinoatrial and atrioventricular nodes are considered as interacting sine-wave oscillators; in sinus rhythm, the sinoatrial oscillator entrains the intrinsically slower atrioventricular oscillator so that they both run at the same frequency; atrioventricular delay results from the phase difference between the two oscillators. We now consider how a change of rate is effected in the model.

Fig. 8.3 shows the P-R intervals generated by the model when the intrinsic rate of the sinoatrial oscillator (SAN) is alternately slowed and quickened during repeated 5 second cycles. The period chosen, 5 seconds, is typical for respiration, so respiratory sinus arrhythmia is simulated. The atrioventricular oscillator (AVN) rate is, for the moment, unchanged. As the sinoatrial oscillator slows, P-P intervals lengthen, and P-R intervals shorten because the resetting stimulus arrives at a later phase of the atrioventricular oscillator's cycle; the interval between the arrival of the resetting stimulus and completion of the atrioventricu-lar oscillator's cycle is thereby shortened. If the rate of the sinoatrial oscillator is now kept constant, but that of the atrioventricular oscillator changed by the same percentage as before, then P-P intervals are constant but P-R intervals lengthen as the atrioventricular oscillator decelerates, and shorten as it accelerates, the

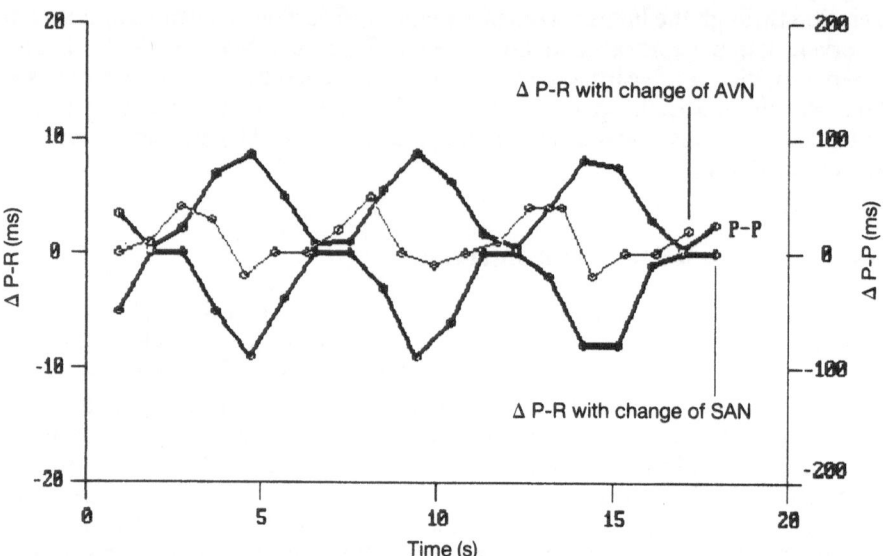

Fig. 8.3. Variation of P-P and P-R intervals with simulated respiratory sinus arrhythmia (respiratory cycle length 5 s, amplitude 10%) in the model of the conducting system. The sinoatrial and atrioventricular cycle lengths are respectively 900 and 1200 milliseconds, and coupling is 2. The dromotropic effect is achieved by varying the intrinsic rate of the atrioventricular oscillator (AVN), and the chronotropic effect by changing that of the sinoatrial oscillator (SAN), which also results in a change of P-P intervals.

opposite effect of a change of rate of the sinoatrial oscillator. However, the plot of P-R intervals with changing rate of sinoatrial oscillator is far from being an exact inversion of the P-R intervals resulting from a changing rate of the atrioventricular oscillator, and the extent of P-R interval shortening and lengthening in the two plots is not equal. Neither are the two curves in phase with each other.

Fig. 8.4 shows what happens when the intrinsic rate of the atrioventricular oscillator is changed at the same time as that of the sinoatrial oscillator, and by the same percentage. The P-R and P-P intervals now lengthen almost in phase with each other, the predominant effect being a lengthening of the P-R interval during deceleration of both nodes. The duration of atrioventricular delay depends on the relative timing of the discharge of sinoatrial and atrioventricular oscillators. If the intrinsic rates, while changing, do not keep perfectly in step, variation of the atrioventricular delay will result. If the sinoatrial oscillator gains on the atrioventricular oscillator then the delay will be shortened, and vice versa.

In the model no allowance is made for any phase difference between the oscillators, nor any difference in their speed of response to vagal stimulation. In reality, the conduction time from sinoatrial to atrioventricular node would effectively result in a phase difference between the nodal oscillators, which are also unlikely to respond with the same speed and to the same extent to changing vagal tone. There are therefore many possible ways in which atrioventricular node function might change during respiration or when autonomic tone is altered. As a first approximation, though, it seems reasonable to assume an equal percentage variation in rate of the two oscillators, in phase with each other and

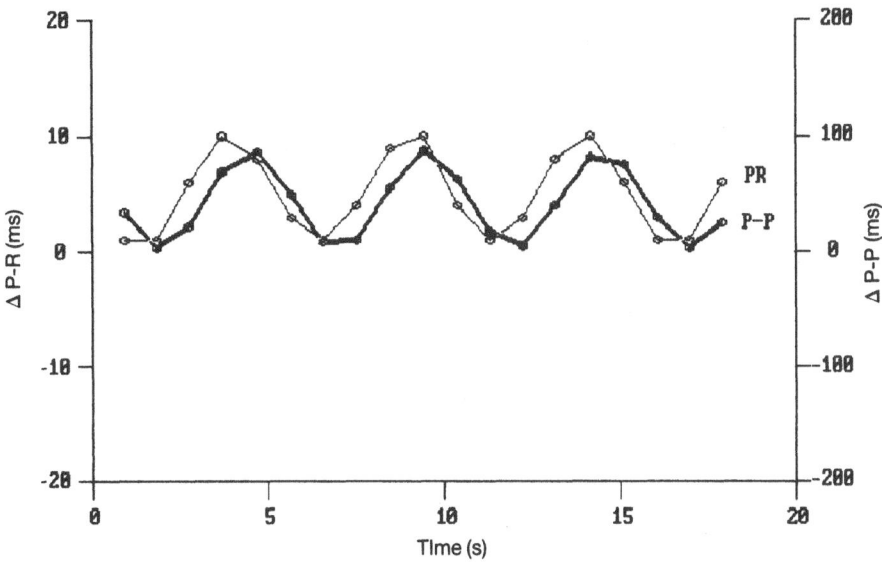

Fig. 8.4. Variation of P-P and P-R intervals with simulated respiratory sinus arrhythmia. The intrinsic rate of both oscillators is altered together.

with respiration, so that dromotropic and chronotropic effects are balanced one against the other. This simulates closely the observed changes in respiratory sinus arrhythmia, where parallel changes in vagal action on the two nodes lead to a small and variable net effect on the P-R interval, which alters in phase with the P-P interval.

The curve showing what actually happens to P-R intervals when both oscillators alter their cycle lengths together contrasts with the sum of the two curves showing the changes of P-R intervals when the oscillators change rate separately (SAN + AVN) (Fig. 8.5). The contrast between these two curves emphasises the point that atrioventricular delay is not an algebraic summation of the dromotropic and chronotropic actions. The model behaves in a markedly non-linear fashion.

In the model it is possible to show the effect on the P-R interval of altering the rate of the atrioventricular oscillator by itself, holding the rate of the sinoatrial oscillator constant. Experimentally it is difficult to isolate the nodes and their nerve supplies, so it is not possible solely to stimulate the vagal supply to the atrioventricular node. But the effect on atrioventricular delay of vagal stimulation of the atrioventricular node alone may be replicated if the atrial rate is held constant by pacing during vagal stimulation. Martin (1977) measured atrioventricular delay in the dog after single vagal stimuli. Vagal effect curves were drawn showing atrioventricular delay as a function of time after a single vagal burst, the net effect of dromotropic and chronotropic vagal action. Repeating the experiment with atrial pacing at a constant rate yielded a vagal effect curve for dromotropic action alone. The recorded sequence of cardiac intervals during vagal stimulation was then used to trigger an atrial pacemaker, without vagal stimulation. This gave the response of the atrioventricular node solely to a change of driving frequency, the chronotropic action.

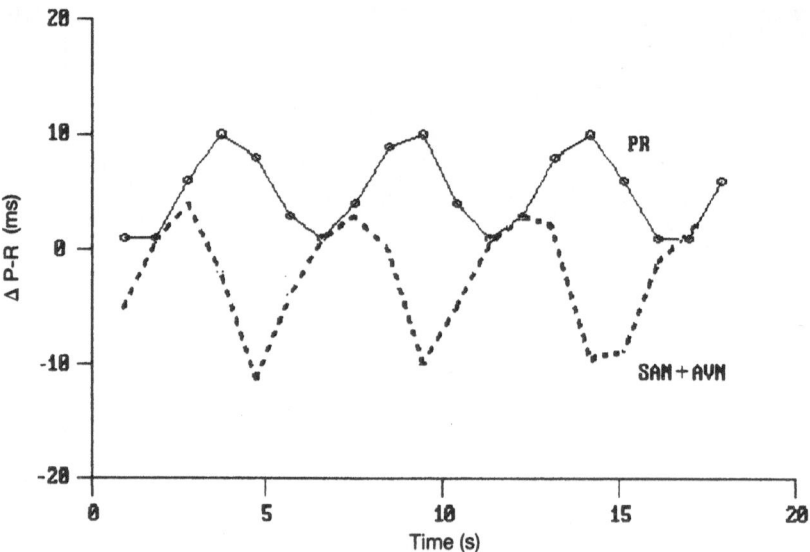

Fig. 8.5. The variation of P-R intervals with simulated sinus arrhythmia achieved by altering both oscillators together (PR) is contrasted with that calculated from the addition of the separate chronotropic and dromotropic effects (SAN + AVN).

Martin (1977) wished to test the hypothesis that atrioventricular delay after vagal stimulation of the heart was the summation of the separate chronotropic and dromotropic responses. His experiments disproved the hypothesis: like the model, the atrioventricular node of the dog behaves in a markedly non-linear manner and the net effect of vagal stimulation in the unpaced heart differed significantly from the sum of the separate chronotropic and dromotropic effects. In his experiments atrioventricular delay after vagal stimulation in the unpaced heart gave shorter delays than the sum of the delays separately. This was interpreted as a paradoxical dynamic interaction of heart period and vagal activity. By contrast, under the particular conditions of testing the model, the interaction between the two effects produces a lengthening of atrioventricular delay greater than the summation of the separate chronotropic and dromotropic effects.

In summary, in the dog and in the model, the atrioventricular node behaves in a non-linear fashion so that the net atrioventricular delay is not the same as the sum of the delays that result from separate chronotropic and dromotropic actions. In order to explain the observed variation of the P-R interval in respiratory sinus arrhythmia, it is necessary to assume a similar phasic variation of the rates of both sinoatrial and atrioventricular nodes.

Respiratory Variation of Ventricular Rate in Atrial Fibrillation

In sinus rhythm vagal tone fluctuates in phase with respiration, resulting in the phenomenon of respiratory sinus arrhythmia which is associated with parallel changes in atrioventricular node function. Demonstration of respiratory variation of ventricular rate in atrial fibrillation would therefore provide circumstantial evidence of autonomic control of heart rate by the vagus nerve acting on the atrioventricular node.

In atrial fibrillation there is a differential of about 5 : 1 between the rate of arrival of stimuli at the atrioventricular node from the atria, and the rate of egress of stimuli to the ventricles (Kirsh et al. 1988). The conducting properties of the atrioventricular node would therefore be expected to have a marked effect on ventricular rate, and it would be surprising if ventricular rate did not alter phasically with respiration. In atrial fibrillation, however, because of the widely variable R-R intervals the background variance is high and the respiratory variation of ventricular rate may be difficult to demonstrate with statistical conviction.

By means of cosinor analysis, respiratory variation of R-R intervals was demonstrated ($p<0.05$) in 7 out of 50 cases with atrial fibrillation, compared with the 2–3 that would have been expected just by chance (Rawles et al. 1989b). When cosinor analyses were done against an arbitrary time-mark rather than respiration, 2 positive ($p<0.05$) results were obtained, as expected. The greater frequency of positive tests related to respiration rather than to an arbitrary time mark suggests the presence of weak respiratory variation of ventricular rate.

The method of cosinor analysis assumes that the variation of R-R intervals around their mean value is sinusoidal in form, which is not necessarily the case, but it is valuable in indicating the phase of respiration when R-R intervals are longest, taking into account all the data points. The phase angles calculated by cosinor analysis against respiration showed a non-random distribution in these 50 subjects with atrial fibrillation. When the analyses were repeated against an arbitrary time mark the distribution was random, as expected. The non-random distribution of phase angles calculated against respiration, compared with the random distribution against a time-mark, also suggests that respiratory variation of ventricular rate is genuinely present in atrial fibrillation.

However, in contrast to sinus rhythm, where shortest P-P, P-R and R-R intervals all occurred in inspiration, in atrial fibrillation the shortest R-R intervals occurred in expiration, the differences in phase angle distribution from that of cardiac intervals in sinus rhythm being statistically highly significant (Fig. 8.6). Thus, in sinus rhythm respiratory variation of heart rate is found in most people, with acceleration in inspiration; in atrial fibrillation respiratory variation of heart rate is uncommon, and when it occurs heart rate accelerates in expiration.

Fig. 8.7 shows 343 beats of atrial fibrillation recorded from a 64 year old subject during spontaneous respiration with an average respiratory cycle length of 5.19 seconds. Each R-R interval is plotted as its angular timing in the respiratory cycle in which it took place, against its duration, expressed as the percentage difference from the mean of all R-R intervals. Each respiratory cycle commences and ends with the end-expiratory point and its duration is normalised to 360°. The best-fit cosine function curve is shown. Cosinor analysis reveals an underlying respiratory modulation of heart rate which has an amplitude of 4.2% and phase of 163°; on average R-R intervals are longer and heart rate is slower in the middle of the respiratory cycle at the end of inspiration.

It might be objected that cosinor analysis forces the data to conform to a sine-wave pattern, however poor the fit. Another method of analysis that avoids this criticism is illustrated in Fig. 8.8, using the same data as in the previous figure. Each respiratory cycle in the 3 minute recording is divided into six equal sectors; the average R-R interval of beats falling into the first to sixth sectors is computed

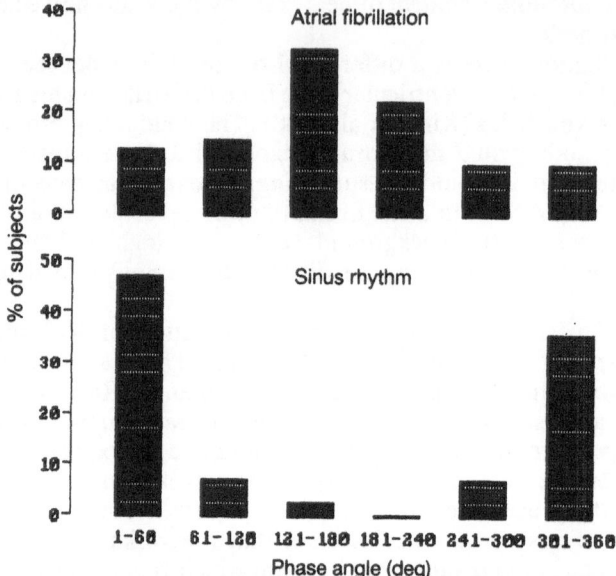

Fig. 8.6. The percentage of subjects with various phase angles for respiratory variation of R-R intervals in atrial fibrillation and sinus rhythm. By phase angle is meant the position in respiration when the intervals are longest. Angles of 0°/360° represent end-expiration.

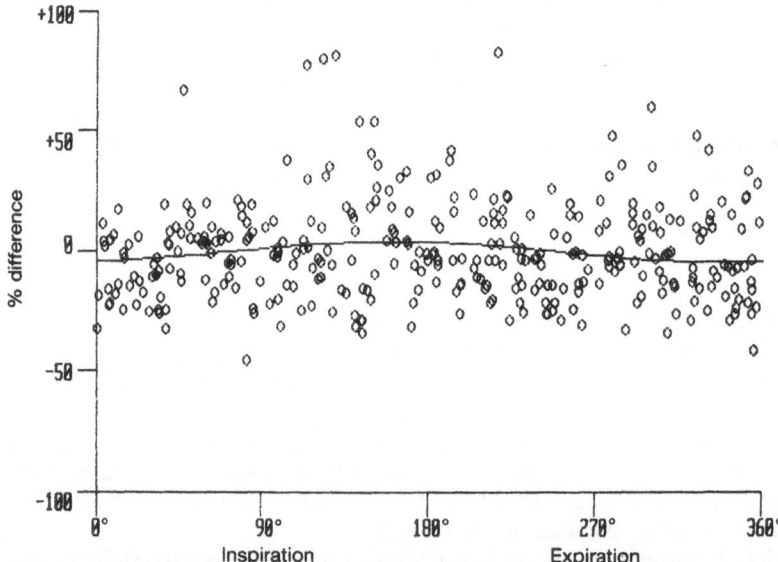

Fig. 8.7. R-R intervals in atrial fibrillation plotted by their timing in the respiratory cycle in which they occurred, against the difference from the mean of all R-R intervals in the 3-minute recording. Each respiratory cycle is demarcated by the end-expiratory point (0°/360°. The best-fit cosine function curve is shown, reaching a maximum amplitude of +4.2% at 163°. $r = 0.15$, $p<0.05$.

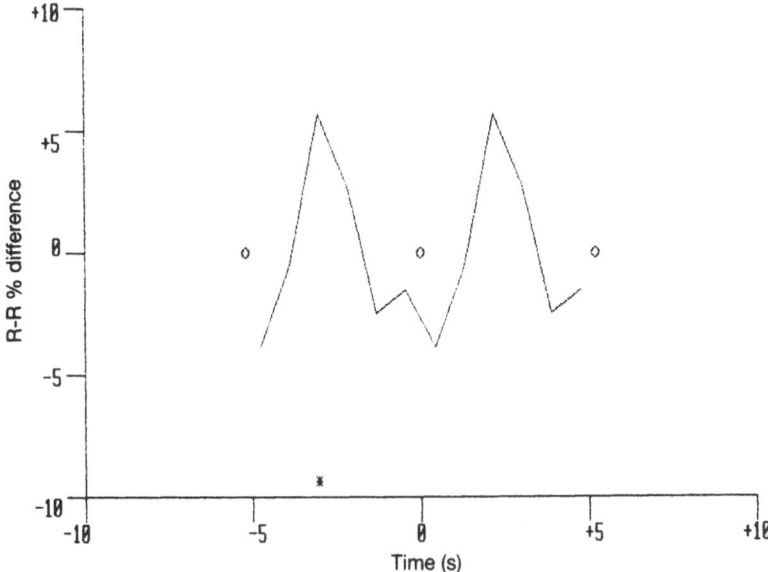

Fig. 8.8. The same data as in Fig. 8.7; mean respiratory cycle length 5191 milliseconds; mean cardiac cycle length 757 milliseconds. The average difference of R-R intervals from their mean value is plotted against their timing in respiration; the respiratory cycle is divided into six equal sectors. For ease of pattern recognition two identical composite respiratory cycles are shown; *small circles* mark end-expiratory points. R-R intervals are longest ($p<0.05$) in mid-respiratory cycle.

and displayed as a percentage difference from the mean of all beats in the recording. Mean R-R intervals from two identical composite respiratory cycles are displayed in order that the pattern may be more readily visualised; the small circles indicate end-expiratory points marking the beginning and end of the cycles. On average, R-R intervals are 5% longer at the end of inspiration ($p<0.05$), and about 4% shorter at end-expiration (NS).

There are several possible explanations for the difference in phase of respiratory variation of heart rate in atrial fibrillation and sinus rhythm. Patients in atrial fibrillation were older, had a greater mean heart rate and respiratory rate than those in sinus rhythm, and in many cases had underlying cardiac disease. However, while the greater age of the patients with atrial fibrillation would be expected to result in a lower amplitude of respiratory variation of heart rate than is seen in sinus rhythm, a difference in phase would not be anticipated. In sinus rhythm neither age, heart rate nor respiratory rate are related to phase angle, and a complete inversion of phase seems unlikely to be explained by these differences even though the factors that determine the phase angle of sinus arrhythmia are not fully known.

In 1920 Kilgore described "Respiratory variations of heart rate in the presence of auricular fibrillation" (Kilgore 1920). "Out of nine cases of auricular fibrillation studied, six show a fairly consistent tendency for shorter heart intervals during late expiration or early inspiration." He commented, "It will be noticed that the tendency to acceleration during late expiration and inspiration is the reverse of the customary relation to respiration in cases of sinus arrhythmia."

Respiratory variation of ventricular rate showing the same relationship to respiration as in sinus rhythm has been demonstrated in anaesthetised dogs (Hoff and Geddes 1965, 1966).

In Cheyne–Stokes respiration associated with cardiac failure, slowing of ventricular rate generally occurs in the hyperpnoeic phase with sinus rhythm, but more often in the apnoeic phase in atrial fibrillation, when the ventricular response is more commonly increased in expiration than in inspiration (Urbach et al. 1970; Flowers et al. 1972).

Inversion of the phase of the respiratory variation of heart rate in atrial fibrillation could be due to changes in the afferent or the efferent arm of the reflex mechanism. The exact afferent stimulus for sinus arrhythmia is unknown but possibly arises from volume receptors on the right side of the heart, where pressure– volume relationships are undoubtedly different in atrial fibrillation compared with sinus rhythm. In the effector arm, the vagus nerve may not always have parallel effects on the sinoatrial and atrioventricular nodes, or its action on the fibrillating atrium may oppose its action on the atrioventricular node.

The paucity of reports of respiratory variation of ventricular rate in atrial fibrillation in man, the difficulty of proving its presence, and the paradoxical nature of the effect reported by Kilgore (1920), Urbach et al. (1970) and ourselves are all noteworthy in the light of the expectation that the effect of respiration on ventricular rate in atrial fibrillation would be an exaggeration of that seen in sinus rhythm. If the atrioventricular node is considered as a slow but direct route for the transmission of stimuli from atria to ventricles, it is difficult to see why variation of vagal tone during respiration does not have a pronounced effect on ventricular rate in atrial fibrillation; a paradoxical action is even more puzzling.

The weak, contradictory effect of respiration on ventricular rate challenges the view of the atrioventricular node as a through conductor, and gives further support to the idea of it being a biological oscillator. But it also raises questions concerning the action of the autonomic nervous system on ventricular rate in atrial fibrillation.

Modelling the Action of the Vagus on the Ventricular Response to Atrial Fibrillation

Study of the respiratory variation of P-R intervals in sinus arrhythmia led us to conclude that most probably there is respiratory variation of the intrinsic rate of the atrioventricular node of similar magnitude and phase to that of the sinoatrial node.

In atrial fibrillation we might also expect to find evidence of respiratory variation of the intrinsic rate of the atrioventricular node. The node is bombarded by impulses from the fibrillating atria so that the ventricular rate is determined by the combination of intrinsic rate and rate of arrival of atrial impulses. Cyclical variation of the atrioventricular rate would be manifest as modulation of the otherwise random beat-to-beat ventricular response, and may be detected by cosinor analysis, using exactly the same methods as in genuine atrial fibrillation. Cosinor analysis does indeed reveal an underlying variation of heart rate, but it has approximately the same amplitude and phase as the modulation of the intrinsic rate of the atrioventricular oscillator that was programmed into the model.

The paradoxical effect of respiration on the ventricular rate in atrial fibrillation therefore poses problems for the oscillator model of the atrioventricular node, as well as for the through-conductor model. Modelling a respiratory effect on the atrioventricular oscillator in simulated atrial fibrillation yields a readily detectable respiratory variation of R-R intervals that has the same phase relationship to respiration as that of P-P intervals in sinus rhythm. The paradoxical ventricular response observed in atrial fibrillation remains paradoxical.

The ventricular response in atrial fibrillation depends on the intrinsic rate of the atrioventricular node, and the rate of arrival at the node of impulses from the atria. In respiratory sinus arrhythmia we believe that there is respiratory modulation of the intrinsic rate of the atrioventricular node, which in atrial fibrillation would be expected to lead to respiratory variation of ventricular rate of similar magnitude and phase to that found in sinus rhythm. The general absence of such an effect, and the occasional occurrence of the very opposite, an acceleration of ventricular rate in expiration, requires an explanation – an explanation which may be found in the effect of vagal stimulation, not on the atrioventricular node, but on the fibrillating atria. As discussed in Chapter 2, vagal stimulation leads to a reduction in the atrial refractory period, shortening of the fibrillation wavelength, and an increase in the fibrillation rate; an increased fibrillation rate in expiration has been reported in the dog (Hoff and Geddes 1966). By this means an increased rate of arrival of impulses at the atrioventricular node occurs at the same time as the intrinsic rate of the node is depressed by vagal action. The resultant ventricular rate will be decreased, unchanged or increased, depending on the balance between these opposing influences.

Fig. 8.9 shows the analysis of simulated atrial fibrillation in which there is

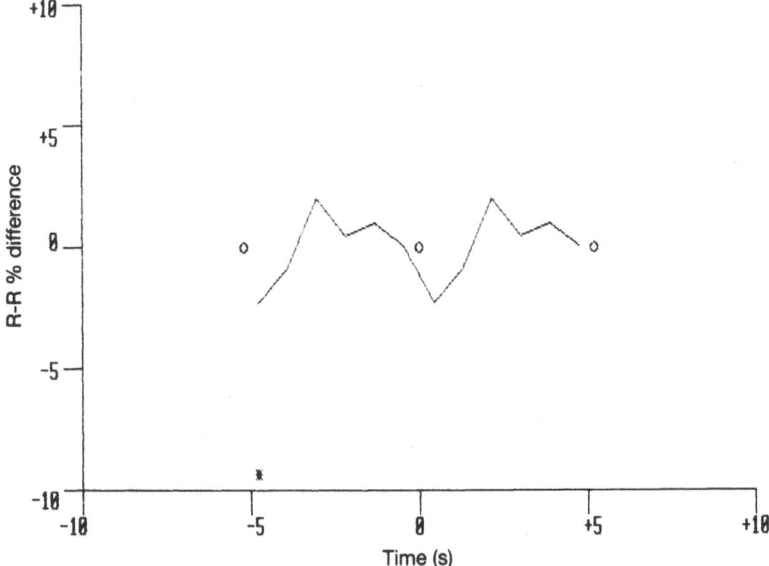

Fig. 8.9. As Fig. 8.8, with simulated atrial fibrillation. R-R intervals are shortest at end-expiration ($p<0.05$). Mean respiratory cycle length 5190 milliseconds, mean cardiac cycle length 798 milliseconds.

simultaneous modulation of the atrioventricular oscillator (1.5%) and of the average rate of generation of atrial impulses (20%), the former being in opposite phase to the latter. The resultant respiratory variation of ventricular rate is of low amplitude with the shortest R-R intervals tending to fall at end-expiration ($p<0.05$), and the longest at end-inspiration (NS), as in the real life example previously given (Fig. 8.8).

The paradoxical ventricular response to respiration observed in atrial fibrillation may be explained in terms of the known effects of vagal action on the atria and on the atrioventricular node, which have opposing consequences for ventricular rate. There remains a further puzzle, however. If vagal action has little or no net effect on ventricular rate, how is it that ventricular rate increases with exercise or with atropine?

The paradoxical response has only been observed by us in patients at rest. At these times vagal tone might be expected to be fairly high, and fluctuating in phase with respiration. With exercise, vagal tone is withdrawn and sympathetic tone increases, altering the intrinsic rate of the atrioventricular node over a wide range, if it behaves like the sinoatrial node. In the atria, the effect of withdrawal of vagal tone is to slow the fibrillation rate, but increasing sympathetic drive does not result in further slowing of the fibrillation rate, since sympathetic and parasympathetic do not have opposing effects on atrial refractory period, and hence, fibrillation rate.

Under conditions of high sympathetic and low vagal tone, a rapid ventricular rate is a consequence of the high discharge rate of the atrioventricular node, while under conditions of high vagal and low sympathetic tone, the lower rate and paradoxical response are brought about by a lower nodal intrinsic rate partially opposed by an increased fibrillation rate (Fig. 8.10).

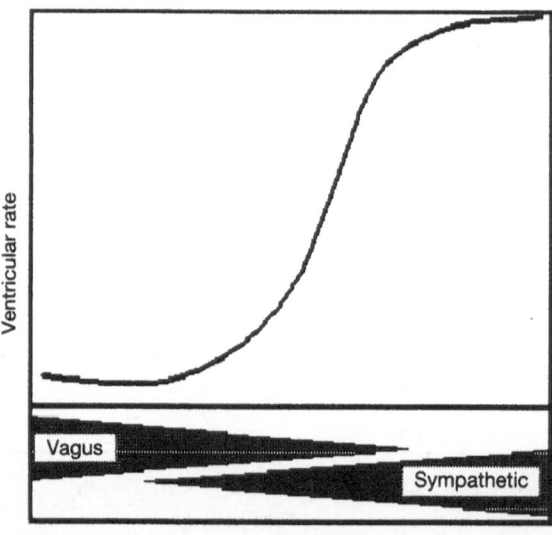

Fig. 8.10. Hypothetical relationship between sympathetic and vagal tone and ventricular rate in atrial fibrillation. A paradoxical effect is seen at the left hand end of the curve.

Measurement of Baroreflex Function

Ramp Method

About half a second after stimulation of the aortic and carotid baroceptors by a change in blood pressure, the neural discharge rate of the vagal supply to the sinoatrial node alters, bringing about a change in heart rate which is maximal between 1 to 2 seconds after the stimulus (Pickering and Davies 1973; Eckberg 1977b; Borst and Karemaker 1983). In the ramp method of assessing the baroreflex an intravenous bolus of a vasoconstrictor such as phenylephrine is given, which causes an acute rise of blood pressure associated with a bradycardia (Smyth et al. 1969; Pickering and Davies 1973). During the period when blood pressure is rising, beat-to-beat intra-arterial systolic pressure is correlated with succeeding R-R intervals (Fig. 8.11).

In assessing the sensitivity of the reflex cardiac intervals are used in place of direct measurement of autonomic nerve firing. The timing of a heart beat is used to sample autonomic activity, which is continually varying. Unfortunately, the sampling interval is long in relation to the speed of operation of the reflex, and the peak response may occur between sampling beats. Moreover the interval between the stimulus (systolic blood pressure rise) and the time of sampling (the ensuing R wave) itself varies as the heart rate changes. The sampling beats do not occur at random times but are temporally linked to the stimulus of the systolic pressure generated by the previous beat. It is quite possible that sampling beats may never occur at the time of maximum response. For these reasons, the method of analysis employed in the ramp method, correlating systolic blood pressures with ensuing R-R intervals, is unsatisfactory. To overcome these objections the use of interpolated estimates of R-R intervals has been described (Smith et al. 1986).

A further problem results from the variation of heart rate with respiration, discussed previously. That problem may be circumvented by collecting or analysing data only during expiration. However, this is only a partial solution since it imposes a severe constraint on the duration of data acquisition and the

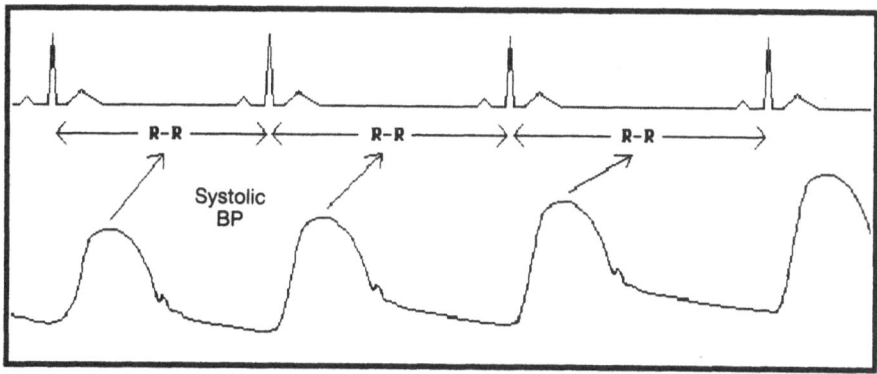

Fig. 8.11. Ramp method of determining baroreflex sensitivity.

result is only applicable to expiration and should not be extrapolated to the subject breathing normally.

Neck-Suction Method

The neck-suction method of assessing baroreflex sensitivity utilises a rigid neck chamber to which suction may be applied (Eckberg et al. 1975; Ludbrook et al. 1975). This results in increased transmural pressure in the carotid arteries, and selective stimulation of the baroceptors there. The method has the advantage of being non-invasive with no possibility of widespread pharmacological action. Further, a brief suction stimulus may be accurately timed to take place at any desired point in the cardiac cycle. On the other hand, the selective stimulation of carotid but not aortic baroceptors has no counterpart in nature. If the duration of neck-suction is brief, the response may be assessed under nearly open-loop conditions before any compensatory adjustments can take place, but if suction is prolonged the response to stimulation of carotid baroceptors may evoke a contrary reaction from aortic baroceptors.

Eckberg and Eckberg (1982) described a protocol in which a train of progressively stronger suction stimuli lasting 0.6 seconds are delivered 50 milliseconds after the R wave of four consecutive beats. When the ventricular rate is less than 75 beats per minute sinoatrial inhibition is maximal in the cardiac cycle in which the stimulus is delivered, affecting the first beat after the stimulus, but for higher heart rates subsequent heart beats are more strongly inhibited. The difficulty of analysing the heart rate response therefore remains and, as in the ramp method, the problem of the interaction with respiration is circumvented by delivering the stimuli during held expiration.

Cowie and Rawles (1989) described a modified method of assessing the carotid baroceptor heart rate reflex which utilises the neck chamber. The stimulus to the baroceptors is cyclical, with equal periods of suction and no-suction. Six 1-minute runs at different suction pressures are employed. The heart rate response to this cyclical stimulus is sinusoidal with the same period as the suction, the amplitude of the response varying in proportion to the suction pressure. The average heart rate response at any pressure is calculated by finding the best fit of a sine-wave to the heart rate data collected during each run, using the technique of cosinor analysis previously mentioned in connection with respiratory sinus arrhythmia. In this way all heart beats contribute to the analysis regardless of when they occur in relation to the varying stimulus, and data may be acquired over an indefinite period. Respiration is continued normally during neck suction, but at a different frequency, so that during each 1 minute run there are 10 complete pressure cycles and 14 complete respiratory cycles, the respiratory cycle moving in and out of phase with the pressure cycle exactly 4 times. The assessment of baroreflex sensitivity is based on some 400 cardiac intervals occurring in 6 minutes, and this large volume of data is reflected in high correlation coefficients between suction and amplitude, and good reproducibility.

Baroreflex sensitivity is log-normally distributed, and steeply and negatively correlated with age, as is the case with other manifestations of autonomic function, such as respiratory sinus arrhythmia. In the 48 subjects studied, baroreflex sensitivity was not independently related to blood pressure.

Integration of Respiratory Sinus Arrhythmia and the Baroreflex

Respiratory sinus arrhythmia and the carotid baroceptor–heart rate reflex are both mediated principally by the vagus nerve acting on the sinus node. Brief stimulation of the carotid baroceptors electrically, or by an acute rise in arterial or transmural pressure, evokes bradycardia during expiration but little or no response when delivered during the inspiratory phase of respiration (Haymet and McCloskey 1975; Davidson et al. 1976; Eckberg and Orshan 1977; Trzebski et al. 1980). This has led to the suggestion that the afferent limb of the baroreflex is gated, stimuli from the baroceptors being prevented from reaching the nucleus ambiguus of the vagus nerve during inspiration (Lopes and Palmer 1976). The all-or-none interaction implied by a gating mechanism has been discredited (Borst and Karemaker 1980), but the interaction has not been precisely quantified. Mostly, the baroreflex in man has been studied in isolation, the confounding interaction with respiration being circumvented by only analysing data collected during expiration, which may be voluntarily prolonged in human experiments (Hainsworth and Al-Shamma 1988).

Previously in this chapter separate methods for quantifying respiratory sinus arrhythmia (Rawles et al. 1989a) and the baroreflex (Cowie and Rawles 1989) have been described. Both methods permit the acquisition of data over an indefinite time period and assume a sinusoidal heart rate response to the cyclical stimulus of respiration and neck-suction respectively. These methods may be combined, to enable the separate heart rate responses to respiration and neck-suction at different frequencies to be determined from simultaneous recordings. By this means the extent of the mutual interactions of respiratory sinus arrhythmia and the baroreflex may be demonstrated. We have found that the heart rate responses to respiration and to neck- suction at different frequencies are independent of each other, the heart rate at any moment resulting from the algebraic summation of the two responses.

In Fig. 8.12 the average amplitude of the heart rate response at five different suction frequencies is plotted against the frequency used; logarithmic scales are employed on both axes. A similar plot of the amplitude of the heart rate response to respiration at the same five frequencies is seen in Fig. 8.13. Over a frequency range equivalent to cycle lengths ranging from 3 to 9.5 seconds the responses of heart rate to neck-suction and respiration are almost identical, declining steeply with increasing frequency of respiration or neck-suction. This suggests that the efferent limb, common to both responses, largely determines the system's behaviour at different frequencies. By comparing in this way the response of the heart rate control system to the same challenge at different frequencies, its behaviour may be described in general terms, independently of the particular challenge employed. Using logarithms of amplitude and frequency, the slopes of the responses are calculated, not in absolute but in relative units of measurement, as decibels (db) per octave; 20 db represents a 10 fold change in amplitude, and 1 octave is a doubling of frequency. Expressed in this way the responses of heart rate to the different challenges of respiration and neck- suction are very similar, with attenuation of 7.6 and 7.0 db per octave respectively.

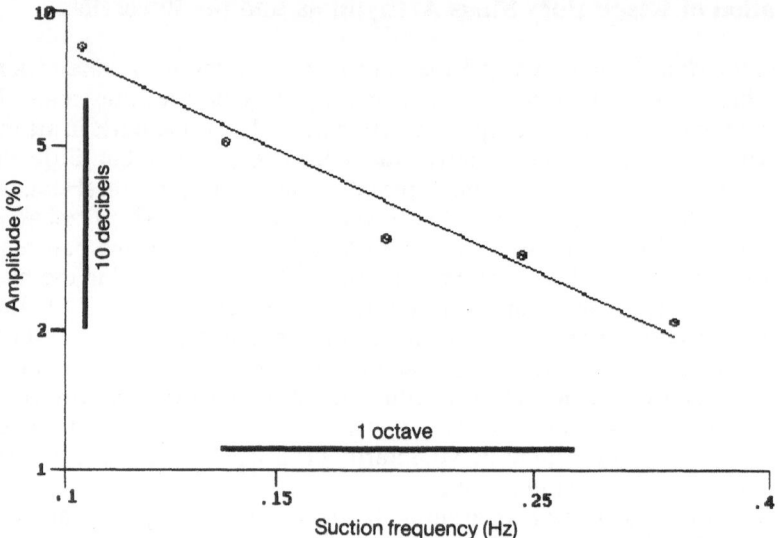

Fig. 8.12. Frequency–response curve showing the amplitude of the heart rate response to cyclical neck-suction at five different frequencies. Note that both axes are logarithmic.

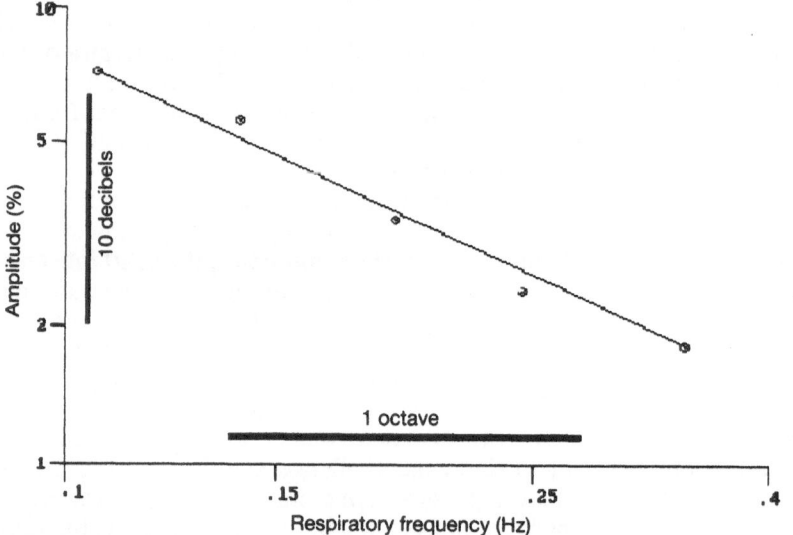

Fig. 8.13. Amplitude of respiratory sinus arrhythmia at five different respiratory frequencies. Compare with Fig. 8.12.

The Frequency Response of the Heart Rate Control System

This method of plotting the results of forcing a system to respond at different frequencies is known as a frequency– response curve. The frequency response of the heart rate control system reveals fundamental properties of the system, properties that may not otherwise be disclosed however fully the system may be

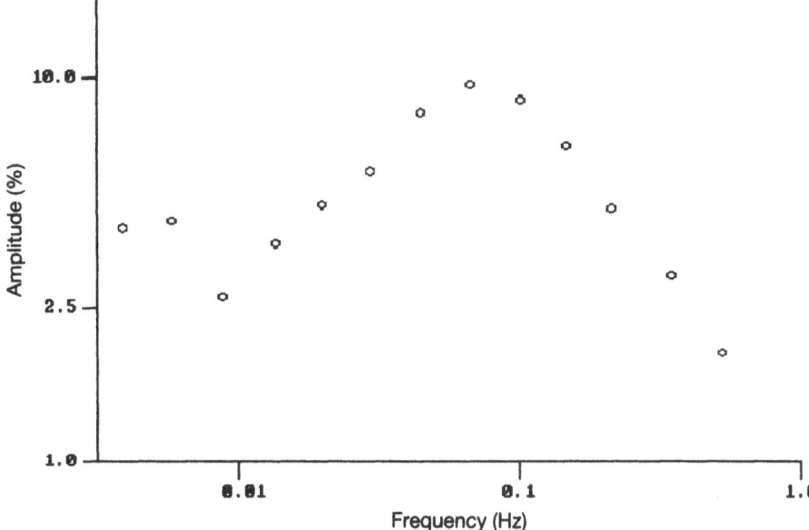

Fig. 8.14. Frequency–response curve of heart rate response to cyclical neck-suction over a wider range of frequencies than shown in Figs. 8.12 and 8.13. Sinus rhythm.

characterised at a single frequency (Warner and Cox 1962; Chess and Calaresu 1971). The knowledge that the effects of respiration and neck-suction on heart rate are additive makes it possible to study the response to cyclical neck-suction over a wide range of frequencies while respiration at a normal rate is maintained. The method of multiple cosinor analysis enables any changes of heart rate to be separately attributed to neck-suction or respiration.

A frequency response curve of the baroreflex obtained over a wider range of frequencies than previously used is shown in Fig. 8.14. It is seen that as the suction frequency is reduced the amplitude of the response increases until it reaches a maximum at 0.07 Hz, equivalent to a cycle length of 15 s. The amplitude then falls, to reach a plateau value of about 6%. The frequency–response curve depicted is characteristic of a complex, high order control system.

A frequency–response curve for a patient with atrial fibrillation is shown in Fig. 8.15. Because of the high background variance the frequency–response curve is less regular than in sinus rhythm, but the overall position and shape of the curve does not appear to differ significantly from that in sinus rhythm. In this patient there is no evidence of any parodoxical action of the baroreflex.

A Systems Approach to Heart Rate Control

The function of the baroreflex is thought to be the maintenance of constant blood pressure; in particular, the reflex may provide an immediate defence against a falling blood pressure. It seems probable that the response of the baroceptors is related not so much to the absolute level of systolic blood pressure, but to pulse pressure. This is suggested by the sinus node response being proportional to

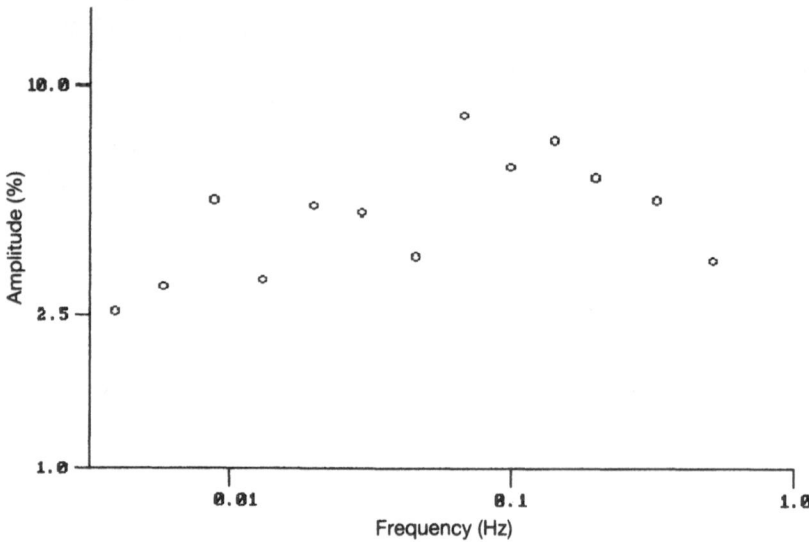

Fig. 8.15. Frequency–response curve of heart rate response to cyclical neck-suction. Atrial fibrillation.

arterial dp/dt (Eckberg 1977a), the rapid adaptation of the baroreflex to constant pressure (Eckberg 1977b), rapid resetting of the baroreflex to a change of blood pressure (Smyth et al. 1969), and the independence of baroreflex gain and the level of blood pressure (Cowie and Rawles 1989).

In sinus rhythm a fall in blood pressure will lead to a rise in heart rate which in turn results in increased blood pressure, tending to restore the *status quo*. The effectiveness of this control system is determined by the gain of the baroreflex, but also, though much less studied, by the influence of heart rate on the controlled variable, blood pressure. There is a negative relationship between blood pressure and heart rate, and a positive relation between heart rate and blood pressure (Fig. 8.16). Because of the opposing signs of the two arms of the control system, the system is intrinsically stable. Any departure of blood pressure from a set point is counteracted by a change of heart rate, and any change of heart rate is counteracted by a change of blood pressure acting through the baroreflex. The heart and its neural control systems may be considered as a black box, whose overall function may be assessed by the heart rate response to a baroceptor challenge (Fig. 8.16).

In atrial fibrillation we have reasons to believe that the baroreflex is intact, operating through the atrioventricular node, though the frequency response of the whole system may be slightly different from that in sinus rhythm. But what of the effector arm of the control system? What is the effect of ventricular rate on blood pressure in atrial fibrillation?

The Effect of Ventricular Rate on Blood Pressure in Atrial Fibrillation

We have seen in Chapters 5 and 6 how left ventricular performance in atrial fibrillation is positively related to the duration of the preceding R-R interval

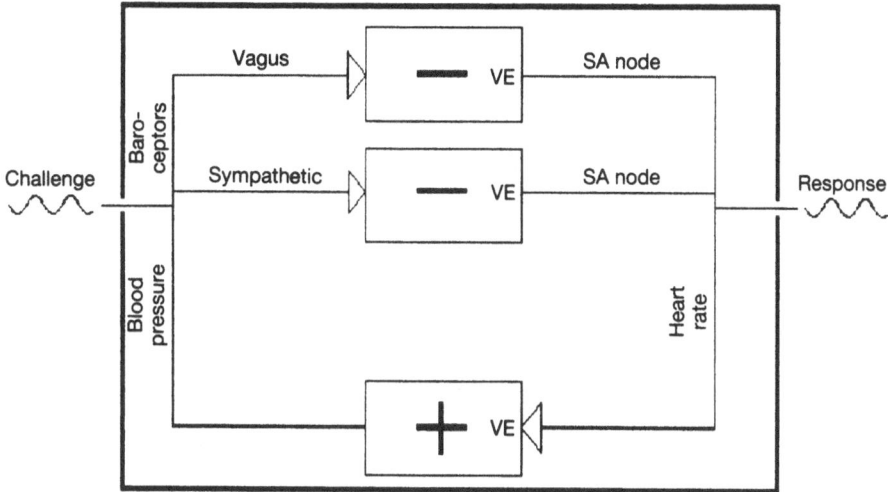

Fig. 8.16. The blood pressure control system in sinus rhythm.

(RR−1), and negatively related to the pre-preceding interval (RR−2). Left ventricular performance may be expressed in many ways, and we chose to study stroke distance, an analogue of stroke volume. But the mathematical model developed in Chapter 5, being physiologically based, is applicable regardless of how left ventricular function is expressed.

Fig. 8.17 shows a haemodynamic profile in the format with which we have

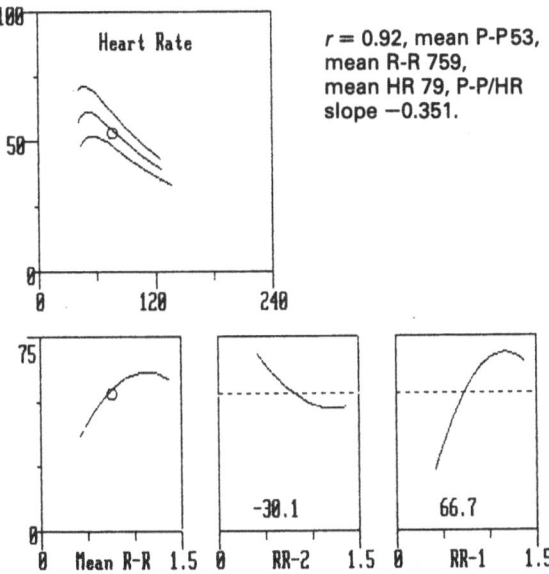

Fig. 8.17. Haemodynamic profile showing relationship between pulse pressure (mmHg) and previous R-R interval (seconds, *bottom right*), pre-preceding R-R interval (*bottom middle*), mean R-R interval (*bottom left*), and heart rate (*top*).

become familiar in earlier chapters. This profile, however, is based not upon stroke distance but on pulse pressure. It was computed from a 100 beat recording of aortic blood pressure taken at the time of cardiac catheterisation. Pulse pressure is regressed against the two preceding R-R intervals according to the equation

$$\text{Pulse pressure} = a + b(\text{RR}-1) + c(\text{RR}-1)^2 + d(\text{RR}-2) + e(\text{RR}-2)^2$$

where a–e are constants, unique to each patient. The multiple correlation coefficient is 0.92, indicating that 85% of the variance of pulse pressure is explained by this regression equation. There is a positive relation between pulse pressure and RR−1 (Fig. 8.17, bottom right), and a negative relation between pulse pressure and RR−2 (Fig. 8.17, bottom middle). As in the case of stroke distance, the positive influence of RR−1 is greater than the negative influence of RR−2, so the slope of the pulse pressure/mean R-R curve (Fig. 8.17, bottom left) is positive. When this curve is translated into that between mean ventricular rate and pulse pressure (Fig. 8.17, top), it is merely transposed into its mirror image, which has a negative slope. Thus, in this patient with a rate of 79 beats per minute there is a steep negative relationship between ventricular rate and pulse pressure with a slope of −0.351 mmHg per beat per minute. A reduction of ventricular rate by 10 beats per minute would lead to an increase of pulse pressure of approximately 3.5 mmHg.

In 19 patients studied in a similar way the mean value of the multiple correlation coefficient was 0.88, the mean ventricular rate was 78, and the mean slope of the pulse pressure/mean heart rate curve was −0.4 mmHg beats per minute, being negative in 18 out of 19 cases. The single patient with a positive slope had the lowest ventricular rate of all: 56.

The Effect of Ventricular Rate on Blood Pressure in Sinus Rhythm

The effect of atrial contractions on the haemodynamic profile for pulse pressure may be surmised. If the atrial rate were to vary from beat to beat as much as does the ventricular rate in atrial fibrillation, the force-frequency effect would result in a negative relation between blood pressure and RR−2 just as in atrial fibrillation. However, the presence of an atrial contraction prior to each ventricular systole would render the filling of the left ventricle largely independent of the duration of RR−1, left ventricular end-diastolic pressure being raised to the same value whatever the length of diastole. The pulse pressure/RR−1 slope would therefore be expected to be zero. The pulse pressure/mean R-R graph would thus be the same as the pulse pressure/RR−2 graph, and would have a negative slope. When transposed into the blood pressure/mean ventricular rate graph it would have a positive slope.

Although there is little direct evidence on the point, it is widely assumed, and we have argued above, that there is a positive relationship between heart rate and blood pressure in sinus rhythm. But in atrial fibrillation the loss of atrial contraction results in the heart rate/blood pressure relationship being negative except at very slow ventricular rates.

Regulation of Ventricular Rate in Atrial Fibrillation

In the system for the control of blood pressure in atrial fibrillation (Fig. 8.18), the sign of the heart rate–pulse pressure loop is negative. So too is the predominant sympathetic branch of the baroreflex loop, the parasympathetic branch playing a lesser role in heart rate control in atrial fibrillation than in sinus rhythm. The system is now intrinsically unstable, the neural control mechanisms exacerbating rather than correcting any departure from a steady state.

When atrial fibrillation starts, the ventricular rate is often extremely fast – 180–200 beats per minute is not exceptional. The initial fall of blood pressure due to loss of the atrial contribution to cardiac output may result in a sympathetic response and withdrawal of vagal tone, leading to tachycardia and a further fall in blood pressure. This has its parallel in an oscillating control system where negative feedback has been commuted into positive feedback.

In spite of the disastrous potential for developing uncontrolled tachycardia, most patients with atrial fibrillation do not have a ventricular rate that is anything like the maximum of which they are capable. One explanation as to why this should be resides in the frequency–response curves depicted in Figs. 8.14 and 8.15. In both sinus rhythm and atrial fibrillation there is a steep attenuation of heart rate response as the frequency of baroceptor stimulation increases. At high ventricular rates, the baroreflex gain is reduced so much that it is, in effect, turned off.

Conclusions

Much is known of the anatomy and physiology of the component parts of the autonomic system that regulates heart rate in sinus rhythm. Less is known about the integrated function of the system acting as a whole.

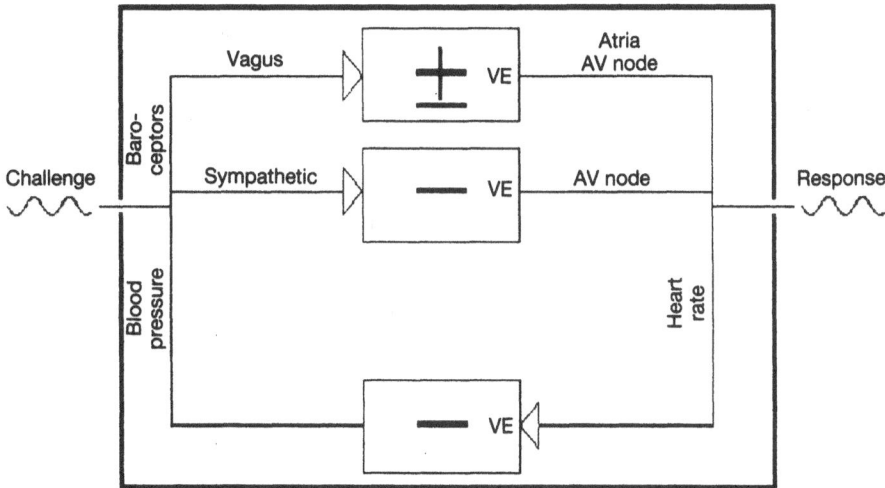

Fig. 8.18. The blood pressure control system in atrial fibrillation.

Very little indeed is known about the disordered operation of the physiological control systems in atrial fibrillation (Waxman et al. 1983), but the potential certainly exists for oscillation and chaos in the interaction of the autonomic nervous system with the circulation (Glass and Mackay 1979). A systems approach to the control of the circulation (Grodins 1963) offers the best hope of understanding how the circulation adapts to atrial fibrillation, or how the compensatory mechanisms themselves may contribute to the disorder. Analysis of the response of the intact circulatory and autonomic system to a challenge may make it possible to define the circumstances under which homeostasis is defended or disrupted.

References

Aberg H, Strom G, Werner I (1972a) Heart rate during exercise in patients with atrial fibrillation. Acta Med Scand 191:315–20

Bainbridge FA (1915). The influence of venous filling upon the rate of the heart. J Physiol 50:65–84

Bainbridge FA (1920) The relation between respiration and the pulse-rate. J Physiol 54:192–202

Balsano FA, Salerno LA, Musca AL, Pitucco GA (1967) The different influence of the right and of the left vagus upon atrial fibrillation. Experimental and clinical research. Cardiologia 50:84–94

Bassan MM (1980) Branham's sign and atrial fibrillation [letter] Am Heart J 100:411–12

Beasley R, Smith DA, McHaffie DJ (1985) Exercise heart rate at different serum digoxin concentrations in patients with atrial fibrillation. Br Med J 290:9–11

Bellet S (1971) Clinical disorders of the heart beat. 3rd edn. Lea & Febiger, Philadelphia

Bleecker ET, Engel BT (1973) Learned control of ventricular rate in patients with atrial fibrillation. Psychosom Med 35:161– 75

Borst C, Karemaker JM (1980) Respiratory modulation of reflex bradycardia evoked by brief carotid sinus nerve stimulation: additive rather than gating mechanism. In: Sleight P (ed) Arterial baroreceptors and hypertension. Oxford University Press, Oxford, pp 276–81

Borst C, Karemaker JM (1983) Time delays in the human baroreceptor reflex. J Auton Nerv Syst 9:399–409

Borst C, Meijler FL (1984) Baroreflex modulation of ventricular rhythm in atrial fibrillation. European Heart Journal 5:870– 5

Brooks CMcC, Lu HH, Lange G, Mangi R, Shaw RB, Geoly K (1966) Effects of localised stretch of the sinoatrial node region of the dog heart. Am J Physiol 211:1197–202

Channer KS, Papouchado M, James MA, Pitcher DW, Rees JR (1987) Towards improved control of atrial fibrillation. Eur Heart J 8:141–7

Chess GF, Calaresu FR (1971) Frequency response model of vagal control of heart rate in the cat. Am J Physiol 220:554–7

Cowie MR, Rawles JM (1989) A modified method of quantifying the carotid baroreceptor–heart rate reflex in man: the effect of age and blood pressure. Clin Sci 77:223–228

Davidson NS, Goldner S, McCloskey DI (1976) Respiratory modulation of baroreceptor and chemoreceptor reflexes affecting heart rate and cardiac vagal efferent nerve activity. J Physiol 259:523–530

Eckberg DL (1977a) Baroreflex inhibition of the human sinus node: importance of stimulus intensity, duration, and rate of pressure change. J Physiol 269:561–77

Eckberg DL (1977b) Adaptation of the human carotid baroreceptor- cardiac reflex. J Physiol 269:579–80

Eckberg DL (1983) Human sinus arrhythmia as an index of vagal cardiac outflow. J Appl Physiol 54:961–6

Eckberg DL, Eckberg MJ (1982) Human sinus node responses to repetitive, ramped carotid baroreceptor stimuli. Am J Physiol 242:H638–44

Eckberg DL, Orshan CR (1977) Respiratory and baroreceptor reflex interactions in man. J Clin Invest 59:780–85

Eckberg DL, Cavanagh MS, Mark AL, Abboud FM (1975) A simplified neck-suction device for activation of carotid baroreceptors. J Lab Clin Med 85:167–73

Ewing DJ, Clarke BF (1982) Diagnosis and management of diabetic autonomic neuropathy. Br Med J 285:916–18

Eyster JAE (1906) Clinical and experimental observations upon Cheyne–Stokes respiration. J Exper Med 8:565–613

Flowers NC, Dawson JE, Horan LG (1972) Modification of the ventricular response in atrial fibrillation during Cheyne–Stokes breathing: a possible source of therapeutic error. J Tenn Med Assoc 65:804–8

Glass L, Mackay MC (1979) Pathological conditions resulting from instabilities in physiological control systems. Ann N Y Acad Sci 316:214

Grodins FS (1963) Control theory and biological systems. Columbia University Press, New York

Hainsworth R, Al-Shamma YMH (1988) Cardiovascular responses to stimulation of carotid baroreceptors in healthy subjects. Clin Sci 75:159–65

Halberg F, Johnson EA, Nelson W, Runge W, Sothern R (1972) Autorhythmometry – Procedures for physiologic self-measurements and their analysis. Physiol Teacher 1(4):1–11

Haymet BT, McCloskey DI (1975) Baroreceptor and chemoreceptor influences on heart rate during the respiratory cycle in the dog. J Physiol 245:699–712

Hellestrand KJ, Nathan AW, Camm AJ (1982) Differential response to carotid sinus pressure during sinus rhythm and atrial fibrillation. Br Heart J 47:504–6

Hellman JB, Stacy RW (1976) Variation of respiratory sinus arrhythmia with age. J Appl Physiol 41:734–8

Hering E (1871) Uber eine reflectorische Beziehung zwischen Lunge und Herz. Sitzber Akad Wiss Wien 64:333–53

Hoff HEH, Geddes LA (1965) The respiratory–heart rate response in atrial fibrillation. Cardiovasc Res Centre Bull 4:54–64

Hoff HEH, Geddes LA (1966) An analysis of the relationship between respiration and heart rate in atrial fibrillation. Cardiovasc Res Centre Bull 4:81–95

Horan LG, Kistler JC (1961) Study of ventricular response in atrial fibrillation. Circ Res 9:305–11

Hrushesky WJM, Fader D, Schmitt O, Gilbertsen V (1984) The respiratory sinus arrhythmia: a measure of cardiac age. Science 224:1001–4

James TN, Urthaler F, Isobe JH (1973) Neurogenic influence on the atrial repolarization (P-Tp) segment. Am J Cardiol 32:799–807

Kilgore ES (1920) Time relations of heart beats. Respiratory variations of heart rate in the presence of auricular fibrillation. Heart 7:81–104

Kirsh JA, Sahakian AV, Baerman JM, Swiryn S (1988) Ventricular response to atrial fibrillation: role of atrioventricular conduction pathways. J Am Coll Cardiol 12:1265–72

Lopes and Palmer JF (1976) Proposed respiratory 'gating' mechanism for cardiac slowing. Nature 264:454–6

Ludbrook J, Mancia G, Ferrari A, Zanchetti A (1975) The variable- pressure neck chamber method for studying the carotid baroreflex in man. Clin Sci Mol Med 85:167–73

Mackay MC, Glass L (1977) Oscillation and chaos in physiological control systems. Science 197:287–9

Martin P (1977) Paradoxical dynamic interaction of heart period and vagal activity on atrioventricular conduction in the dog. Circ Res 40:81–9

Matthews E, Wood WB (1940) Cardiac arrhythmia during Cheyne– Stokes respiration. Bull Johns Hopkins Hosp 66:335–52

Mears EJ (1956) The association of Adams-Stokes attacks and Cheyne-Stokes respiration. Am Heart J 52:935–9

Melcher A (1976) Respiratory sinus arrhythmia in man: A study of heart rate regulating mechanisms. Acta Physiol Scand 97 (Suppl 435):1–31

Myerburg RJ (1972) Electrocardiographic diagnosis of sinus node rhythm variations and SA block. In: Schlant RC, Hurst JW (eds) Advances in electrocardiography. Grune & Stratton, New York

O'Brien IAD, O'Hare P, Corrall RJM (1986) Heart rate variability in healthy subjects: effects of age and the derivation of normal ranges for tests of autonomic function. Br Heart J 55:348–54

Pickering TG, Davies J (1973) Estimation of the conduction time of the baroceptor–cardiac reflex in man. Cardiovasc Res 7:213–19

Rawles JM, Pai GR, Reid SR (1989a) A method of quantifying sinus arrhythmia: parallel effect of respiration on P-P and P-R intervals. Clinical Science 76:103–8

Rawles JM, Pai GR, Reid SR (1989b) Paradoxical effect of respiration on ventricular rate in atrial fibrillation. Clin Sci 76:109–12

Resnik WH, Lathrop FW (1925) Changes in the heart rhythm associated with Cheyne-Stokes respiration. Arch Intern Med 36:229–38

Smith SA (1982) Reduced sinus arrhythmia in diabetic autonomic neuropathy: diagnostic value of an age-related normal range. Br Med J 285:1599–601

Smith SA, Stallard TJ, Littler WA (1986) Estimation of sinoaortic baroceptor heart rate reflex sensitivity and latency in man: a new microcomputer assisted method of analysis. Cardiovasc Res 20:877–82

Smyth HS, Sleight P, Pickering GW (1969) Reflex regulation of arterial pressure during sleep; a quantitative method of assessing baroreflex sensitivity. Circ Res 24:109–21

Soloff LA, Zatuchni J (1954) The hyperactive carotid sinus reflex of the cardioinhibitory type in individuals with auricular fibrillation. Am J Med Sci 226:281–9

Trzebski A, Raczkowska M, Kubin L (1980) Influence of respiratory activity and hypocapnia on the carotid baroreceptor reflex in man. In: Sleight P (ed) Arterial baroreceptors and hypertension. Oxford University Press, Oxford, pp 282–90

Urbach JR, Grauman JJ, Straus SH (1970) Effects of inspiration, expiration, and apnea upon pacemaking and block in atrial fibrillation. Circulation 42:261–9

Warner HR, Cox A (1962) A mathematical model of heart rate control by sympathetic and vagus efferent information. J Appl Physiol 17:349–55

Waxman MB, Wald RW, Cameron D (1983) Interactions between the autonomic nervous system and tachycardias in man. Cardiol Clin 1:143–85

Chapter 9

Clinical Aspects of Atrial Fibrillation

Epidemiology

Incidence

Although chronic atrial fibrillation may be found in young people, its incidence rises steeply with age; incidence does not differ between the sexes. In the Framingham study, the population incidence was 2% over a 20-year period (Kannel et al. 1982), with an age-specific incidence rising from 0.2 per 1000 at ages 30–39, to 39.0 per 1000 at ages 80–89 years (Wolf et al. 1987).

Prevalence

Godtfredsen (1982) suggests that, as a guideline, the prevalence rates for atrial fibrillation are 0.4% in the general population, 4% in cardiac patients at large, and 40% in patients with manifest heart failure.

The prevalence of atrial fibrillation in the adult population of Tecumseh was 0.43%, and 16 of the 22 people affected were over 60 (Ostrander et al. 1965). In 18 403 male civil servants the prevalence in those aged 40–49 was 0.2%, aged 50–59 0.4%, and aged 60–64 1.1% (Rose et al. 1978); in 9067 subjects aged 32–64 it was 0.3% (Onundarson et al. 1987). In older people the prevalence is much higher. Campbell et al. (1974) recorded electrocardiograms in 2254 people over 64 years old living at home in East Kilbride. Atrial fibrillation was present in 2% of those under 75, and 5% of those over 75. Hill et al. (1987) screened their general practice patients over 65 years of age; atrial fibrillation was found in 30 of 819 patients (3.7%), and 20 of the cases were already known to them; the prevalence was 3.4% in those aged 65–69, and 5.9% in those aged over 85. Of 1770 people over 60 in Busselton, Western Australia, 40 had atrial fibrillation (2.3%), and 47 developed it during a 20 year follow-up (Lake et al. 1989). Atrial fibrillation was recorded in 14% of people over 84 in the community in Tampere, Finland (Rajala et al. 1987), but the prevalence was even higher at 15% in 1171 patients over the age of 65 in a geriatric institution (Vaidya et al. 1976), and 17% in a group of 396 hospital and clinic patients over 74 (Wasserburger 1975).

Prognosis

In the Framingham study, chronic atrial fibrillation was amongst the risk factors for development of cardiac failure, and for sudden death (Kannel et al. 1988), and was associated with a two-fold increase in mortality (Kannel et al. 1982).

In Busselton, in unselected subjects 40–79 years of age the 13- year mortality from cardiovascular disease was significantly higher in those whose initial electrocardiogram showed atrial fibrillation or flutter (Cullen et al. 1982). In those over 60, atrial fibrillation was associated with a relative mortality of 1.92 for all causes, 1.82 for cardiovascular causes, and 3.78 for deaths from stroke; the excess mortality associated with atrial fibrillation declined with age (Lake et al. 1989).

Thus, in a case control study of atrial fibrillation in 25 younger subjects aged 32–64, the relative risks of cardiovascular and cerebrovascular death were higher at 6.1 and 12.2 respectively (Onundarson et al. 1987).

Rajala et al. (1985) related electrocardiographic abnormalities in 559 people over 84 years of age to their 5-year survival. The lowest mortality was in those with no codable ECG abnormality, and the highest was in those with atrial fibrillation or first degree heart block.

Gajewski and Singer (1981) reported the mortality in 3099 life insurance applicants with atrial fibrillation, and compared it with that derived from standard insurance mortality tables. Mortality was standard in those with paroxysmal atrial fibrillation provided the condition was not associated with other cardiovascular disease, but it was increased in the presence of mitral stenosis or coronary heart disease. Chronic atrial fibrillation carried a worse prognosis than paroxysmal atrial fibrillation, the risk being greatest when it occurred with mitral stenosis.

Atrial fibrillation occurring in the absence of heart disease or thyroid toxaemia was called "lone" by Evans and Swann (1954), and was said to have a benign prognosis. A favourable outcome for most patients with lone atrial fibrillation followed up for an average of 7.5 years was reported by Close et al. (1979). Kopecky et al. (1987) conducted a population-based study of lone atrial fibrillation over three decades. Their patients were under 61 at entry to the study, with atrial fibrillation but no overt cardiovascular disease or precipitating illness. After an average follow-up of 15 years survival was 94%, and only 1.3% (4 patients) had strokes. Such a low incidence of embolic strokes led the authors to conclude that routine anticoagulation may not be warranted for lone atrial fibrillation. However, only a minority (22%) of patients in this study had chronic atrial fibrillation; in the remainder atrial fibrillation was paroxysmal (58%), or an isolated incident (21%).

In the Framingham study, the association of chronic lone atrial fibrillation with a four-fold increased risk of stroke, but not of death, has been reported by Brand et al. (1985). And Kulbertus et al. (1982) found that elderly individuals with no apparent sign of cardiac disease other than atrial fibrillation had a higher mortality than matched controls in sinus rhythm. Lone atrial fibrillation can no longer be considered entirely benign.

In a study of patients admitted to hospital with atrial fibrillation, Godtfredsen (1982) concludes that a poor prognosis, with 5-year survival less than 50%, is associated with the diagnoses of arteriosclerotic and hypertensive heart disease, aortic valve disease, and the presence of congestive heart failure NYHA class 2 or

3. Conversely, a good prognosis – 5-year survival better than 50% – is associated with an age at onset of less than 70 years, mitral valve disease, thyrotoxic heart disease, lone atrial fibrillation, and absence of congestive heart failure before onset of atrial fibrillation.

A weak association between atrial fibrillation and subsequent manifestation of coronary heart disease has been shown by Rose et al (1978), and in patients known to have ischaemic heart disease the occurrence of atrial fibrillation increases overall mortality independently of its association with age (Kitchin and Milne 1977).

In 16% of patients with acute myocardial infarction atrial fibrillation occurred during the hospital stay; patients with atrial fibrillation had a higher fatality rate as in-patients (27.6% versus 16.6%) as well as during follow-up (Goldberg et al. 1990). However, opinion is divided as to whether atrial fibrillation has an independent effect on survival when potentially confounding prognostic factors are examined by multivariate analysis (Kitchin and Pocock 1977; Goldberg et al. 1990).

Alt et al. (1985) compared the survival after pacemaker implantation of patients with sick sinus syndrome, complete heart block and atrial fibrillation. The 10 year survival rates were 55%, 34% and 25% respectively. Survival in the sick sinus syndrome was normal, but in heart block and atrial fibrillation it was significantly decreased.

Aetiology

Published studies of atrial fibrillation based on hospital experience include a much higher proportion of patients with underlying cardiac pathology than do community-based studies. In a review of 463 cases of atrial fibrillation from Philadelphia General Hospital from 1955 to 1965, Aberg (1968) reported that arteriosclerotic and hypertensive heart diseases were responsible in almost 65% of the cases; atrial fibrillation was considered to be idiopathic in only 6%. There was a male to female preponderance of 4 : 1.

In a study of 704 patients with atrial fibrillation presenting at the emergency room, the associated diagnoses were ischaemic heart disease 55%, rheumatic heart disease 23%, chronic obstructive airways disease 3%, Wolff–Parkinson–White syndrome 3%, thyrotoxicosis 3%, and idiopathic causes 5% (Davidson et al. 1989).

Godtfredsen (1982) studied 1212 cases of atrial fibrillation admitted to hospital between 1940 and 1967; 45% were men, 55% women. During that time the proportion of patients with rheumatic heart disease fell from 27% to 15%, while the proportion with arteriosclerotic heart disease rose from 23% to 38%. The declining importance of rheumatic heart disease as a cause of atrial fibrillation led to an increase in the median age of onset, from 59 years in 1940–1948 to 70 years in 1958–1967. As consequence the 5-year survival fell from 50% to 29%.

A very different pattern of underlying heart disease emerges in a community-based study of elderly ambulatory patients with atrial fibrillation (Kulbertus et al. 1982). Hypertension was present in 39%, but the prevalence of hypertension and ischaemic heart disease was not significantly different in a group of matched subjects in sinus rhythm; 26% had no evidence of cardiovascular disease; a higher proportion of patients with atrial fibrillation than with sinus rhythm had cardiomegaly.

In children, atrial fibrillation is associated with a wide variety of conditions, congenital heart disease predominates, though idiopathic atrial fibrillation and the sick sinus syndrome are not unknown (Radford and Izukawa 1977).

Coronary Artery Disease

Before the introduction of coronary angiography, atrial fibrillation was probably attributed to "arteriosclerotic heart disease" more often than was warranted. Cheng (1974) reported that coronary artery disease was an uncommon cause of chronic atrial fibrillation. Only 5% of patients with angiographically proven coronary artery disease had atrial fibrillation, and only 5% of patients with chronic lone atrial fibrillation showed significant coronary artery disease. An even lower prevalence was reported by Haddad et al. (1978); of 496 patients with coronary artery disease only 1 had atrial fibrillation, and of 45 with atrial fibrillation only 1 had coronary artery disease.

In the Framingham study, about 8% of atrial fibrillation in men and 3% in women was attributable to overt coronary heart disease (Kannel et al. 1983). Hypertensive cardiovascular disease, cardiac failure, and rheumatic heart disease were more powerful predictors than was coronary heart disease for subsequent chronic or transient atrial fibrillation, and when these other predictors were excluded, coronary heart disease was a statistically significant predictor only for transient atrial fibrillation, and only in men.

Of 18 343 patients with proven coronary artery disease on the CASS Registry, atrial fibrillation was present in 116 (0.6%). The presence of atrial fibrillation was negatively related to the number of diseased arteries, but was associated with increasing age, male sex, mitral regurgitation and congestive cardiac failure (Cameron et al. 1988). In these patients with coronary artery disease, atrial fibrillation was an independent predictor associated with a doubling of mortality risk. This comes about because atrial fibrillation is a marker for abnormal left ventricular function in coronary artery disease (Kramer et al. 1982).

Atrial fibrillation is common after coronary artery bypass grafting (Rubin et al. 1987), an incidence of 28% being reported by Fuller et al. (1989). Its occurrence is unpredictable, but age is the most important determinant (Leitch et al. 1990); it does not adversely affect survival but is associated with postoperative stroke (Taylor et al. 1987).

Myocardial Infarction

Atrial fibrillation complicates acute myocardial infarction in 10%–20% of cases (Fluck et al. 1967; Cristal et al. 1976; Liem et al. 1976; Hunt et al. 1978; Sugiura et al. 1985). When it occurs within 3 hours of the onset of symptoms it may specifically indicate ischaemia of the left atrium and the atrioventricular node due to occlusion of the left circumflex artery proximal to the left atrial branch (Hod et al. 1987); otherwise it may be associated with interruption of the supply to the sinoatrial node (James 1961).

Old age, a history of heart failure, the development of heart failure, low systolic blood pressure and pericarditis are just some of the factors that predispose to the development of atrial fibrillation after acute myocardial infarction (Liem et al. 1976; Sugiura et al. 1985; Flugelman et al. 1986, 1987). Pericarditis occurs in 15% of patients with acute myocardial infarction, and is associated with large, transmural infarcts and atrial fibrillation (Liem et al. 1975). However, atrial fibrillation may reflect ventricular dysfunction rather than pericarditis, since the dysrhythmia is uncommon in acute infective pericarditis (Mallet and Letac 1975), and when it does occur in pericarditis it is usually in patients with pre-existing heart disease (Spodick 1976).

Valvular Heart Disease

Atrial fibrilation is the most common complication of rheumatic heart disease, occurring at some time in 40%–50% of cases (Selzer and Cohn 1972; Shimada 1986). Initially atrial fibrillation may be paroxysmal, but later it becomes chronic; its onset often results in disability where there was none in sinus rhythm. In mitral regurgitation it occurs early in the course of the disorder, and is almost always present if the patient is seriously disabled (Ellis and Ramirez 1969). In mitral stenosis the occurrence of atrial fibrillation is not related to the severity of mitral valve disease, the mitral valve area at surgery being similar in those with and without atrial fibrillation, though fibrosis of atrial myocardium is more extensive if atrial fibrillation is present (Unverferth et al. 1984). The onset of atrial fibrillation in mitral stenosis is associated with clinical deterioration by at least one NYHA class, and with reduction of cardiac output; pulmonary artery wedge pressure is usually but not always increased (Dubin et al. 1971).

The left atrium is enlarged in mitral stenosis with sinus rhythm, and it is larger still if atrial fibrillation is present (Wade et al 1952). In a group of 265 patients with mitral or aortic valve disease or hypertrophic obstructive cardiomyopathy, atrial fibrillation was rare (3%) if left atrial dimension was less than 40 mm, but common (54%) when it was more than 40 mm (Henry et al. 1976). The right atrium, however, is of normal size in the presence of mitral stenosis and sinus rhythm, but enlarges once atrial fibrillation is established (Keren et al. 1987). The left atrial dimension in patients with atrial fibrillation of recent onset was found to be less than that in patients who had had atrial fibrillation for more than a year. But when the measurements were repeated 6 months later, the left atrial dimensions in the two groups were not significantly different (Petersen et al. 1987). This evidence supports the contention that atrial enlargement is a consequence as well as a cause of atrial fibrillation, most of the enlargement occurring in the first 6 months after the onset of atrial fibrillation (Probst et al. 1973; Keren et al. 1987; Sosa-Suarez et al. 1989; Sanfilippo et al. 1990).

Atrial fibrillation has been reported in 12% of patients with mitral valve prolapse (Beton et al. 1983).

In aortic stenosis atrial fibrillation is uncommon; when it occurs it carries a very poor prognosis (Myler et al. 1968; Gorentsvit 1981). Atrial fibrillation is uncommon, too, in patients with idiopathic hypertrophic subaortic stenosis (Glancy et al. 1970). It occurs in 10% of patients, late in the course of the disease, and is unrelated to the severity of outflow tract obstruction or to mitral regurgitation. Loss of atrial systole in the presence of a poorly compliant ventricle leads to reduction of cardiac output and symptomatic deterioration (Robinson et al. 1990).

Thyrotoxicosis

Haslett et al. (1983) noted the frequent occurrence of thyroid dysfunction in patients with rheumatic heart disease. When atrial fibrillation occurs in rheumatic heart disease, the possibility of coexisting thyrotoxicosis should therefore be remembered. Occult thyrotoxicosis should also be considered as a cause of atrial fibrillation in the elderly, in whom the somatic signs of thyrotoxicosis may not be very evident (Tibaldi et al. 1986). It has been suggested that occult thyrotoxicosis is a correctable cause of "idiopathic" atrial fibrillation, and, in the presence of a normal T_4 concentration, may only be diagnosed consistently by an abnormal thyrotrophin-stimulating hormone(TSH) response to thyrotrophin-releasing hormone (TRH) (Symons et al. 1978; Forfar et al 1979). However, when Davies et al. (1985) examined 75 elderly patients with atrial fibrillation and compared them with 73 patients in sinus rhythm, reduced or absent TSH responses to TRH were common in both groups. An absent response is thus an uncertain marker of hyperthyroidism; diagnosis of thyrotoxicosis should be based on levels of free T_3 or T_4.

Atrial fibrillation occurs in about 20% of cases of thyrotoxicosis (Hurley et al. 1981; Bar-Sela et al. 1981; Iwasaki et al. 1989), often in patients with coexisting heart disease (Gayet et al. 1987). Age over 40 is a better predictor of atrial fibrillation in thyrotoxicosis than is an enlarged left atrium; in patients over 40, serum T_3 and T_4 concentrations were higher in patients with atrial fibrillation than in those with sinus rhythm (Iwasaki et al. 1989). Hyperthyroidism predisposes to atrial fibrillation because it is associated with accelerated atrial repolarisation and a shorter atrial refractory period (Olsson 1981).

Alcohol

There are many published case reports of atrial fibrillation occurring in association with acute or chronic abuse of alcohol (Lutterotti 1967; Zuber and Kalasz 1967; Zuber and Doroszew 1969; Szczerbinski 1969; Vainshtein et al. 1970; Watt 1985; Rich et al. 1984; Sipila 1985; Garcia Pascual 1986). Thornton (1984) describe atrial fibrillation after alcoholic binges in four people who usually drank little or no alcohol; all spontaneously reverted to sinus rhythm within a day. And in 40 cases of new-onset atrial fibrillation, alcohol intoxication was a causal or contributory factor in 14 (35%), the proportion increasing to 63% in those under 65 years old; most reverted to sinus rhythm within a day (Lowenstein et al. 1983).

Rich et al. (1985) reported a case-control study of 64 patients with acute idiopathic atrial fibrillation. Sixty-two per cent of patients and 33% of controls were heavy users of alcohol. Of 100 patients with new-onset atrial fibrillation, 65 had underlying conditions, of which the commonest were ischaemic and hypertensive heart disease; 35 cases were considered idiopathic. The alcohol intake of patients with idiopathic atrial fibrillation was higher than in those where an underlying condition was identified (Koskinen et al. 1987).

The precipitation of arrhythmias, including atrial fibrillation, by an alcoholic binge has been nicknamed the "holiday heart" syndrome (Nissen and Lemberg 1984), and was replicated in 10 out of 14 patients with such a history by administration of 90 ml of 80°-proof whisky (Greenspon and Schaal 1983). Mechanisms suggested to explain the induction of atrial fibrillation by alcohol in occasional drinkers are release of catecholamines, depression of conduction, prolongation of refractoriness, change of electrolyte balance, and release of catecholamines on withdrawal of alcohol (Anonymous 1985). However, in anaesthetised normal dogs, alcohol appears to have a mild antiarrhythmic effect on the atria, raising the threshold for the induction of atrial fibrillation (Kostis et al. 1977), which is sustained for shorter periods (Nguyen et al. 1987).

Metabolic Causes of Atrial Fibrillation

In patients undergoing cardiac surgery, the potassium concentration of atrial myocardium was highest in those who were in sinus rhythm throughout, intermediate in those who developed atrial fibrillation perioperatively, and lowest in those who were in atrial fibrillation preoperatively (Ebert 1970). Olsson (1981) showed that decreased atrial muscle potassium leads to accelerated atrial repolarisation, predisposing to atrial fibrillation. Hypokalaemia leading to atrial fibrillation has been reported with chronic diarrhoea (Sarma et al. 1966), after a drinking spree (Kounis and Kenmure 1976), and with thiazide treatment for hypertension (Emara and Saadet 1986).

However, in acute myocardial infarction, while serum potassium concentration is inversely related to the occurrence of ventricular arrhythmias, it is unrelated to the development of atrial fibrillation and other atrial arrhythmias (Nordrehaug and von der Lippe 1986).

Magnesium is another electrolyte a deficiency of which may result in ventricular or supraventricular arrhythmias, including atrial fibrillation refractory to digoxin (Sheehan and White 1982); the administration of magnesium sulphate may suppress arrhythmias or bring the ventricular rate under control (Singh et al. 1976; Iseri 1990).

Atrial fibrillation may be precipitated by hypercalcaemia (Drucker 1981), which may counteract the effect of the calcium antagonist verapamil (Bar-Or and Gasiel 1981; Santo et al. 1982).

Alstrup and Sorensen (1975) were unable to identify any abnormalities in the metabolism of lipid, insulin or glucose that accompanied the onset of atrial fibrillation following thoracotomy for lung cancer.

The Role of the Autonomic Nervous System in Precipitating Atrial Fibrillation

Although atrial fibrillation has been reported in the denervated transplanted heart (Berke et al. 1973; Harrison et al. 1978), the autonomic nervous system plays a major role in the initiation of atrial fibrillation in many cases. The vagus is closely implicated in the induction of atrial fibrillation or the conversion of atrial flutter to atrial fibrillation by carotid sinus pressure (Franke 1968; Anbe 1969; El-Sherif 1972), or the onset of atrial fibrillation in association with severe nausea and vomiting (Wilson and Davis 1978), or hiccough (Thorne 1969). Afferent and efferent arms of the autonomic nervous system are undoubtedly involved in the induction of atrial fibrillation by swallowing, in which there is incoordinate peristaltic activity of the oesophagus (Kalloor et al. 1977; Keidar et al. 1984; Morady et al. 1987).

Direct causation of atrial fibrillation by catecholamines is seen in phaeochromocytoma (Bucher et al. 1969; Slavina et al. 1973), and raised catecholamines may be responsible for atrial fibrillation after hypoglycaemia (Collier et al. 1987). Psychological stress, in which catecholamines are increased, has long been recognised as a cause of atrial fibrillation (Nakamoto 1965). In US navy personnel, atrial fibrillation related to emotional strain was observed in patients whose normal heart rate was slow and who often had ventricular premature beats (Master and Eichert 1946; Peter et al. 1968). In monkeys and rabbits, stress-induced arrhythmias are associated with raised concentrations of adrenaline in blood and myocardium (Ul'ianinskii 1981). Moderate motor activity at the time of stress leads to decreased adrenaline in myocardium, and reduction of arrhythmias.

The frequency and duration of bouts of paroxysmal atrial fibrillation vary widely within and between patients (Takahashi et al. 1981), and in many patients no pattern in the occurrence of bouts can be discerned (Greer et al. 1989). The ventricular rate at the onset of a paroxysm is higher than that of the preceding sinus rate, and in attacks lasting less than 2 hours the ventricular rate at the end is no different from that at the beginning; ventricular rate decelerates during attacks lasting more than 2 hours (Kawakubo et al. 1986). Gabathuler and Adamec (1985) classified paroxysmal atrial fibrillation by whether bradycardia or normal rate immediately preceded its onset, and by associated sinoatrial node dysfunction. In the first category attacks are more frequent by night than by day, and in the second and third categories no attacks occur when the immediately preceding sinus rate is more than 94 beats per minute.

Coumel et al. (1978) emphasised the role of the parasympathetic nervous system in precipitating attacks of paroxysmal atrial fibrillation or flutter. Attacks of vagal origin occur predominantly in middle aged males with no underlying heart disease; they are usually nocturnal but also occur after meals or alcohol (Coumel 1990b). Because of its vagotonic effect, digoxin may increase the number of attacks in these subjects (Coumel et al 1982). Changes in heart rate prior to and at the onset of the attack are said to reflect the autonomic state, a falling ventricular rate that is less than 60, or atrial fibrillation alternating with atrial flutter, indicating vagal dominance, while a ventricular rate over 75 and accelerating, or atrial tachycardia, indicates sympathetic action (Balducelli et al. 1989; Coumel 1990a). The distinction between adrenergic and vagally induced attacks is not clear-cut, however, and intermediate states are described. In Coumel's experience, attacks of vagal origin, typically nocturnal, are commoner than adrenergic atrial fibrillation in patients with lone fibrillation. However, Rawles et al. (1990) reported the more frequent occurrence of attacks of atrial fibrillation by day than by night in 72 patients with paroxysmal atrial fibrillation, suggesting that, in these patients, attacks are more often precipitated by sympathetic than vagal activity.

There is no doubt, however, that the autonomic nervous system plays a major role in the onset of atrial fibrillation, especially in many of the unusual and painful circumstances cited below.

Physical Factors Precipitating Atrial Fibrillation

Atrial fibrillation has been reported following a lightning strike, which also caused inferior myocardial ischaemia (Gupta et al 1988). Reports of atrial fibrillation following electric shock are more commonplace (Merle d'Aubigne and Saint-Maurice 1967; Kernohan 1967; Wachtel and Rothfeld 1967; Sharma 1971; Cotoi and Dragulescu 1974).

Exposure to smoke, and poisoning by carbon monoxide, may precipitate atrial fibrillation in association with reversible signs of myocardial damage (Bass and Hildreth 1979; Carnevali et al 1987).

Atrial fibrillation is commonly seen in moderate or severe accidental hypothermia in which the core temperature is below 32 °C; the dysrhythmia itself appears benign and makes no difference to the mortality rate (Sgobba et al. 1982; Okada 1984; Rankin and Rae 1984). Atrial fibrillation may also occur after drowning (Rivers et al. 1970).

Fighter pilots in the Japanese Air Self Defense Force undergo anti-gravity training in a human centrifuge. More than half of the trainees develop a variety of arrhythmias, including premature ventricular and supraventricular contractions, atrioventricular dissociation, sinoatrial block, and atrial fibrillation. A pilot with the Wolff–Parkinson–White syndrome developed atrial fibrillation and lost consciousness (Sekiguchi et al. 1986). Continuous electrocardiogram monitoring of civil air crews during flight operations revealed one case of atrial fibrillation associated with ST-segment depression while walking up a stairway after a flight (Sekiguchi et al. 1977). Exertion is not a potent cause of atrial fibrillation; in patients known to have paroxysmal atrial fibrillation, exercise testing by bicycle ergometry precipitated atrial fibrillation in only 16% of subjects (Buslo et al. 1982).

Blunt injury to the chest may rarely result in atrial fibrillation (Borgeat and Grbic 1986; Vesterby and Gregersen 1980), but atrial fibrillation following head injury has been reported more frequently (Bonofiglio et al. 1967; Palma et al. 1969; Marshall 1976). Presumably, cardiac dysrhythmias following head injury involve similar mechanisms to those found in spontaneous subarachnoid haemorrhage or subdural haematoma, in which a wide variety of dysrhythmias commonly occur (Wong and Cooper 1969; Van der Ark 1975; Dimant and Grob 1977; Di Pasquale et al. 1987). Yamour et al. (1980) reported that intracerebral haemorrhage into the frontal lobes results in prolongation of the Q-T interval and T-wave changes, while brainstem haemorrhage leads to non-cardiogenic pulmonary oedema and atrial fibrillation. They suggest that both types of response are mediated by the autonomic nervous system.

Other traumatic experiences followed by atrial fibrillation include anaphylaxis after a wasp sting (Brasher and Sanchez 1974), snake bite (Gupta et al. 1987), and the bite of a black widow spider (Weitzman et al. 1977).

Iatrogenic Atrial Fibrillation

Atrial fibrillation has been reported as an occasional complication of many diagnostic and therapeutic procedures in daily use. In cardiac electrophysiological studies, atrial fibrillation requiring treatment occurred in 10 out of 1000 consecutive cases. Apart from severe proarrhythmic events in patients undergoing investigation of ventricular arrhythmias, atrial fibrillation was the commonest complication seen (Horowitz et al. 1987). There are case reports of atrial fibrillation triggered by placement of Swan–Ganz and central venous catheters (Teramoto et al. 1967; Geha et al. 1973; Nakajima et al. 1987), the use of radiological contrast media (Sandrasegram and Kumar 1986), and by the injection of cold fluids into the right atrium during measurement of cardiac output by thermodilution (Todd 1983).

Long-term ventricular pacing is implicated as a cause of atrial fibrillation (Langenfeld et al. 1988), due to atrial enlargement and retrograde conduction of ventricular stimuli during the vulnerable period of the atria. Markewitz et al. (1986) reported that 55% of patients with ventricular demand pacemakers developed atrial fibrillation after 5 years, while Rosenqvist et al. (1986) recounted that 30% of patients with sinus node disease developed atrial fibrillation with ventricular pacing compared with 4% with atrial pacing. Comparable figures of 22% for ventricular pacing versus 4% for atrial pacing were obtained by Sutton and Kenny (1986) after a follow up period of 2.5 years in patients with the sick sinus syndrome.

Atrial fibrillation has been precipitated by the application of a test-magnet to a demand pacemaker; atrial stimulation in the vulnerable period presumably resulted from retrograde conduction of fixed-rate pacing stimuli (Staller 1984).

Jordaens et al. (1985) reported the induction of atrial fibrillation by the operation of an automatic defibrillator which had been implanted for refractory ventricular tachycardia. More importantly, 20% of patients developed new-onset atrial fibrillation after implantation of these devices (Gartman et al. 1990).

Gastric endoscopy, or gastric hypothermia, may precipitate atrial fibrillation, perhaps by eliciting a strong vagal response (Fujita and Kumura 1975; Allaz et al. 1983; Rose and Harrell 1967), while Switz et al. (1976) attributed atrial fibrillation and death to electrical malfunction of a transformer-powered flexible fibre gastroscope.

Other procedures which have resulted in atrial fibrillation by misadventure are transtracheal aspiration (McCartney and McMurtry 1973; Pitts et al. 1977), maintenance haemodialysis (Goodwin et al. 1969), hydrotherapy (Greentree 1986), general anaesthesia with enflurane (Pratila and Pratilas 1977), epidural anaesthesia with bupivacaine (Pratila and Pratilas 1982), the use of suxamethonium during electroconvulsive therapy (O'Melia 1970), intravenous fluorescein for ophthalmic examination (Kirson and Wilson 1987), hexoprenaline for premature labour (Frederiksen et al. 1983), and physostigmine for a patient who was slow in recovering from a general anaesthetic (Maister 1983).

Atrial fibrillation may be a manifestation of acute cardiotoxicity associated with anthracyclines used in the treatment of malignancy (Oster and Rakowski 1981; Fuster Siebert et al. 1983; Okuma et al. 1984), but it may also follow the administration of other chemotherapeutic agents (Eskilsson et al. 1988), or the immunosuppressant azothioprine (Dodd et al. 1985).

Atrial fibrillation has been reported after the use of antidepressants (Gorelick et al. 1981; Whiteford et al. 1984; White and Wong 1985), ophthalmic atropine (Merli et al. 1986), inhaled albuterol (Breeden and Safirstein 1990), theophylline (Sessler and Cohen 1990), digitalis (Maljar et al. 1968; Agarwal and Agarwal 1971; Cosma and Munteanu 1975), adenosine triphosphate (Belhassen et al. 1984), atenolol (Rasmussen et al. 1982), phenylbutazone (Ford and Prescott 1982) and bile salts (Patton 1970).

Self-Induced Atrial Fibrillation

Bartall et al. (1979) described "push-up palpitations" – paroxysmal atrial fibrillation and shortness of breath following press-up exercises in a middle aged man, resulting in rupture of the chordae tendineae. Another unlikely recreational activity associated with atrial fibrillation is the game of bowls (Louis et al. 1986).

Paroxysmal atrial fibrillation has been triggered by the consumption of excessive quantities of nicotine chewing gum (Stewart and Catterall 1985; Rigotti and Eagle 1986), by the abuse of a breath spray containing ethyl alcohol (Ridker et al. 1989), by occupational exposure to trichloroethylene (Panov and Dolmatov 1988), and by consumption of tyramine containing foods (Jacob and Carron 1987).

Atrial fibrillation is a well-recognised complication of heroin abuse (Labi 1969; Gann et al. 1971; Kaufman et al. 1972; Glauser et al 1977), and has followed self-poisoning with digitalis (Ranquin and Parizel 1975), sodium azide (Albertson et al. 1986), desipramine (Colvard 1969), buprenorphine (Gorgolas Hernandez-Mora 1988), alkylphosphate (Himmel and Sterz 1968), and carbon tetrachloride (Kennaugh 1975).

Congenital and Familial Causes of Atrial Fibrillation

Atrial fibrillation has been recorded *in utero*, during both pregnancy and labour (Komaromy et al. 1977). Belhassen et al. (1982) reported fetal atrial fibrillation at the thirty- second week of gestation; a baby with the Wolff–Parkinson–White syndrome was delivered. Intrauterine atrial fibrillation was associated with fetomaternal haemorrhage in a case described by Bacevice et al (1985). Idiopathic atrial fibrillation may present at birth (Makarova and Ageeva 1982) or in infancy (Zaldivar et al. 1973). Both idiopathic atrial fibrillation and the sick sinus syndrome may be familial, and may present as atrial fibrillation in childhood (Derrida et al. 1976; Ardiaca Capell et al. 1987; Yan et al 1983; Ector and van der Hauwaert 1980). Vacuolar degeneration and early atrial cell necrosis were observed in the members of one such family (Amat-Y-Lyon et al. 1974), while right atrial dilatation was observed in another (Jenni et al. 1981).

Atrial fibrillation was present in 6 out of 32 (19%) adults with atrial septal defect. Its occurrence in these patients was related to left atrial pressure, left atrial enlargement and age (Tikoff et al. 1968).

In patients with the Wolff–Parkinson–White syndrome, atrial fibrillation is seen more frequently than might be expected considering that the syndrome is not usually associated with overt heart disease. Atrial fibrillation is documented in 12%–39% of reported series (Gallagher et al. 1978). The electrophysiological mechanisms underlying this association are discussed in Chapter 2 and reviewed by Fujimura et al. (1990).

Atrial fibrillation presenting in midlife may be a manifestation of cardiac involvement in several familial neuromuscular disorders, including Friedreich's ataxia (Malo et al. 1976), Emery–Dreifuss muscular dystrophy (Miller et al. 1985), and some non-eponymous conditions (Graber et al. 1986; Waters et al. 1975). Cardiac involvement resulting in atrial fibrillation is common in haemochromatosis (Marty et al. 1973; Passa et al. 1975; Chernii and Shmidt 1980), and amyloidosis (Hodkinson and Pomerance 1977), including the familial form which is associated with polyneuropathy (Juillet and Grosgogeat 1978; Eriksson and Olofsson 1984). Atrial fibrillation and other dysrhythmias may occur unusually in Wilson's disease (Wolf et al. 1983; Kuan 1987).

Atrial fibrillation has been reported in thalassaemia (Conley and Wintrobe 1976), Tangier disease (Plancher et al. 1984) and alcaptonuria (Tsunashima et al. 1976).

Infection and Infestation

Atrial fibrillation is commonplace in elderly patients with chest infection, especially if heart failure is present. Published case reports, on the other hand, emphasise atrial fibrillation as an unusual complication, or as the result of an uncommon infection. In these categories are accounts of atrial fibrillation in mild viral hepatitis (Fedi et al. 1968; Agarwal et al. 1982), Rocky Mountain spotted fever (Atwater et al. 1968), brucellosis (Axon and Rimmer 1969), murine typhus (Grand and Gallet 1972), filariasis (Gerbaux et al. 1973), ornithosis (Reid et al. 1982), infection with *Nocardia asteroides* (Susens et al. 1967), Lyme carditis (Ballmer and Hany 1988), and Mediterranean boutonneuse fever (Scaffidi et al. 1981).

Cardiac involvement by coxsackie virus infection is well recognised, and atrial fibrillation may recur with each recrudescence (Bacca et al. 1967; Sainani et al. 1968; Dawson and Rogen 1970). Less well known are the electrocardiographic consequences of influenza, documented by Verel et al. (1976). These include transient ST-segment deviation, T-wave inversion, flattening of the T wave, sinus bradycardia, sinus tachycardia, nodal rhythm and atrial fibrillation. A more serious infection involving the heart is leptospirosis, reported from New Caledonia; cardiovascular collapse with tachyarrhythmia features prominently (Dussarat et al. 1988). Two papers describe toxoplasmic infection as an explanation for obscure cardiomyopathy, atypical chest pain and arrhythmias (Leak and Meghji 1979; Novikov and Gracheva 1982).

Soto-Rojas et al. (1984) reported that electrocardiographic abnormalities, including asymptomatic atrial fibrillation, are common in apparently healthy subjects with positive serological tests for Chagas' disease.

Atrial fibrillation was present in 18 out of 69 cases (26%) of subacute bacterial endocarditis described by Eisinger (1971). It was more common in the elderly, and was associated with a high mortality, partly because of the age of the subjects and partly because underlying cardiac function was poorer in those with the arrhythmia.

Malignancy

Malignant infiltration of the atria may predispose to fibrillation because of dispersion of refractoriness, reduction of conduction velocity, and atrial enlargement. Frequent electrocardiographic abnormalities in patients with leukaemia are reported by Kafkas et al. (1973). Case reports are available for atrial fibrillation in multiple myeloma (Atkinson et al. 1974), lymphocytic leukaemia with mucor mediastinitis (Connor et al. 1979), acute myeloblastic leukaemia (Jaeger and Rivier 1975), and metastatic bronchogenic carcinoma; in the latter an ante-mortem diagnosis was based on a distinctive PR segment and the occurrence of atrial arrhythmias (Goldberger and Ludwig 1978). Non-malignant fatty infiltration of myocardium may have a similar end-result (Balsaver et al. 1967). Atrial fibrillation has been reported in left atrial myxoma (Guillet et al. 1981), and with a lipoma of the inter-atrial septum (Klein and Schaefer 1973).

Connective Tissue Disorders

Conduction disturbances and supraventricular arrhythmias are common in progressive systemic sclerosis (Clements et al. 1981; Cozzi et al. 1983) and ankylosing spondylitis (Nitter-Hauge and Otterstad 1981; Bergfeldt et al. 1982), and atrial fibrillation has been reported in Reiter's syndrome (De Mestral et al. 1978).

The Acute Abdomen

Atrial fibrillation is not uncommon in the patient with an acute abdomen (Tullio et al. 1981). An obvious mechanism is a strong vagal response to severe pain, seen most clearly in renal colic due to acute ureterolithiasis (Vargo 1985). Reaction to pain may be the explanation for atrial fibrillation complicating acute pancreatitis (Barbezat and Waterworth 1978; Gullo et al. 1988) and acute cholangitis (Sandu and Popescu 1975). Direct involvement of the left atrium by a fistula arising from an ulcer in the lower oesophagus was the explanation in cases reported by Cunnane (1978) and Snyder et al. (1990). The diagnosis is suggested by the triad of chronic dysphagia, haematemesis and acute neurological signs; there may be pericarditis, atrial fibrillation or shock.

In some cases the causal connection between atrial fibrillation and the acute abdomen is reversed, fibrillation causing visceral infarction by embolism from the left atrium.

Neurological Conditions

Migraine attacks may be associated with atrial fibrillation, perhaps because of vagal participation in nausea and vomiting (Shuaib et al. 1987), or by direct involvement of the nucleus of the vagus nerve in the brainstem in basilar migraine (Peterson et al 1977). Holter electrocardiogram recordings in patients with cluster headaches showed rhythm disturbances in a fifth of them (Russell and Storstein 1983).

Atrial fibrillation associated with seizures in a case of frontal meningioma was reported by Mathew et al. (1970), and paroxysmal atrial fibrillation occurred in a young woman with an acute right pontine lesion, later shown by magnetic resonance imaging to be due to multiple sclerosis (Chagnac et al. 1986).

Respiratory and Sundry Conditions

Atrial fibrillation is common after pneumonectomy, occurring in 22%–28% of cases, usually in the first 3 post-operative days (Stougard 1969). There is poor correlation with preoperative pulmonary function, but atrial fibrillation often follows intrapericardial dissection (Krowka et al. 1987).

Although Davies and Pomerance (1972) reported that atrial fibrillation in just the last 2 weeks of life was commonly associated with pulmonary embolism in a prospective study of 90 patients with pulmonary embolism, none had atrial flutter or fibrillation (Stein et al. 1975).

Edwards and Wilkins (1987) described atrial fibrillation precipitated by acute hypovolaemia in 6 patients with acute cardiorespiratory failure, all of whom had low pulmonary artery pressure. Atrial fibrillation reverted to sinus rhythm within 30 minutes of plasma volume expansion.

In 102 patients with chronic pulmonary heart disease, atrial fibrillation was present in only 6 (Thomas and Valabhji 1969).

There are case reports of atrial fibrillation in spontaneous pneumothorax (Saidi et al. 1967), idiopathic pulmonary haemosiderosis (Roberts et al. 1972), giant cell myocarditis (Sundstrom 1973), pulmonary sarcoidosis (Capritti 1976), severe anaemia (Buxbaum and Furgerson 1970), psoriasis (Gavrilova 1978), and a bronchogenic cyst that was compressing the left atrium (Volpi et al. 1988).

Spontaneous Termination of Atrial Fibrillation

There are several case reports of spontaneous termination of atrial fibrillation which had been present continuously for a decade or more (Zimmerman et al. 1973; Chevalier 1979; Cecchetti 1980; Khan 1980; Gardner and Dunn 1982; Kaul et al. 1983). Most patients had had rheumatic mitral valve disease, and it was exceptional for the resumption of sinus rhythm to be associated with any clinical improvement (Ciaccheri et al 1981). Where investigated, the left atrium was shown to be mechanically and electrically inactive, generating a low voltage P wave that lacked the left atrial component (Froment et al. 1976; Olsson et al 1980). Histological examination showed severe loss of myocardial cells in the atria (Yoneda et al 1978).

Reversion of atrial fibrillation to sinus rhythm has been reported in association with hyperkalaemia (Manchester and Lamberti 1970; Mathew 1979; Rothschild et al. 1985). Quite another mechanism may have been involved in a conversion from atrial fibrillation to sinus rhythm that took place in a prayer meeting (Kowey et al. 1986).

Thromboembolism

Prevalence of Thromboembolism in Atrial Fibrillation

Atrial fibrillation owes much of its clinical importance to its association with thromboembolism and, in particular, cerebral embolism resulting in stroke.

In a prospective study endocardial thrombus was present in 17% of 1000 consecutive autopsies, but the proportion was 35% in patients with atrial fibrillation at the time of death (Burry and Row 1969).

Aberg (1969) studied 642 patients who had atrial fibrillation when they died. At autopsy 21% had atrial thrombi, and 44% had evidence of embolism, the proportion with emboli being higher where atrial fibrillation had been of long duration, and in those with valvular and congenital heart disease. In another autopsy study of atrial fibrillation embolism was present in 41% of patients with mitral valve disease, and 35% of patients with ischaemic heart disease. By contrast, embolism was only seen in 7% of patients with ischaemic heart disease who were in sinus rhythm when they died (Hinton et al. 1977).

Cabin et al. (1990) followed up 272 patients with atrial fibrillation without mitral stenosis for a mean period of 33 months. Twenty-seven (10%) had an embolic event, cerebral in 23 and peripheral in 4. The risk of such an event was increased in females, by underlying heart disease, and by an enlarged left atrium, but not by age, hypertension, or type of atrial fibrillation, whether paroxysmal or chronic. On the other hand, chronic atrial fibrillation was the only significant predictor of embolism in a similar study reported by Wiener et al. (1987).

In the Copenhagen AFASAK study, previous myocardial infarction was the only significant predictor of thromboembolic events in patients with non-rheumatic atrial fibrillation (Petersen et al. 1990). Petersen (1990) put the annual risk of thromboembolism in chronic non-rheumatic atrial fibrillation at 3%–6%, some 5–7 times that in sinus rhythm.

Embolic events cluster around the time of onset of paroxysmal atrial fibrillation, the incidence increasing several-fold with transition to chronic atrial fibrillation, when clustering of embolic events takes place during the first year (Petersen and Godtfredsen 1986).

Fairfax et al. (1976) compared the prevalence of systemic embolism in chronic sinoatrial disorder (16%), chronic complete heart block (1.3%) and chronic ventricular bradycardia with atrial fibrillation or flutter (7.3%). They concluded that impaired atrial function appears to be a key factor in predisposing to intracardiac thrombosis and systemic embolism.

Systemic embolism is not uncommon in thyrotoxic atrial fibrillation (Staffurth et al. 1977; Yuen et al. 1979; Bar-Sela et al 1981; Hurley et al. 1981; Petersen and Hansen 1988), and out of 79 thromboembolic episodes in these studies 51 (65%) were cerebral. Alone among these studies, that of Petersen and Hansen (1988) indicates that atrial fibrillation is not an independent risk factor for thromboembolic events in thyrotoxicosis. However, Presti and Hart (1989) advised that this study should not be viewed in isolation from the others which together suggest that the rate of embolism in thyrotoxic atrial fibrillation actually exceeds that for non-rheumatic atrial fibrillation in euthyroid subjects. The question raised by Petersen and Hansen's (1988) results is not whether anticoagulants are indicated for patients with thyrotoxicosis and atrial fibrillation, but whether they are also indicated for patients with thyrotoxicosis in sinus rhythm.

Prevalence of Atrial Fibrillation in Thromboembolism

Atrial fibrillation is commonly present in patients with thromboembolism. Of 174 patients with peripheral arterial embolism described by MacGowan and Mooneeram (1973), 107 (61%) had atrial fibrillation. As reported by Andersson et al. (1989), 50% of 106 patients with arterial embolism had chronic atrial fibrillation, and 21 of 28 patients with sinus rhythm on admission to hospital had previous atrial arrhythmias. Thirty- seven per cent of 130 patients with peripheral arterial embolism reported by Lorentzen et al. (1980) had atrial fibrillation; their 5 year survival rate was only 39%, death being due primarily to the underlying cardiovascular disease.

A case of acute myocardial infarction due to coronary embolism from left atrial thrombus and atrial fibrillation has been reported (Noto et al. 1990).

Prevalence of Stroke in Atrial Fibrillation

In clinical as well as pathological series of patients with systemic embolism, cerebral emboli are more numerous than peripheral and visceral emboli, by a factor of about 4 to 1 (Fisher 1982). Since the brain receives 20% of the cardiac output, fewer cerebral than peripheral emboli would be expected, in a ratio of 1 to 4 – the reverse of what is found. A likely explanation is that small emboli that cause rapid and extensive infarction in the cerebral circulation are harmless elsewhere (Fisher 1982).

In the Framingham study, chronic atrial fibrillation in the absence of rheumatic heart disease was associated with a greater than 5-fold increase in stroke independently of associated congestive cardiac failure and coronary heart disease, while atrial fibrillation with rheumatic heart disease had a 17-fold increase (Wolf et al. 1978). The age-adjusted rate for stroke in lone atrial fibrillation was 4 times that of the control group (Brand et al. 1985). The greater risk of stroke in atrial fibrillation associated with rheumatic heart disease has been confirmed by Petersen and Godtfredsen (1988).

While in 1978 there was no evidence of a particularly vulnerable period for stroke, in a later publication from Framingham there was said to be a clustering of strokes at the time of onset of atrial fibrillation; the recurrence of stroke within 6 months was more than twice as common in atrial fibrillation than in sinus rhythm (Wolf et al. 1983). Most cerebral infarcts in patients with atrial fibrillation are large, disabling, and unheralded by transient cerebral ischaemic attacks. A quarter of the recurrences take place within 2 weeks of the initial stroke (Sherman et al. 1984).

In a Canadian study of 91 patients with non-rheumatic atrial fibrillation, there were 7.9 new strokes per 100 person-years; patients with previous embolic events were 2.3 times more likely to have another event (Flegel and Hanley 1989). The risk of recurrent stroke in patients with atrial fibrillation and non- valvular heart disease remained at approximately 20% per year throughout a 9 year observation period in a study reported by Sage and van Uitert (1983).

A longitudinal study of a Chinese population aged 60 and over showed that subjects who initially had a history of transient cerebral ischaemic attacks and atrial fibrillation had a greater than 10-fold increase in the risk of stroke in the subsequent 30 months (Woo and Lau 1990).

In geriatric patients admitted to hospital, the prevalence of stroke was 42% in those with chronic atrial fibrillation, 27% in those with transient atrial fibrillation, and 19% in patients in sinus rhythm (Treseder et al. 1986).

However, in a prospective study of 1804 residents over 50 years old in Rochester, Minnesota, atrial fibrillation was not found to be an independent risk factor for stroke when examined by multivariate analysis (Davis et al. 1987). Phillips (1990) questioned whether atrial fibrillation is an independent risk factor for stroke, pointing out that the increased probability of stroke attributed to this arrhythmia might be due to its association with other risk factors such as hypertension, diabetes and atherosclerosis. However, the relationship between stroke and atrial fibrillation is generally much stronger than that between stroke and other risk factors (Woo and Lau 1990).

Flegel et al. (1987) compared the risk of stroke in three studies. Non-rheumatic atrial fibrillation conferred a relative risk for stroke of 6.9 in the Whitehall study of civil servants, and 5.6 in Framingham. In the British Region Heart Study only 1 stroke had occurred during the follow-up period. The authors point out that the absolute risk of stroke is less in British studies than in Framingham, and this bears on the potential benefit from prophylaxis.

Patients with chronic atrial fibrillation have evidence of previous silent cerebral infarction more often than do control subjects in sinus rhythm (Kempster et al. 1988; Guidotti et al. 1990), but this is not the case in paroxysmal atrial fibrillation (Petersen et al. 1989). ·

The Prevalence of Atrial Fibrillation in Stroke

Cardiac arrhythmias are common in acute stroke, and are of a variety that could have given rise to cerebral embolism in 13%–25% of cases (Norris et al. 1978; Britton et al. 1979; Lowe et al. 1983; Sandercock et al. 1989). A higher detection rate of potentially embolising arrhythmias is achieved with Holter monitoring (Abdon et al. 1982).

Harrison and Marshall (1984) reported the presence of atrial fibrillation in 5.6% of completed strokes and 1.6% of transient cerebral ischaemic attacks. Transient ischaemic attacks in the presence of atrial fibrillation last longer than 60 minutes, except when there is coexistent carotid disease that might be a source of microemboli (Weinberger et al. 1988); atrial fibrillation itself is only rarely responsible for brief transient cerebral ischaemic attacks.

Infarction due to cerebral embolism was present in 24% of all patients with stroke reported by Caplan et al. (1983). Coronary heart disease, atrial fibrillation, valvular heart disease, mitral calcification and cardiomyopathy were the commonest underlying conditions giving rise to emboli; the prognosis was worse for embolic than for thrombotic stroke. Because of the association between atrial fibrillation and embolic stroke, the presence of atrial fibrillation would be expected to carry a worse prognosis. This was reported by Lowe et al. (1983) for patients aged 60–79 years, but in a study of 1484 stroke cases by Abu-Zeid et al. (1978), the presence of atrial fibrillation did not significantly affect survival.

In a black urban population stroke was due to embolism in 14% of stroke victims, in half of whom the presumed cause of embolism was atrial fibrillation (Rosman 1986). This is in line with the estimate of Fisher (1982) that 10% of all strokes are due to cerebral embolism from atrial fibrillation.

Van Merwijk et al. (1990) determined the prevalence of atrial fibrillation in patients with strokes due to primary intracerebral haemorrhage, lacunar infarcts, and cortical infarcts. Atrial fibrillation was as common in lacunar infarcts as in haemorrhagic infarcts, but much commoner in cortical infarcts, with an odds ratio of 5.6. Knowing the frequency of atrial fibrillation in haemorrhagic infarcts, the authors suggest that as many as 30% of cortical infarcts associated with atrial fibrillation do not result from the arrhythmia.

Intracardiac Thrombi in Atrial Fibrillation

In post-mortem studies thrombus is present more often in the left atrial appendage than in the left atrial chamber, and more often in the left atrium than the right atrium (Fisher 1982). Intracardiac thrombus is commoner in rheumatic than in ischaemic heart disease, and in atrial fibrillation than in sinus rhythm (Aberg 1969). In patients with severe mitral valve disease, left atrial thrombi were present in 36% (Wallach et al. 1953).

Beppu et al. (1984) emphasised the dynamic nature of intracardiac thrombi – they grow or shrink spontaneously, or sometimes become detached (Sogaard 1981) and embolise (Kushwaha and Jepson 1990). But the reasons for these changes, and for some patients developing atrial thrombi while others do not, are obscure. Coagulation studies show that in mitral valve disease there is a thrombotic tendency in blood in the left heart, which is particularly pronounced during exercise (Toy et al. 1980). In mitral stenosis antithrombin III levels are reduced due to increased consumption (Fukuda et al. 1980), and platelet function is augmented with increased thromboglobulin (Fukuda and Nakamura 1984). Chronic atrial fibrillation itself is associated with raised concentrations of fibrin degradation products indicative of increased intracardiovascular clotting (Kumagai et al. 1990). The fibrinolytic system is inhibited if congestive cardiac failure is present (Rawles et al. 1973), so that any thrombi formed would be more likely to persist.

Stasis of blood in the left atrium is another factor predisposing to thrombosis there; blood flow velocities in the left atrial appendage are less in atrial fibrillation than in sinus rhythm, especially in the presence of mitral stenosis (Suetsugu et al. 1988). Cardiac output is often reduced in patients with mitral stenosis and systemic embolism, further contributing to intracardiac stasis (Casella et al. 1964).

In vivo, intra-atrial thrombus may be detected by cine- angiography; most commonly, the presence of clot has been an incidental finding during the preoperative investigation of rheumatic heart disease, open heart surgery being an opportunity to validate the diagnosis (Parker et al. 1965; Fisher et al. 1965; Lewis et al. 1965).

Eriksson et al. (1984) used pulmonary artery cine-angiography for detection of cardiac sources of cerebral embolism. Among 33 patients with non-haemorrhagic stroke and chronic atrial fibrillation, an atrial thrombus was detected in 8 (24%). About 1 in 3 patients was estimated to have remaining intracardiac thrombus after stroke. At present, angiography is more sensitive than echocardiography in detecting thrombus (Eriksson et al. 1988), but computed tomography is an alternative non-invasive technique with promise (Tomoda et al. 1980).

References

Abdon NJ, Zettervall O, Carlson J et al. (1982) Is occult atrial disorder a frequent cause of non-hemorrhagic stroke? Long-term ECG in 86 patients. Stroke 13:832–7

Aberg H (1968) Atrial fibrillation. A review of 463 cases from Philadelphia General Hospital from 1955 to 1965. Acta Med Scand 184:425–31

Aberg H (1969) Atrial fibrillation. I. A study of atrial thrombosis and systemic embolism in a necropsy material. Acta Med Scand 185:373–9

Abu-Zeid AH, Choi NW, Hsu PH, Maini KK (1978) Prognostic factors in the survival of 1484 stroke cases observed for 30 to 48 months. II. Clinical variables and laboratory measurements. Arch Neurol 35:213–18

Agarwal BD, Gahlaut DS, Tandon RN, Seth ON, Sikka KK (1982) Atrial fibrillation in mild viral hepatitis [letter]. J Assoc Physicians India 30:122

Agarwal BL, Agarwal BV (1971) Atrial fibrillation: a manifestation of digitalis intoxication. J Assoc Physicians India 19:719–24

Albertson TE, Reed S, Siefkin A (1986) A case of fatal sodium azide ingestion. J Toxicol Clin Toxicol 24:339–51

Allaz AF, Guillaume P, Male PJ, Dubas J, Adamec R (1983) [Continuous electrocardiography monitoring during esophago- gastro-duodenoscopy.] Schweiz Med Wochenschr 113:1188–91

Alstrup P, Sorensen HR (1975) Metabolic studies following thoracotomy for lung cancer with particular reference to postoperative atrial fibrillation. Scand J Thorac Cardiovasc Surg 9:149–53

Alt E, Volker R, Wirtzfeld A (1985) Survival and follow-up after pacemaker implantation: A comparison of patients with sick sinus syndrome, complete heart block, and atrial fibrillation. PACE 8:849–55

Amat-Y-Leon F, Racki AJ, Denes P et al. (1974) Familial atrial dysrhythmia with A-V block. Intracellular microelectrode, clinical electrophysiologic, and morphologic observations. Circulation 50:1097–1104

Anbe DT, Rubenfire M, Drake EH (1969) Conversion of atrial flutter to atrial fibrillation with carotid sinus pressure. J Electrocardiol 2:377–80

Andersson B, Abdon NJ, Hammarsten J (1989) Arterial embolism and atrial arrhythmias. Eur J Vasc Surg 3:261–6

Anonymous (1985) Alcohol and atrial fibrillation [Editorial]. Lancet i:1374

Ardiaca Capell A, Ibanez Regales M, Rubio Caballero M (1987) [Familial idiopathic auricular fibrillation.] Rev Esp Cardiol 40:65–8

Atkinson K, McElwain TJ, Mackay AM (1974) Myeloma of the heart. Br Heart J 36:309–12

Atwater JS, Markle CD, Stubbs W (1968) Rocky Mountain spotted fever with thrombocytopenia and atrial fibrillation: a case report. J Med Assoc Ga 57:210–2

Axon AT, Rimmer DM (1969) Brucellosis presenting with arrhythmia. Br Med J ii:695

Bacca F, Lovreglio V, Brindicci G, Rizzon P (1967) [Parosymal atrial fibrillation in the course of recurrences of viral pericarditis.] Cuore Circ 51:17–25

Bacevice AE Jr, Dierker LJ Jr, Wolfson RN (1985) Intrauterine atrial fibrillation associated with fetomaternal hemorrhage. Am J Obstet Gynecol 153:81–2

Balducelli M, Capucci A, Boriani G, Magnani B (1989) Paroxysmal atrial fibrillation: study of the pattern of onset using dynamic electrocardiography. Cardiologia 34:713–20

Ballmer PE, Hany A (1988) [Lyme carditis.] Schweiz Med Wochenschr 118:358–62

Balsaver AM, Morales AR, Whitehouse FW (1967) Fat infiltration of myocardium as a cause of cardiac conduction defect. Am J Cardiol 19:261–5

Barbezat GO, Waterworth MW (1978) Atrial fibrillation in acute pancreatitis. A report of two cases. S Afr Med J 53:554–5

Bar-Or D, Gasiel Y (1981) Calcium and calciferol antagonise effect of verapamil in atrial fibrillation. Br Med J 282:1585–6

Bar-Sela S, Ehrenfeld M, Eliakim M (1981) Arterial embolism in thyrotoxicosis with atrial fibrillation. Arch Intern Med 141:1191–2

Bartall H, Brown S, Benchimol A, Desser KB, Sheasby C (1979) "Push-up palpitations": unusual presentation of ruptured chordae tendineae: a case report. Angiology 30:347–50

Bass HN, Hildreth BF (1979) Paroxysmal atrial fibrillation and exposure to smoke [letter]. Lancet i:1036

Belhassen B, Pauzner D, Blieden L et al. (1982) Intrauterine and postnatal atrial fibrillation in the Wolff–Parkinson–White syndrome. Circulation 66:1124–8

Belhassen B, Pelleg A, Shoshani D, Laniado S (1984) Atrial fibrillation induced by adenosine triphosphate. Am J Cardiol 53:1405–6

Beppu S, Park YD, Sakakibara H, Nagata S, Nimura Y (1984) Clinical features of intracardiac thrombosis based on echocardiographic observation. Jpn Circ J 48:75–82

Bergfeldt L, Edhag O, Vallin H (1982) Cardiac conduction disturbances, an underestimated manifestation in ankylosing spondylitis. A 25-year follow-up study of 68 patients. Acta Med Scand 212:217–23

Berke DK, Graham AF, Schroeder JS, Harrison DC (1973) Arrhythmias in the denervated transplanted human heart. Circulation 48(Suppl 3):112–15

Beton DC, Brear SG, Edwards JD, Leonard JC (1983) Mitral valve prolapse: an assessment of clinical features, associated conditions and prognosis. Q J Med 52:150–64

Bonofiglio A, Bugaro L, Pantaleoni A (1967) [On a case of atrial fibrillation during cranial trauma.] Boll Soc Ital Cardiol 12:859–65

Borgeat A, Grbic M (1986) [Syncope in a 16-year-old young man during a football match.] Schweiz Med Wochenschr 116:1419–21

Brand FN, Abbott RD, Kannel WB, Wolf PA (1985) Characteristics and prognosis of lone atrial fibrillation. 30-year follow-up in the Framingham Study. JAMA 254:3449–53

Brasher GW, Sanchez SA (1974) Reversible electrocardiographic changes associated with wasp sting anaphylaxis. JAMA 229:1210–1

Breeden CC, Safirstein BH (1990) Albuterol and spacer-induced atrial fibrillation. Chest 98:762–3

Britton M, de Faire U, Helmers C, Miah K, Ryding C, Wester PO (1979) Arrhythmias in patients with acute cerebrovascular disease. Acta Med Scand 205:425–8

Bucher HW, Aepli R, Zenger F, Stucki P (1969) [Pheochromocytoma with unusual cardiac arrhythmias and their treatment with alpha- and beta-receptor-blocking agents.] Schweiz Med Wochenschr 99:956–60

Burry AF, Row PG (1969) Endocardial thrombosis: a prospective study of 1000 autopsies. Pathology 1:141–8

Buslo EA, Koltunova MI, Dobrotvorskaia TE (1982) [Diagnostic efficacy of the graded exercise test in various arrhythmias.] Kardiologiia 22:37–41

Buxbaum J, Furgerson W (1970) Atrial fibrillation in severe anemia. JAMA 212:1958–9

Cabin HS, Clubb KS, Hall C, Perlmutter RA, Feinstein AR (1990) Risk for systemic embolization of atrial fibrillation without mitral stenosis. Am J Cardiol 65:1112–6

Cameron A, Schwartz MJ, Kronmal RA, Kosinski AS (1988) Prevalence and significance of atrial fibrillation in coronary artery disease (CASS Registry). Am J Cardiol 61:714–7

Campbell A, Caird FI, Jackson TF (1974) Prevalence of abnormalities of electrocardiogram in old people. Br Heart J 36:1005–11

Caplan LR, Hier DB, D'Cruz I (1983) Cerebral embolism in the Michael Reese Stroke Registry. Stroke 14:530–6

Capritti AG (1976) [Electrocardiographic findings in a case of pulmonary sarcoidosis.] Minerva Cardioangiol 24:297–302

Carnevali R, Omboni E, Rossati M, Villa A, Checchini M (1987) [Electrocardiographic changes in acute carbon monoxide poisoning.] Minerva Med 78:175–8

Casella L, Abelmann WH, Ellis LB (1964) Patients with mitral stenosis and systemic emboli. Hemodynamic and clinical observations. Arch Intern Med 114:773–81

Cecchetti G (1980) [Spontaneous return of sinus rhythm after 15 years of uninterrupted atrial fibrillation in a patient with serious rheumatic valvulopathy. Subsequent intermittent reappearance of the arrhythmia.] Minerva Cardioangiol 28:805–12

Chagnac Y, Martinovits G, Tadmor R, Goldhammer Y (1986) Paroxysmal atrial fibrillation associated with an attack of multiple sclerosis. Postgrad Med J 62:385–7

Cheng TO (1974) Coronary artery disease as an uncommon cause of chronic atrial fibrillation. Clin Res 22:268A

Chernii IaM, Shmidt AD (1980) [Myocardial lesion in hemochromatosis.] Vrach Delo 25:19–22

Chevalier H (1979) Spontaneous resumption of sinus rhythm in an elderly patient after 13 years of permanent atrial fibrillation. Am Heart J 98:361–5

Ciaccheri M, Dolara A, Cecchi F et al. (1981) Spontaneous reversion to normal sinus rhythm after prolonged atrial fibrillation. Acta Cardiol 36:125–9

Clements PJ, Furst DE, Cabeen W, Tashkin D, Paulus HE, Roberts N (1981) The relationship arrhythmias and conduction disturbances to other manifestations of cardiopulmonary disease in progressive systemic sclerosis (PSS). Am J Med 71:38–46

Close JB, Evans DW, Bailey SM (1979) Persistent lone atrial fibrillation – its prognosis after clinical diagnosis. J R Coll Gen Pract 29:547–9

Collier A, Matthews DM, Young RJ, Clarke BF (1987) Transient atrial fibrillation precipitated by hypoglycaemia: two case reports. Postgrad Med J 63:895–7

Colvard C (1969) Overdosage of desipramine hydrochloride with marked electrocardiographic abnormalities. South Med J 61:1218

Conley CL, Wintrobe MM (1976) Thalassemia in the "D" family: case presentation: Mr. M.D. Johns Hopkins Med J 139:201–4

Connor BA, Anderson RJ, Smith JW (1979) *Mucor* mediastinitis. Chest 75:525–6

Cosma M, Munteanu R (1975) [Atrial fibrillation and flutter caused by digitalis.] Rev Med Chir Soc Med Nat Iasi 79:379– 84

Cotoi S, Dragulescu SI (1974) Idiopathic persistent atrial fibrillation precipitated by electrocution in a 40-year-old man. G Ital Cardiol 4:80–3

Coumel P (1990a) Clinical approach to paroxysmal atrial fibrillation. Clin Cardiol 13:209–12

Coumel P (1990b) Role of the autonomic nervous system in paroxysmal atrial fibrillation. In: Touboul P, Waldo AL (eds) Atrial arrhythmias. Current concepts and management. Mosby Year Book, St Louis, pp 248–61

Coumel P, Attuel P, Lavallee J, Flammang D, Leclercq JF, Slama R (1978) [The atrial arrhythmia syndrome of vagal origin.] Arch Mal Coeur 71:645–56

Coumel P, Leclercq J-F, Attuel P (1982) Paroxysmal atrial fibrillation. In: Kulburtus HE, Olsson SB, Schlepper M (eds) Atrial fibrillation. AB Hassle, Molndal, pp 158–75

Cozzi F, Tessier R, Glorioso S, Peserico A, Todesco S (1983) Electrocardiogram in progressive systemic sclerosis. Analysis of 73 cases. Acta Cardiol 38:27–34

Cristal N, Peterburg I, Szwarcberg J (1976) Atrial fibrillation developing in the acute phase of myocardial infarction. Prognostic implications. Chest 70:8–11

Cullen K, Stenhouse NS, Wearne KL, Cumpston GN (1982) Electrocardiograms and 13-year cardiovascular mortality in Busselton study. Br Heart J 47:209–12

Cunnane K (1978) Esophagoatrial fistula. Can J Surg 21:466–7

Davidson E, Weinberger I, Rotenberg Z, Fuchs J, Agmon J (1989) Atrial fibrillation. Cause and time of onset. Arch Intern Med 149:457–9

Davies AB, Williams I, John R, Hall R, Scanlon MF (1985) Diagnostic value of thyrotrophin releasing hormone tests in elderly patients with atrial fibrillation. Br Med J 291:773–6

Davies MJ, Pomerance A (1972) Pathology of atrial fibrillation in man. Br Heart J 34:520–5

Davis PH, Dambrosia JM, Schoenberg BS et al. (1987) Risk factors for ischemic stroke: a prospective study in Rochester, Minnesota. Ann Neurol 22:319–27

Dawson KP, Rogen AS (1970) Cardiac complications of Coxsackie virus group B infection. Practitioner 205:333–5

De Mestral E, Katchaluba M, Gerster JC, Saudan Y (1978) [Auricular fi fibrillation and acute aortic insufficienncy in Reiter's syndrome.] Schweiz Med Wochenschr 108:1886–7

Derrida JP, Gaudeau S, Vernant P (1976) [Familial auricular fibrillation.] Nouv Presse Med 5:1297–9

Di Pasquale G, Pinelli G, Andreoli A, Manini G, Grazi P, Tognetti F (1987) Holter detection of cardiac arrhythmias in intracranial subarachnoid hemorrhage. Am J Cardiol 59:596–600

Dimant J, Grob D (1977) Electrocardiographic changes and myocardial damage in patients with acute cerebrovascular accidents. Stroke 8:448–55

Dodd HJ, Tatnall FM, Sarkany I (1985) Fast atrial fibrillation induced by treatment of psoriasis with azothioprine. Br Med J 291:706

Drucker D (1981) Atrial fibrillation after administration of calcium and pentagastrin [letter]. N Engl J Med 304:1427–8

Dubin AA, March HW, Cohn K, Selzer A (1971) Longitudinal hemodynamic and clinical study of mitral stenosis. Circulation 44:381–9

Dussarat GV, Cointet F, Capdevielle P, Le Bris H, Brethes B (1988) [Cardiac manifestations in leptospirosis. Apropos of 15 cases observed in New Caledonia.] Ann Cardiol Angeiol (Paris) 37:449–53

Ebert PA (1970) Relationship of myocardial potassium content and atrial fibrillation. Circulation 41(Suppl 2):137–41

Ector H, Van der Hauwaert LG (1980) Sick sinus syndrome in childhood. Br Heart J 44:684–91

Edwards JD, Wilkins RG (1987) Atrial fibrillation precipitated by acute hypovolaemia. Br Med J 294:283–4

Eisinger AJ (1971) Atrial fibrillation in bacterial endocarditis. Br Heart J 33:739–41

Ellis LB, Ramirez A (1969) The clinical course of patients with severe "rheumatic" mitral insufficiency. Am Heart J 78:406– 18

El-Sherif N (1972) Paroxysmal atrial flutter and fibrillation. Induction by carotid sinus compression and prevention by atropine. Br Heart J 34:1024–8

Emara MK, Saadet AM (1986) Transient atrial fibrillation in hypertensive patients with thiazide induced hypokalaemia. Postgrad Med J 62:1125–7

Eriksson P, Olofsson BO (1984) Pacemaker treatment in familial amyloidosis with polyneuropathy. PACE 7:702–6

Eriksson S, Osterman G, Asplund K et al. (1984) Pulmonary-artery cineangiocardiography to demonstrate cardiac thrombi in patients with cerebral infarction. Acta Neurol Scand 69:27–33

Eriksson S, Backman C, Osterman G (1988) Pulmonary-artery cineangiocardiography and echo-cardiography for detection of cardiac sources of cerebral embolism. Acta Med Scand 223:27–33

Eskilsson J, Albertsson M, Mercke C (1988) Adverse cardiac effects during induction chemotherapy treatment with cis-platin and 5-fluorouracil. Radiother Oncol 13:41–6

Evans W, Swann P (1954) Lone auricular fibrillation. Br Heart J 16:189–94

Fairfax AJ, Lambert CD, Leatham A (1976) Systemic embolism in chronic sinoatrial disorder. N Engl J Med 295:190–2

Fedi A, Scalise G, Boggiano CA (1968) [Electrocardiographic changes during viral hepatitis.] Rass Int Clin Ter 48:383–9

Fisher CM (1982) Embolism in atrial fibrillation. In: Kulburtus HE, Olsson SB, Schlepper M (eds) Atrial fibrillation. AB Hassle, Molndal, pp 192–207

Fisher DL, Brent LB, Kent EM, Magovern GJ (1965) The preoperative detection of atrial thrombi by selective left atriography. J Thorac Cardiovasc Surg 50:473–81

Flegel KM, Hanley J (1989) Risk factors for stroke and other embolic events in patients with nonrheumatic atrial fibrillation. Stroke 20:1000–4

Flegel KM, Shipley MJ, Rose G (1987) Risk of stroke in non- rheumatic atrial fibrillation. Lancet i:526–9

Fluck DC, Olsen E, Pentecost BL et al. (1967) Natural history and clinical significance of arrhythmias after acute cardiac infarction. Br Heart J 29:170–89

Flugelman MT, Flugelman AA, Rozenman J et al. Prediction of atrial and ventricular fibrillation complicating myocardial infarction from admission data: A prospective study. Clin Cardiol 10:503–5

Flugelman MY, Hasin Y, Shefer A, Sebbag D, Freiman I, Gotsman MS (1986) Atrial fibrillation in acute myocardial infarction. Isr J Med Sci 22:355–9

Ford MJ, Prescott LF (1982) Phenylbutazone pericarditis. Scott Med J 27:252–3

Forfar JC, Miller HC, Toft AD (1979) Occult thyrotoxicosis: a correctable cause of "idiopathic" atrial fibrillation. Am J Cardiol 44:9–12

Franke H (1968) [Heart rhythm disorders in hyperactive carotid sinus reflex.] Internist (Berlin) 9:289–96

Frederiksen MC, Toig RM, Depp R (1983) Atrial fibrillation during hexoprenaline therapy for premature labor. Am J Obstet Gynecol 145:108–9

Froment R, Touboul P, Gallavardin L, Porte J, Dufour R (1976) [Significance of spontaneous reduction of long term auricular fibrillation. Apropos of a case followed by subtotal auricular paralysis.] Arch Mal Coeur 69:315–20

Fujimura O, Klein GJ, Yee R, Sharma AD, Boahene KA (1990) Atrial fibrillation in the Wolff–Parkinson–White syndrome. In: Touboul P, Waldo AL (eds) Atrial arrhythmias. Current concepts and management. Mosby Year Book, St Louis, pp 262–9

Fujita R, Kumura F (1975) Arrhythmias and ischemic changes of the heart induced by gastric endoscopic procedures. Am J Gastroenterol 64:44–8

Fukuda Y, Nakamura K (1984) The incidence of thromboembolism and the hemocoagulative background in patients with rheumatic heart disease. Jpn Circ J 48:59–66

Fukuda Y, Kuroiwa Y, Okumiya K et al. (1980) Hypercoagulability in patients with mitral stenosis – from the viewpoint of the behavior of plasma antithrombin III and alpha2-plasmin inhibitor. Jpn Circ J 44:867–74

Fuller JA, Adams GG, Buxton B (1989) Atrial fibrillation after coronary artery bypass grafting. Is it a disorder of the elderly? Thorac Cardiovasc Surg 97:821–5

Fuster Siebert M, Fernandez Iglesias A, Couselo Sanchez JM, Gallego Garcia D, Iglesias Diz JL, Alvez Gonzalez F (1983) [Daunomycin and atrial fibrillation.] An Esp Pediatr 19:415–6

Gabathuler J, Adamec R (1985) [Triggering of paroxysmal auricular fibrillation. Study using continuous electrocardiographic recording (Holter system).] Arch Mal Coeur 78:1255–62

Gajewski J, Singer RB (1981) Mortality in an insured population with atrial fibrillation. JAMA 245:1540–4

Gallagher JJ, Pritchett ELC, Sealy WC, Kasell J, Wallace AG (1978) The preexitation syndromes. Prog Cardiovasc Dis 20:285–327

Gann D, Mansour N, Crosby DJ (1971) Atrial fibrillation and pulmonary edema in acute heroin intoxication. Ariz Med 28:672–4

Garcia Pascual J (1986) [Paroxysmal auricular fibrillation and acute alcoholic intoxication (letter)]. Med Clin (Barc) 87:82

Gardner JD, Dunn M (1982) Spontaneous conversion of long-standing atrial fibrillation. Chest 81:429–32

Gartman DM, Bardy GH, Allen MD, Misbach GA, Ivey TD (1990) Short- term morbidity and mortality of implantation of automatic implantable cardioverter-defibrillator. J Thorac Cardiovasc Surg 100:353–7

Gavrilova LV (1978) [Heart lesion in psoriasis.] Vrach Delo 1978 Mar(3):60–2

Gayet C, Wilner C, Orgiazzi J, Tourniaire J, Berthezene F, Matagrin C (1987) [Auricular fibrillation in hyperthyroidism. Incidence of associated cardiopathy and of dilatation of the left auricle.] Arch Mal Coeur 80:1278–82

Geha DG, Davis NJ, Lappas DG (1973) Persistent atrial arrhythmias associated with placement of a Swan–Ganz catheter. Anesthesiology 39:651–3

Gerbaux A, Dubost C, Maurice P et al. (1973) [Endomyocardial fibrosis during filariasis. Apropos of a surgically treated case.] Ann Med Interne (Paris) 124:471–82

Glancy DL, O'Brien KP, Gold HK, Epstein SE (1970) Atrial fibrillation in patients with idiopathic hypertrophic subaortic stenosis. Br Heart J 32:652–9

Glauser FL, Downie RL, Smith WR (1977) Electrocardiographic abnormalities in acute heroin overdosage. Bull Narc 29:85–9

Godtfredsen J (1982) Atrial fibrillation: course and prognosis – a follow-up study of 1212 cases. In: Kulburtus HE, Olsson SB, Schlepper M (eds) Atrial fibrillation. AB Hassle, Molndal, pp 134–145

Goldberg RJ, Seeley D, Becker RC et al. (1990) Impact of atrial fibrillation on the in-hospital and long-term survival of patients with acute myocardial infarction: a community-wide perspective. Am Heart J 119:996–1001

Goldberger AL, Ludwig M (1978) Metastatic atrial tumor: case report with electrocardiographic-pathologic correlation. J Electrocardiol 11:297–300

Goodwin NJ, Castronuovo JJ, Friedman EA (1969) Recurrent septic pulmonary embolism complicating maintenance hemodialysis. Ann Intern Med 71:29–38

Gorelick DA, Marder SR, Sack D, Marks H (1981) Atrial flutter/fibrillation associated with tranylcypromine treatment. J Clin Psychopharmacol 1:402–4

Gorentsvit IE (1981) [Nature of the heart rhythm, size of the left atrium and cardiac volume in aortic stenosis.] Kardiologiia 21:72–5

Gorgolas Hernandez-Mora P, Marcos Sanchez F, Aparicio Martinez J, de Juana Velasco P (1988) [Paroxysmal auricular fibrillation after the ingestion of buprenorphine (letter).] Rev Clin Esp 183:99–100

Graber HL, Unverferth DV, Baker PB, Ryan JM, Baba N, Wooley CF (1986) Evolution of a hereditary cardiac conduction and muscle disorder: a study involving a family with six generations affected. Circulation 74:21–35

Grand A, Gallet J (1972) [Murine typhus revealed by pericarditis.] Arch Mal Coeur 65:611–20

Greenspon AJ, Schaal SF (1983) The "holiday heart": electrophysiologic studies of alcohol effects in alcoholics. Ann Intern Med 98:135–9

Greentree LB (1986) Auricular fibrillation: an unusual experience [letter]. N Engl J Med 315:1680–1

Greer GS, Wilkinson WE, McCarthy EA, Pritchett EL (1989) Random and nonrandom behavior of symptomatic paroxysmal atrial fibrillation. Am J Cardiol 64:339–42

Guidotti M, Tadeo G, Zanasi S, Pellegrini G (1990) Silent cerebral ischemia in patients with chronic atrial fibrillation–a case–control study. Ir J Med Sci 159:96–7

Guillet P, Baconnet C, Labrousse A et al. (1981) Left atrial myxoma in the elderly: diagnosis by M-mode and bidimensional echocardiography. J Am Geriatr Soc 29:453–9

Gullo L, Labriola E, Di Benedetto S, Priori P, Iannacci P, Pezzilli R (1988) Acute pancreatitis associated with paroxysmal atrial fibrillation. A case report. Panminerva Med 30:111–3

Gupta OP, Mewar SH, Kalantri SP, Jain AP, Jajoo UN (1987) Reversible atrial fibrillation following snake bite. J Assoc Physicians India 35:535–6

Gupta GB, Gupta SR, Somani PN, Agrawal BV (1988) Atrial fibrillation with inferior wall myocardial ischaemia following lightning [letter]. J Assoc Physicians India 36:354–5

Haddad AH, Prchkov VK, Dean DC (1978) Chronic atrial fibrillation and coronary artery disease. J Electrocardiol 11:67–9

Harrison DC, Mason JW, Schroeder JS et al. (1978) Effects of cardiac denervation on cardiac arrhythmias and electrophysiology. In: Prevention of arrhythmias. Br Heart J 40 (Suppl):17–23

Harrison MJ, Marshall J (1984) Atrial fibrillation, TIAs and completed strokes. Stroke 15:441–2

Haslett C, Douglas JG, Munro JF (1983) Rheumatic heart disease and thyroid status. Scott Med J 28:17–20

Henry WL, Morganroth J, Pearlman AS et al. (1976) Relation between echocardiographically determined left atrial size and atrial fibrillation. Circulation 53:273–9

Hill JD, Mottram EM, Killeen PD (1987) Study of the prevalence of atrial fibrillation in general practice patients over 65 years of age. J R Coll Gen Pract 37:172–3

Himmel G, Sterz H (1968) [Auricular fibrillation in alkylphosphate poisoning.] Wien Klin Wochenschr 80:350–1

Hinton RC, Kistler JP, Fallon JT, Friedlich AL, Fisher CM (1977) Influence of etiology of atrial fibrillation on incidence of systemic embolism. Am J Cardiol 40:509–13

Hod H, Lew AS, Keltai M et al. (1987) Early atrial fibrillation during evolving myocardial infarction: a consequence of impaired left atrial perfusion Circulation 75:146–50

Hodkinson HM, Pomerance A (1977) The clinical significance of senile cardiac amyloidosis: a prospective clinico-pathological study. Q J Med 46:381–7

Horowitz LN, Kay HR, Kutalek SP et al. (1987) Risks and complications of clinical cardiac electrophysiologic studies: a prospective analysis of 1,000 consecutive patients. J Am Coll Cardiol 9:1261–8

Hunt D, Sloman G, Penington C (1978) Effects of atrial fibrillation on prognosis of acute myocardial infarction. Br Heart J 40:303–7

Hurley DM, Hunter AN, Hewett MJ, Stockigt JR (1981) Atrial fibrillation and arterial embolism in hyperthyroidism. Aust N Z J Med 11:391–3

Iseri LT (1990) Role of magnesium in cardiac tachyarrhythmias. Am J Cardiol 65:47K–50K

Iwasaki T, Naka M, Hiramatsu K et al. (1989) Echocardiographic studies on the relationship between atrial fibrillation and atrial enlargement in patients with hyperthyroidism of Graves' disease. Cardiology 76:10–7

Jacob LH, Carron DB (1987) Atrial fibrillation precipitated by tyramine containing foods. Br Heart J 57:205–6

Jaeger M, Rivier JL (1975) [Development of persistent atrial paralysis secondary to acute leukemia.] Arch Mal Coeur 68:1217–24

James TN (1961) Myocardial infarction and atrial arrhythmias. Circulation 24:761–76

Jenni R, Goebel N, Schneider L, Krayenbuhl HP (1981) [Idiopathic familial right atrial dilatation.] Schweiz Med Wochenschr 111:1565–72

Jordaens L, Hamerlynck R, Clement DL (1985) Efficacy of an implanted automatic defibrillator which had induced atrial fibrillation. Br Heart J 54:605–8

Juillet Y, Grosgogeat Y (1978) [Cardiac amyloidosis: clinical, x-ray, electric, hemodynamic, developmental and pathological aspects. Apropos of 85 cases collected from French cardiology departments.] Arch Mal Coeur 71:361–70

Kafkas P, Papaevangelou G, Kanaginis T, Iordanoglou J, Hatzioannou J (1973) Frequency of electrocardiographic alterations in patients with leukaemia. Statistical analysis of 480 cases. Ann Clin Res 5:23–6

Kalloor GJ, Singh SP, Collis JL (1977) Cardiac arrhythmias on swallowing. Am Heart J 93:235–8

Kannel WB, Abbott RD, Savage DD, McNamara PM (1982) Epidemiologic features of chronic atrial fibrillation: the Framingham study. N Engl J Med 306:1018–22

Kannel WB, Abbott RD, Savage DD, McNamara PM (1983) Coronary heart disease and atrial fibrillation: the Framingham Study. Am Heart J 106:389–96

Kannel WB, Plehn JF, Cupples LA (1988) Cardiac failure and sudden death in the Framingham Study. Am Heart J 115:869–75

Kaufman DM, Hegyi T, Duberstein JL (1972) Heroin intoxication in adolescents. Pediatrics 50:746–53

Kaul U, Ramachandran P, Bhatia ML (1983) Spontaneous conversion of long standing atrial fibrillation to sinus rhythm – an unusual pre-terminal phenomenon. Indian Heart J 35:241–4

Kawakubo K, Murayama M, Itai T et al. (1986) Heart rate at onset and termination of paroxysmal atrial fibrillation in apparently healthy subjects. Jpn Heart J 27:645–51

Keidar S, Grenadier E, Fleischman P, Palant A (1984) Swallowing induced atrial tachycardia and fibrillation in a patient with a Wolff–Parkinson–White syndrome. Am J Med Sci 288:32–4

Kempster PA, Gerraty RP, Gates PC (1988) Asymptomatic cerebral infarction in patients with chronic atrial fibrillation. Stroke 19:955–7

Kennaugh RC (1975) Carbon tetrachloride overdosage. A case report. S Afr Med J 49:635–6

Keren G, Etzion T, Sherez J et al. (1987) Atrial fibrillation and atrial enlargement in patients with mitral stenosis. Am Heart J 114:1146–55

Kernohan RJ (1967) Electric shock induced atrial fibrillation. Ir J Med Sci 6:509–10

Khan AH (1980) Spontaneous resumption of sinus rhythm after prolonged AF [letter]. Am Heart J 100:409

Kirson LE, Wilson ME (1987) Atrial fibrillation associated with intravenous fluorescein [letter]. Anesth Analg 66:283

Kitchin AH, Milne JS (1977) Longitudinal survey of ischaemic heart disease in randomly selected sample of older population. Br Heart J 39:889–93

Kitchin AH, Pocock SJ (1977) Prognosis of patients with acute myocardial infarction admitted to a coronary care unit. II. Survival after hospital discharge. Br Heart J 39:1167–71

Klein PJ, Schaefer HE (1973) [Cardiac lipoma of the interatrial septum.] Z Krebsforsch 79:11–8

Komaromy B, Gaal J, Lampe L (1977) Fetal arrhythmia during pregnancy and labour. Br J Obstet Gynaecol 84:492–6

Kopecky SL, Gersh BJ, McGoon MD et al. (1987) The natural history of lone atrial fibrillation. A population based study over three decades. New Engl J Med 317:669–74

Koskinen P, Kupari M, Leinonen H, Luomanmaki K (1987) Alcohol and new onset atrial fibrillation: a case–control study of a current series. Br Heart J 57:468–73

Kostis JB, Goodkind MJ, Skvaza H, Gerber NH Jr, Kuo PT (1977) Effect of alcohol on the atrial fibrillation threshold in dogs. Angiology 28:583–7

Kounis NG, Kenmure AC (1976) Micturition syncope, hypokalemia, and atrial fibrillation. JAMA 236:954

Kowey PR, Friehling TD, Marinchak RA (1986) Prayer-meeting cardioversion [letter]. Ann Intern Med 104:727–8

Kramer RJ, Zeldis SM, Hamby RI (1982) Atrial fibrillation – a marker for abnormal left ventricular function in coronary heart disease. Br Heart J 47:606–8

Krowka MJ, Pairolero PC, Trastek VF, Payne WS, Bernatz PE (1987) Cardiac dysrhythmia following pneumonectomy. Clinical correlates and prognostic significance. Chest 91:490–5

Kuan P (1987) Cardiac Wilson's disease. Chest 91:579–83

Kulbertus HE, de Leval-Rutten F, Bartsch P, Petit J-M (1982) Atrial fibrillation in elderly, ambulatory patients. In: Kulbertus HE, Olsson SB, Schlepper M (eds) Atrial fibrillation. AB Hassle, Molndal, pp 148–55

Kumagai K, Fukunami M, Ohmori M, Kitabatake A, Kamada T, Hoki N (1990) Increased intracardiovascular clotting in patients with chronic atrial fibrillation. J Am Coll Cardiol 16:377–80

Kushwaha S, Jepson EM (1990) Resolution of right atrial thrombus following anticoagulation. Int J Cardiol 27:269–71

Labi M (1969) Paroxysmal atrial fibrillation in heroin intoxication. Ann Intern Med 71:951–9

Lake FR, Cullen KJ, de Klerk NH, McCall MG, Rosman DL (1989) Atrial fibrillation and mortality in an elderly population. Aust NZ J Med 19:321–6

Langenfeld H, Grimm W, Maisch B, Kochsiek K (1988) Atrial fibrillation and embolic complications in paced patients. PACE 11:1667–72

Leak D, Meghji M (1979) Toxoplasmic infection in cardiac disease. Am J Cardiol 43:841–9

Leitch JW, Thomson D, Baird DK, Harris PJ (1990) The importance of age as a predictor of atrial fibrillation and flutter after coronary artery bypass grafting. J Thorac Cardiovasc Surg 100:338–42

Lewis KB, Criley JM, Ross RS (1965) Detection of left atrial thrombus by cineangiocardiography. Am Heart J 70:612–19

Liem KL, Durrer D, Lie KI, Wellens HJ (1975) Pericarditis in acute myocardial infarction. Lancet ii:1004–6

Liem KL, Lie KI, Durrer D, Wellens HJ (1976) Clinical setting and prognostic significance of atrial fibrillation complicating acute myocardial infarction. Eur J Cardiol 4:59–62

Lorentzen JE, Rder OC, Hansen HJ (1980) Peripheral arterial embolism. A follow-up of 130 consecutive patients submitted to embolectomy. Acta Chir Scand (Suppl) 502:111–16

Louis DN, Casper DS, Armenia DC (1986) Bowlogenic dysrhythmias [letter]. Am J Med 81:949
Lowe GD, Jaap AJ, Forbes CD (1983) Relation of atrial fibrillation and high haematocrit to mortality in acute stroke. Lancet i:784-6
Lowenstein SR, Gabow PA, Cramer J, Oliva PB, Ratner K (1983) The role of alcohol in new-onset atrial fibrillation. Arch Intern Med 143:1882-5
Lutterotti A von (1967) [Chronic alcoholism as a cause of arrhythmia.] Z Kreislaufforsch 56:275-80
MacGowan WA, Mooneeram R (1973) A review of 174 patients with arterial embolism. Br J Surg 60:894-8
Maister AH (1983) Atrial fibrillation following physostigmine. Can Anaesth Soc J 30:419-21
Makarova LF, Ageeva RM (1982) [Idiopathic form of atrial fibrillation in a newborn infant.] Pediatriia 69:62-3
Maljar L, Romero RL, Cuneo J, Marchetti B (1968) [Unusual arrhytmias caused by digitalis intoxication.] Rev Clin Esp 109:309-404
Mallet E, Letac B (1975) [Acute pericarditis. Study of 50 consecutive cases.] Nouv Presse Med 4:2853-7
Malo S, Latour Y, Cote M, Geoffroy G, Lemieux B, Barbeau A (1976) Electrocardiographic and vectocardiographic findings in Friedreich's ataxia. Can J Neurol Sci 3:323-8
Manchester JH, Lamberti JJ (1970) Reversion of atrial fibrillation following hyperkalemia. Chest 58:399-402
Markewitz A, Schad N, Hemmer W, Bernheim C, Ciavolella M, Weinhold C (1986) What is the most appropriate stimulation mode in patients with sinus node dysfunction? PACE 9:1115-20
Marshall AJ (1976) Transient atrial fibrillation after minor head injury. Br Heart J 38:984-5
Marty C, Elghozi D, Kulas A, Reboul M, Lainee J, Aubert P (1973) [Heart failure in hemochromatosis. Apropos of a case.] Ann Med Interne (Paris) 124:507-11
Master AM, Eichert H (1946) Functional paroxysmal auricular fibrillation. Am J Med Sci 211:336-45
Mathew PK (1979) Hyperkalemia-induced conversion of chronic atrial fibrillation to normal sinus rhythm; a case report. Angiology 30:143-6
Mathew NT, Taori GM, Mathai KV, Chandy J (1970) Atrial fibrillation associated with seizure in a case of frontal meningioma. Neurology 20:725-8
McCartney RD, McMurtry RJ (1973) Complications of transtracheal aspiration. N Engl J Med 289:1094
Merle d'Aubigne R, Saint-Maurice (1967) [Cardiac fibrillation caused by electrocution.] Mem Acad Chir (Paris) 93:57-62
Merli GJ, Weitz H, Martin JH et al. (1986) Cardiac dysrhythmias associated with ophthalmic atropine. Arch Intern Med 146:45-7
Miller RG, Layzer RB, Mellenthin MA, Golabi M, Francoz RA, Mall JC (1985) Emery-Dreifuss muscular dystrophy with autosomal dominant transmission. Neurology 35:1230-3
Morady F, Krol RB, Nostrant TT, De Buitleir M, Cline W (1987) Supraventricular tachycardia induced by swallowing: a case report and review of the literature. PACE 10:133-8
Myler RK, Sanders CA (1968) Aortic valve disease and atrial fibrillation. Report of 122 patients with electrographic, radiographic, and hemodynamic observations. Arch Intern Med 121:530-3
Nakajima Y, Arai T, Hatano Y (1987) [Persistent atrial fibrillation during insertion of central venous catheter with the catheter-over-guidewire device.] Masui 36:1123-5
Nakamoto K (1965) Psychogenic paroxysmal cardiac arrhythmias. Contents of mental events, age and patterns of arrhythmias. Jpn Circ J 29:701-17
Nguyen TN, Friedman HS, Mokraoui AM (1987) Effects of alcohol on experimental atrial fibrillation. Alcoholism 11:474-6
Nissen MB, Lemberg L (1984) The "holiday heart" syndrome. Heart Lung 13:89-92
Nitter-Hauge S, Otterstad JE (1981) Characteristics of atrioventricular conduction disturbances in ankylosing spondylitis (Mb. Bechterew). Acta Med Scand 210:197-200
Nordrehaug JE, von der Lippe G (1986) Serum potassium concentrations are inversely related to ventricular, but not to atrial, arrhythmias in acute myocardial infarction. Eur Heart J 7:204-9
Norris JW, Froggatt GM, Hachinski VC (1978) Cardiac arrhythmias in acute stroke. Stroke 9:392-6
Noto N, Osaka T, Yamanaka O, Kobayashi S, Ozaki H, Kanoh T (1990) [A case of acute myocardial infarction due to coronary embolism from left atrial thrombus with atrial fibrillation.] Kokyu To Junkan 38:483-7
Novikov FE, Gracheva LI (1982) [Congestive cardiomyopathy of toxoplasmic etiology.] Kardiologiia 22:106-8
Okada M (1984) The cardiac rhythm in accidental hypothermia. J Electrocardiol 17:123-8
Okuma K, Furuta I, Ota K (1984) [Acute cardiotoxicity of anthracyclines - analysis by using Holter ECG.] Gan To Kagaku Ryoho 11:902-11

Olsson SB (1981) Nature of cardiac arrhythmias and electrolyte disturbances. Role of potassium in atrial fibrillation. Acta Med Scand Suppl 647:33–7

Olsson SB, Orndahl G, Ernestrom S et al. (1980) Spontaneous reversion from long-lasting atrial fibrillation to sinus rhythm. Acta Med Scand 207:5–20

O'Melia J (1970) A case of atrial fibrillation following the use of suxamethonium during ECT. Br J Psychiatry 117:718

Onundarson PT, Thorgeirsson G, Jonmundsson E, Sigfusson N, Hardarson T (1987) Chronic atrial fibrillation – epidemiologic features and 14 year follow-up: A case control study. Eur Heart J 8:521–7

Oster MW, Rakowski TJ (1981) Myocardial injury immediately following adriamycin administration. Med Pediatr Oncol 9:463– 5

Ostrander LD, Brandt RL, Kjelsberg MO, Epstein FH (1965) Electrocardiographic findings among the adult population of a total natural community, Tecumseh, Michigan. Circulation 31:888–98

Palma A, Pisapia A, Guglielmotti A, Di Costanzo F (1969) [Paroxystic atrial fibrillation due to contusive craniocerebral injury.] Rass Int Clin Ter 49:335–43

Panov BV, Dolmatov AA (1988) [Subacute trichloroethylene poisoning with myocardial lesion and atrial fibrillation.] Sov Med 33:119–20

Parker BM, Friedenberg MJ, Templeton AW, Burford TH (1965) Preoperative angiocardiographic diagnosis of left atrial thrombi in mitral stenosis. N Engl J Med 273:136–40

Passa P, Gourgon R, Cazor JL, Masquet C (1975) [Cardiac hemochromatosis. Anatomic, clinical and hemodynamical data.] Nouv Presse Med 4:1017–22

Patton RD (1970) Sodium dehydrocholate hypersensitivity in man. N Y State J Med 70:682–6

Peter RH, Gracey JG, Beach TB (1968) A clinical profile of idiopathic atrial fibrillation. A functional disorder of atrial rhythm. Ann Intern Med 68:1288–95

Petersen P (1990) Thromboembolic complications in atrial fibrillation. Stroke 21:4–13

Petersen P, Godtfredsen J (1986) Embolic complications in paroxysmal atrial fibrillation. Stroke 17:622–6

Petersen P, Godtfredsen J (1988) Risk factors for stroke in chronic atrial fibrillation. Eur Heart J 9:291–4

Petersen P, Hansen JM (1988) Stroke in thyrotoxicosis with atrial fibrillation. Stroke 19:15–8

Petersen P, Kastrup J, Brinch K, Godtfredsen J, Boysen G (1987) Relation between left atrial dimension and duration of atrial fibrillation. Am J Cardiol 60:382–4

Petersen P, Pedersen F, Johnsen A et al. (1989) Cerebral computed tomography in paroxysmal atrial fibrillation. Acta Neurol Scand 79:482–6

Petersen P, Kastrup J, Helweg-Larsen S, Boysen G, Godtfredsen J (1990) Risk factors for thromboembolic complications in chronic atrial fibrillation. The Copenhagen AFASAK study. Arch Intern Med 150:819–21

Peterson J, Scruton D, Downie AW (1977) Basilar artery migraine with transient atrial fibrillation. Br Med J ii:1125–6

Phillips SJ (1990) Is atrial fibrillation an independent risk factor for stroke? Can J Neurol Sci 17:163–8

Pitts JC, Brantigan CO, Hopeman AR (1977) Myocardial ischemia associated with transtracheal aspiration. JAMA 237:2527–8

Plancher AC, Maccari S, Tramaloni C, Vigotti G, Giudici GA (1984) [Cardiologic aspects of a case of Tangier disease.] G Clin Med 65:189–97

Pratila MG, Pratilas V (1977) A case of tachydysrhythmia. Refractory to propranolol and responsive to neostigmine. Anaesthesia 32:1017–9

Pratila MG, Pratilas V (1982) Dysrhythmia occurring during epidural anesthesia with bupivacaine. Mt Sinai J Med 49:130– 2

Presti CF, Hart RG (1989) Thyrotoxicosis, atrial fibrillation, and embolism revisited. Am Heart J 117:976–7

Probst P, Goldschlager N, Selzer A (1973) Left atrial size and atrial fibrillation in mitral stenosis. Factors influencing their relationship. Circulation 48:1282–7

Radford DJ, Izukawa T (1977) Atrial fibrillation in children. Pediatrics 59:250–6

Rajala S, Haavisto M, Kaltiala K, Mattila K (1985) ECG findings and survival in very old people. Eur Heart J 6:247–52

Rajala S, Haavisto M, Kaltiala K, Mattila K (1987) Electrocardiographic findings and 5-year cardiovascular mortality in very old people. Ann Clin Res 19:324–7

Rankin AC, Rae AP (1984) Cardiac arrhythmias during rewarming of patients with accidental hypothermia. Br Med J 289:874–7

Ranquin R, Parizel G (1975) Massive digoxin intoxication. Report of a case with serum digoxin level correlation. Acta Cardiol 30:375–82

Rasmussen K, Andersen K, Wang H (1982) Atrial fibrillation induced by atenolol. Eur Heart J 3:276–
    81
Rawles JM, Ogston D, Douglas AS (1973) Studies in the fibrinolytic system in congestive cardiac
    failure. Clin Science Mol Med 45:65–76
Rawles JM, Metcalfe MJ, Jennings K (1990) Time of occurrence, duration and ventricular rate of
    paroxysmal atrial fibrillation: the effect of digoxin. Br Heart J 63:225–7
Reid JM, Kennedy JF, McArthur J (1982) Unusual cause of atrial fibrillation. Br Med J 284:237–8
Rich EC, Siebold CM, Campion BC (1984) Alcohol-related acute atrial fibrillation. Minn Med
    67:687–8
Rich EC, Siebold C, Campion B (1985) Alcohol-related acute atrial fibrillation. A case–control study
    and review of 40 patients. Arch Intern Med 145:830–3
Ridker PM, Gibson CM, Lopez R (1989) Atrial fibrillation induced by breath spray [letter]. N Engl J
    Med 320:124
Rigotti NA, Eagle KA (1986) Atrial fibrillation while chewing nicotine gum [letter]. JAMA 255:1018
Rivers JF, Orr G, Lee HA (1970) Drowning. Its clinical sequelae and management. Br Med J ii:157–
    61
Roberts LN, Montessori G, Patterson JG (1972) Idiopathic pulmonary hemosiderosis. Case report
    with pulmonary function tests and cardiac catheterization data. Am Rev Respir Dis 106:904–8
Robinson K, Frenneaux MP, Stockins B, Karatasakis G, Poloniecki JD, McKenna WJ (1990) Atrial
    fibrillation in hypertrophic cardiomyopathy: a longitudinal study. J Am Coll Cardiol 15:1279–85
Rose G, Baxter PJ, Reid DD, McCartney P (1978) Prevalence and prognosis of electrocardiographic
    findings in middle-aged men. Br Heart J 40:636–43
Rose H, Harrell R (1967) Cardiovascular effects of gastric hypothermia. South Med J 60:152–5
Rosenqvist M, Brandt J, Schuller H (1986) Atrial versus ventricular pacing in sinus node disease: a
    treatment comparison study. Am Heart J 111:292–7
Rosman KD (1986) The epidemiology of stroke in an urban black population. Stroke 17:667–9
Rothschild AH, Sridharan MR, Gondi B, Sohi GS, Flowers NC (1985) Conversion of atrial
    fibrillation to normal sinus rhythm during hyperkalemia. J Ky Med Assoc 83:295–8
Rubin DA, Nieminski KE, Reed GE, Herman MV (1987) Predictors, prevention, and long-term
    prognosis of atrial fibrillation after coronary artery bypass graft operations. J Thorac Cardiovasc
    Surg 94:331–5
Russell D, Storstein L (1983) Cluster headache: a computerized analysis of 24 h Holter ECG
    recordings and description of ECG rhythm disturbances. Cephalalgia 3:83–107
Sage JI, Van Uitert RL (1983) Risk of recurrent stroke in patients with atrial fibrillation and non-
    valvular heart disease. Stroke 14:537–40
Saidi M, Nazem I, Angers B, Leblond S (1967) [Spontaneous pneumothorax with auricular
    fibrillation.] Union Med Can 96:281– 5
Sainani GS, Krompotic E, Slodki SJ (1968) Adult heart disease due to the Coxsackie virus B
    infection. Medicine (Baltimore) 47:133– 47
Sandercock PA, Warlow CP, Jones LN, Starkey IR (1989) Predisposing factors for cerebral
    infarction: the Oxfordshire community stroke project. Br Med J 298:75–80
Sandrasegram M, Kumar CM (1986) Radiological contrast media and atrial fibrillation [letter].
    Anaesthesia 41:771–2
Sandu L, Popescu I (1975) [Uremigenic cholangitis.] Rev Chir 24:229–33
Sanfilippo AJ, Abascal VM, Sheehan M et al. (1990) Atrial enlargement as a consequence of atrial
    fibrillation. A prospective echocardiographic study. Circulation 82:792–7
Santo M, Aderka D, Pinkhas J (1982) Rapid atrial fibrillation during hypercalcemia despite verapamil
    therapy [letter]. Am Heart J 104:320
Sarma RN, Naik BK, Rao SH (1966) Atrial fibrillation in chronic diarrhoea corrected with oral
    potassium. Indian Heart J 18:167–70
Scaffidi A, Furitano G, Scaffidi L (1981) [Cardiovascular involvement and complications in
    Mediterranean boutonneuse fever.] Minerva Med 72:2097–108
Sekiguchi C, Yamaguchi O, Kitajima T, Ueda Y (1977) Continuous ECG monitoring on civil air
    crews during flight operations. Aviat Space Environ Med 48:872–6
Sekiguchi C, Iwane M, Oshibuchi M (1986) Anti-G training of Japanese Air Self Defense Force
    fighter pilots. Aviat Space Environ Med 57:1029–34
Selzer A, Cohn KE (1972) Natural history of mitral stenosis: a review. Circulation 45:878–90
Sessler CN, Cohen MD (1990) Cardiac arrhythmias during theophylline toxicity. A prospective
    continuous electrocardiographic study. Chest 98:672–8
Sgobba G, Nassisi G, Giannelli F et al. (1982) [Electrocardiographic changes in accidental
    hypothermia.] G Ital Cardiol 12:147–50

Sharma NG (1971) Electric shock, a rare cause of atrial fibrillation. (A case report). Indian Heart J 23:306–7

Sheehan J, White A (1982) Diuretic-associated hypomagnesaemia. Br Med J 285:1157–9

Sherman DG, Goldman L, Whiting RB, Jurgensen K, Kaste M, Easton JD (1984) Thromboembolism in patients with atrial fibrillation. Arch Neurol 41:708–10

Shimada S (1986) A 13-year follow-up study of rheumatic valvular diseases. Jpn Circ J 50:1304–8

Shuaib A, Klein G, Dear R (1987) Migraine headache and atrial fibrillation. Headache 27:252–3

Singh RB, Manmohan MD, Dube KP, Singh VP (1976) Serum magnesium concentrations in atrial fibrillation. Acta Cardiol 31:221–6 Sipila R (1985) Atrial fibrillation precipitated by alcohol [letter]. Lancet i:391–2

Slavina LS, Kazeev KN, Kilinskii EL, Kertsman GI (1973) [Changes in the electrocardiogram in pheochromocytoma.] Probl Endokrinol (Mosk) 19:42–5

Snyder RW, Dumas PR, Kolts BE (1990) Esophagoatrial fistula with previous pericarditis complicating esophageal ulceration. Report of two cases and a review of the literature. Chest 98:679–81

Sogaard PE (1981) Free ball thrombus of the left atrium. Eur J Cardiol 12:177–9

Sosa-Suarez G et al. (1989) Changes in left atrial size due to chronic atrial fibrillation. J Am Coll Cardiol 13:206A

Soto-Rojas G, Cortes JM, Medrano GA (1984) [Electrocardiographic changes in 29 apparently healthy subjects with positive serological tests for Chagas' disease.] Arch Inst Cardiol Mex 54:579–83

Spodick DH (1976) Arrhythmias during acute pericarditis. A prospective study of 100 consecutive cases. JAMA 235:39–41

Staffurth JS, Gibberd MC, S Ng Tang Fui (1977) Arterial embolism in thyrotoxicosis with atrial fibrillation. Br Med J ii:688–90

Staller BJ (1984) Atrial fibrillation induced by testing magnet applied to VVI pacemaker. PACE 7:293–5

Stein PD, Dalen JE, McIntyre KM, Sasahara AA, Wenger NK, Willis PW (1975) The electrocardiogram in acute pulmonary embolism. Prog Cardiovasc Dis 17:247–57

Stewart PM, Catterall JR (1985) Chronic nicotine ingestion and atrial fibrillation. Br Heart J 54:222–3

Stougard J (1969) Cardiac arrhythmias following pneumonectomy. Thorax 24:568–72

Suetsugu M, Matsuzaki M, Toma Y et al. (1988) Detection of mural thrombi and analysis of blood flow velocities in the left atrial appendage using transesophageal two-dimensional echocardiography and pulsed Doppler flowmetry. J Cardiol 18:385–94

Sugiura T, Iwasaka T, Ogawa A et al. (1985) Atrial fibrillation in acute myocardial infarction. Am J Cardiol 56:27–9

Sundstrom C (1973) A case of giant cell myocarditis. Ups J Med Sci 78:83–8

Susens GP, al-Shamma A, Rowe JC, Herbert CC, Bassis ML, Coggs GC (1967) Purulent constrictive pericarditis caused by Nocardia asteroides. Ann Intern Med 67:1021–32

Sutton R, Kenny RA (1986) The natural history of sick sinus syndrome. PACE 9:1110–4

Switz DM, Clarke AM, Longacher JW (1976) Electrical malfunction at endoscopy. Possible cause of arrhythmia and death. JAMA 235:273–5

Symons C, Myers A, Kingstone D, Boss M (1978) Response to thyrotrophin-releasing hormone in atrial dysrhythmias. Postgrad Med J 54:658–62

Szczerbinski A (1969) [Transient atrial fibrillation after alcohol abuse.] Wiad Lek 22:2153–5

Takahashi N, Seki A, Imataka K, Fujii J (1981) Clinical features of paroxysmal atrial fibrillation. An observation of 94 patients. Jpn Heart J 22:143–9

Taylor GJ, Malik SA, Colliver JA et al. (1987) Usefulness of atrial fibrillation as a predictor of stroke after isolated coronary artery bypass grafting. Am J Cardiol 60:905–7

Teramoto S, Katsumura T, Nakanishi S, Hayashi S, Doi T (1967) [3 cases of accidental complications during heart catheterization.] Kyobu Geka 20:460–6

Thomas AJ, Valabhji P (1969) Arrhythmia and tachycardia in pulmonary heart disease. Br Heart J 31:491–5

Thorne MG (1969) Hiccup and heart block. Br Heart J 31:397–9

Thornton JR (1984) Atrial fibrillation in healthy non-alcoholic people after an alcoholic binge. Lancet ii:1013–5

Tibaldi JM, Barzel US, Albin J, Surks M (1986) Thyrotoxicosis in the very old. Am J Med 81:619–22

Tikoff G, Schmidt AM, Hecht HH (1968) Atrial fibrillation in atrial septal defect. Arch Intern Med 121:402–5

Todd MM (1983) Atrial fibrillation induced by the right atrial injection of cold fluids during thermodilution cardiac output determination: a case report. Anesthesiology 59:253–5

Tomoda H, Hoshiai M, Tagawa R et al. (1980) Evaluation of left atrial thrombus with computed tomography. Am Heart J 100:306–10

Toy JL, Lederer DA, Tulpule AT, Tandon AP, Taylor SH, McNicol GP (1980) Coagulation studies in rheumatic heart disease. Br Heart J 43:301–5

Treseder AS, Sastry BS, Thomas TP, Yates MA, Pathy MS (1986) Atrial fibrillation and stroke in elderly hospitalized patients. Age Ageing 15:89–92

Tsunashima T, Arima T, Tsuboi S et al. (1976) A case of alcaptonuria with fatal cardiovascular disturbance. Acta Med Okayama 30:87–94

Tullio D, Staniscia GC, Di Bartolomeo A, Paolucci A (1981) [Electrocardiographic changes in acute abdomen.] Minerva Med 72:1951–8

Ul'ianinskii LS, Urmancheeva TG, Stepanian EP, Fufacheva AA, Gritsak AV (1981) [Effect of motor activity on the development of arrhythmia in experimental emotional stress.] Kardiologiia 21:64–7

Unverferth DV, Fertel RH, Unverferth BJ, Leier CV (1984) Atrial fibrillation in mitral stenosis: histologic, hemodynamic and metabolic factors. Int J Cardiol 5:143–54

Vaidya PN, Bhosley PN, Rao DB, Luisada AA (1976) Tachyarrhythmias in old age. J Am Geriatr Soc 24:412–4

Vainshtein SG, Danovskii LV, Klimova NA (1970) [Auricular fibrillation due to alcoholic intoxication.] Sov Med 33:140–1

Van der Ark GD (1975) Cardiovascular changes with acute subdural hematoma. Surg Neurol 3:305–8

Van Merwijk G, Lodder J, Bamford J, Kester AD (1990) How often is non-valvular atrial fibrillation the cause of brain infarction? J Neurol 237:205–7

Vargo BP (1985) Transient atrial fibrillation in a patient with acute ureterolithiasis. J Emerg Med 2:169–73

Verel D, Warrack AJ, Potter CW, Ward C, Rickards DF (1976) Observations on the A2 England influenza epidemic: a clinicopathological study. Am Heart J 92:290–6

Vesterby A, Gregersen M (1980) Atrial fibrillation resulting from cardiac trauma. Z Rechtsmed 85:153–7

Volpi A, Cavalli A, Maggioni AP, Pieri-Nerli F (1988) Left atrial compression by a mediastinal bronchogenic cyst presenting with paroxysmal atrial fibrillation. Thorax 43:216–7

Wachtel FW, Rothfeld D (1967) Transient atrial fibrillation following accidental electric shock. A case report. J Med Soc N J 64:27–9

Wade G, Werko L, Eliasch H, Gidlund A, Lagerlof H (1952) The haemodynamic basis of the symptoms and signs in mitral valvular disease. Q J Med 21:361–83

Wallach JB, Lukash L, Angrist AA (1953) An interpretation of the incidence of mural thrombi in the left auricle and appendage with particular reference to mitral commissurotomy. Am Heart J 45:252–4

Wasserburger RH (1975) An electrocardiographic survey of the aged. Postgrad Med 58:147–50

Waters DD, Nutter DO, Hopkins LC, Dorney ER (1975) Cardiac features of an unusual x-linked humeroperoneal neuromuscular disease. N Engl J Med 293:1017–22

Watt AH (1985) Atrial fibrillation and alcohol [letter]. Lancet i:162

Weinberger J, Rothlauf E, Materese E, Halperin J (1988) Noninvasive evaluation of the extracranial carotid arteries in patients with cerebrovascular events and atrial fibrillations. Arch Int Med 148:1785–8

Weitzman S, Margulis G, Lehmann E (1977) Uncommon cardiovascular manifestations and high catecholamine levels due to "black widow" bite. Am Heart J 93:89–90

White WB, Wong SH (1985) Rapid atrial fibrillation associated with trazodone hydrochloride [letter]. Arch Gen Psychiatry 42(4):424

Whiteford H, Klug P, Evans L (1984) Disturbed cardiac function possibly associated with mianserin therapy. Med J Aust 140:166–7

Wiener I, Hafner R, Nicolai M, Lyons H (1987) Clinical and echocardiographic correlates of systemic embolization in nonrheumatic atrial fibrillation. Am J Cardiol 59:177

Wilson CL, Davis SJ (1978) Recurrent atrial fibrillation with nausea and vomiting. Aviat Space Environ Med 49:624–5

Wolf A, Schomerus H, Risler T, Muller GA (1983) [Unusual complications in Wilson's disease.] Internist (Berlin) 24:721–4

Wolf PA, Dawber TR, Thomas HE Jr, Kannel WB (1978) Epidemiologic assessment of chronic atrial fibrillation and risk of stroke: the Framingham study. Neurology 28:973–7

Wolf PA, Kannel WB, McGee DL, Meeks SL, Bharucha NE, McNamara PM (1983) Duration of atrial fibrillation and imminence of stroke: the Framingham study. Stroke 14:664–7

Wolf PA, Abbott RD, Kannel WB (1987) Atrial fibrillation: a major contributor to stroke in the elderly. The Framingham Study. Arch Intern Med 147:1561–4

Wong TC, Cooper ES (1969) Atrial fibrillation with ventricular slowing in a patient with spontaneous subarachnoid hemorrhage. Am J Cardiol 23:473–7

Woo J, Lau EM (1990) Risk factors predisposing to stroke in an elderly Chinese population – a longitudinal study. Neuroepidemiology 9:131–4

Yamour BJ, Sridharan MR, Rice JF, Flowers NC (1980) Electrocardiographic changes in cerebrovascular hemorrhage. Am Heart J 99:294–300

Yan WT, Liu ZM, Zhang YR, Li JZ, Fan ZH, Yan XF (1983) Clinical study of 9 familial atrial fibrillation cases. Chin Med J [Engl] 96:441–4

Yoneda S, Murata M, Ueda A et al. (1978) Persistent atrial standstill developed in a patient with rheumatic heart disease: electrophysiological and histological study. Clin Cardiol 1:43–7

Yuen RW, Gutteridge DH, Thompson PL, Robinson JS (1979) Embolism in thyrotoxic atrial fibrillation. Med J Aust 1:630–1

Zaldivar N, Gelband H, Tamer D, Garcia O (1973) Atrial fibrillation in infancy. J Pediatr 83:821–2

Zimmerman TJ, Basta LL, January LE (1973) Spontaneous return of sinus rhythm in older patients with chronic atrial fibrillation and rheumatic mitral valve disease. Description of three patients. Am Heart J 86:676–80

Zuber E, Doroszew J (1969) [A case of alcohol-induced atrial fibrillation.] Wiad Lek 22:2203–5

Zuber E, Kalasz J (1967) [2 cases of postalcoholic atrial fibrillation.] Pol Tyg Lek 22:917–18

Chapter 10

The Management of Atrial Fibrillation

Presentation

Symptoms

In many cases of atrial fibrillation any symptoms due to the dysrhythmia itself are greatly overshadowed by the circumstances of its discovery. Thus, atrial fibrillation may be found in patients presenting with stroke, thyrotoxicosis, or pulmonary oedema due to severe mitral stenosis. Or, atrial fibrillation may be completely asymptomatic, and discovered only during systematic electrocardiographic examination of the community. Holter recordings in patients with paroxysmal atrial fibrillation may also reveal many brief episodes of atrial fibrillation which pass unnoticed. Some people, however, are acutely aware of the abnormal rhythm during atrial fibrillation, with anxiety, palpitations, precordial discomfort, shortness of breath and dizziness being the most common complaints. By dizziness is meant not vertigo, but a feeling of unsteadiness and impending syncope, which usually never befalls (Hertzeanu and Aron 1985). In fact, patients with atrial fibrillation have improved tolerance to orthostatic stress, and are less likely to suffer from postural hypotension (Abelman and Fareeduddin 1969). Syncope, or near- syncope, when it does occur in patients with atrial fibrillation is associated with ventricular standstill which, if it exceeds 2–3 seconds, may be indicative of heart block, and respond to pacing (Rebello and Brownlee 1987).

On the other hand, the development of stable atrial fibrillation in patients with the sick sinus syndrome may provide a natural cure of symptoms caused by bradycardia (Vera et al. 1977). Before the development of electrophysiological investigation and precise treatment of supraventricular tachycardia, conversion to atrial fibrillation by rapid atrial stimulation was a therapeutic option, the ventricular rate being more readily controlled in atrial fibrillation than in atrial flutter or tachycardia (Wiener and Dwyer 1968; Chamberlain et al. 1971). Chevalier (1966) made the case that, for some patients with paroxysmal atrial fibrillation, the treatment – at that time cardioversion and maintenance therapy with quinidine – may have been worse than the disease. He argued that for these patients a pacemaker that recognises sinus rhythm and restores atrial fibrillation might be desirable.

In the Wolff–Parkinson–White syndrome atrial fibrillation is common and may be the presenting arrhythmia (Robinson et al. 1988). If conduction is predominantly over the accessory pathway the ventricular rate may be very rapid; syncope or sudden death due to ventricular fibrillation may result (Dreifus et al. 1971, 1976; Papa et al. 1978; Yee and Klein 1984).

Signs

The physical signs of atrial fibrillation are well known: the apparently completely irregular pulse, the absence of an "a" wave in the jugular venous pulsation, and the varying intensity and timing of heart sounds and murmurs. Some beats, audible with a stethoscope, may not produce a palpable peripheral pulse, so a pulse deficit results. The presystolic murmur of mitral stenosis disappears, but there may be presystolic augmentation of heart sounds as the muscular component of the first sound is intensified due to contraction of an underfilled ventricle (Bonner et al. 1976). There may also be diastolic sounds from uncoordinated atrial contractions (Neporent and Da Silva 1967).

Investigations

In the electrocardiogram of atrial fibrillation "P waves are absent and the baseline consists of irregular waveforms which continuously change in shape, duration, amplitude, and direction...the ventricular response is totally irregular (random)" (WHO/ISC Task Force 1978). The QRS complexes are usually narrow unless there is coexisting bundle branch block, or aberration, but in the Wolff–Parkinson–White syndrome there is an irregular broad-complex tachycardia.

Fibrillating contractions of the atria may be visible on echocardiography (Drinkovic 1982), and the rapid blood flow through the mitral valve associated with atrial contraction is lost on Doppler examination.

The irregularity of the heart rate in atrial fibrillation poses problems in measurement. A 20 second sample at rest is required to give an average heart rate that is within 5% of the true value in 92.5% of patients (Watt et al. 1984), and a 20 second sample is necessary in exercise (Atwood et al. 1989a).

The presence of atrial fibrillation increases the variability of blood pressure measurement within and between observers (Sykes et al. 1990). It has been suggested that haemodynamic assessment of patients with atrial fibrillation at cardiac catheterisation would be helped by temporary right ventricular pacing, in order to regularise the rhythm and reduce the variability of pressure measurements (Miller and Carleton 1968).

In spite of the variable R-R intervals in atrial fibrillation, radionuclide ventriculography gated to the electrocardiogram gives a reliable estimate of left ventricular ejection fraction, provided ventricular rate is slow (Bacharach et al. 1981).

Although the variation in heart rate in atrial fibrillation may make haemodynamic assessment more difficult, it may be exploited for the measurement of mitral valve area in mitral stenosis (Davies et al. 1990), or for determining the compliance of the aorta under transient conditions (Yin and Liu 1989). Chapters

6 and 7 give other examples of the uses to which the variation in heart rate may be put.

Patients with atrial fibrillation, like those in sinus rhythm, show a learning effect with treadmill exercise, the performamce of submaximal exercise improving with repeated tests (Kraemer et al. 1989). Compared with normal subjects in sinus rhythm, the ventricular rate of patients with atrial fibrillation is abnormally high at rest, at all levels of exercise, and immediately after exercise (Hornsten and Bruce 1968; Corbelli et al. 1990). The mean maximal ventricular rate in atrial fibrillation is about 20 beats per minute greater than the age-predicted value for sinus rhythm. However, the duration of exercise achieved in atrial fibrillation is no less than in matched subjects in sinus rhythm, and the maximal exercise capacity of patients with lone atrial fibrillation is normal (Atwood et al. 1988a).

General Management

In many cases of acute atrial fibrillation it is not the arrhythmia but the underlying medical problem that gives most cause for concern. Of 97 patients with new-onset atrial fibrillation, 82% reverted to sinus rhythm during their hospital stay (Shlofmitz et al. 1986); the authors concluded that emergency admission to hospital is unnecessary for the arrhythmia itself. Acute myocardial infarction rarely presents as atrial fibrillation in the absence of chest pain, and cardiac arrest is very uncommon in acute atrial fibrillation (Friedman et al. 1987). It is therefore wasteful of resources to admit patients with acute atrial fibrillation to a coronary care unit (Thibault 1985).

The management of atrial fibrillation will be considered under the following headings: control of ventricular rate, cardioversion, prevention of recurrences, and anticoagulant therapy.

Control of Ventricular Rate

Digitalis

Mechanism of Action

Digoxin remains the drug of first choice for the reduction of a rapid ventricular rate in atrial fibrillation. Preferably, treatment should be guided by measurement of cardiac output, which may be made non-invasively by Doppler ultrasound (Chapter 4). In the absence of haemodynamic guidance, the target ventricular rate at rest is 90 beats per minute (Chapter 6). Digoxin exerts its effects both directly on the specialised conducting system of the heart and indirectly through the autonomic nervous system. The relative importance of the direct and the autonomically mediated electrophysiological actions of digoxin has been studied in transplanted, denervated hearts.

Acutely, digoxin has little direct action on either sinoatrial or atrioventricular nodes, or on atrial myocardium of the denervated heart (Goodman et al.

1975a,b). Chronic administration of digoxin, however, leads to lengthening of atrioventricular delay and the refractory period of the atrioventricular node – a direct depressant effect on atrioventricular conduction (Ricci et al. 1978).

In normally innervated hearts the sinoatrial rate falls slightly, the atrial refractory period is reduced, the refractory period of the atrioventricular node is increased, and atrioventricular delay lengthens (Hoffman and Singer 1964). These changes are largely mediated by the vagus, digoxin having a vagomimetic action. However, digoxin also has an anti-adrenergic action which contributes to these electrophysiological changes, particularly in the atrioventricular node (Mendez et al. 1961a,b; Nadeau and James 1963).

Thus, administration of digoxin in atrial fibrillation brings about a reduction of ventricular rate by its several actions on the atrioventricular node. In addition, since digoxin is vagomimetic, a vagally mediated increase in fibrillation rate has been adduced. It has been argued that this indirect effect of digoxin on the atria also results in a reduced ventricular rate. As a greater number of impulses arrive at, and are conducted into, the atrioventricular node, it is rendered refractory for a greater proportion of the time, with consequent slowing of the ventricular rate (Meijler 1985). As an electrophysiological phenomenon, repetitive concealed conduction has been well documented and quantified (Moe et al. 1964; Fujiki et al. 1990), and the through-conductor model of the atrioventricular node based on this phenomenon (Moe and Abildskov 1964) is widely acknowledged to have made a major contribution to our understanding of this topic. The model predicts that ventricular rate will be negatively related to the rate of atrial input, and the paper describing the model contains an illustration showing a faster ventricular rate during a period of slower atrial activity. But, as discussed in Chapter 3, that model fails to account for the observed distribution of R-R intervals in atrial fibrillation, and furthermore requires the assumption that the atrioventricular node consists of two separate sequential compartments with different refractory periods. The authors themselves concede that when the atrial input frequency is rapid and fractionated the impulses may be less capable of penetrating the node, so this could be an alternative explanation for a slower ventricular response.

Moreover, although it is well established that vagal activity results in a reduced atrial refractory period, and that digitalis has a vagomimetic effect, the direct effect of digitalis on the atrial refractory period is to increase it, thus opposing the indirect effect (Farah and Loomis 1950; Mendez and Mendez 1957; Hoffman and Singer 1964; Engel and Gonzalez 1978). The net effect of digitalis on the atrial refractory period, and hence on the fibrillation rate, is likely to be small and variable.

Lewis et al. (1921) reported a rise in the atrial rate of about 6% after digitalis, and were of the opinion that "the rise of auricular rate being relatively small and occurring at very high rates...probably does not influence the ventricular rate materially". Aberg and Nordgren (1970) could not demonstrate any change in the frequency of f waves following digoxin. Prinzmetal et al. (1952) made direct observations of atrial fibrillation with high-speed cinephotography and oscillography, and wrote that digitalis seemed to decrease the amplitude and increase the rate of atrial oscillations in experimental atrial fibrillation in dogs; this observation was not quantified. In man, digitalisation resulted in marked reduction of the amplitude of the baseline oscillations of the electrocardiogram. "The diminution in electrical 'bombardment' of the auriculo-ventricular node which apparently occurs following digitalization may be one of the factors contributing to the slowing of the ventricular rate observed when auricular fibrillation is treated with digitalis" (Prinzmetal et al. 1952, p. 314).

The direct and autonomically mediated actions of digoxin on the atrioventricular node may be readily understood if the node is considered as an oscillator, or pacemaker (Chapters 3 and 8). Digoxin is then seen to have qualitatively similar actions on sinoatrial and atrioventricular nodes, slowing both their rates (Hoffman and Singer 1964). Reduction of the rate of the atrioventricular oscillator leads to increased atrioventricular delay and refractory period (Chapter 3). If digoxin does increase the fibrillation rate in any individual, the effect is likely to be of small magnitude but would tend to counteract, rather than assist, the drug's beneficial actions on the atrioventricular node.

Mackenzie (1911) recognised that digitalis slows the ventricular rate in atrial fibrillation by means of the vagus nerve, as evidenced by the antagonistic action of atropine. As time passes after the start of treatment with digitalis, and the cumulative dose increases, the relative effect of its direct and indirect actions was thought to increase, until with full digitalisation the direct action was considered to be responsible for almost all of the effect (Cushny 1925, p. 223). At this point the ventricular rate after atropine is unchanged, and is unresponsive also to cholinergic manoeuvres. Greenspan and Lord (1973), however, showed that vagal depression of ventricular rate was enhanced by digitalis right up to toxic doses, thus demonstrating that the heart rate response to atropine is no guide to the adequacy of the dose.

The continuing importance of the indirect, vagal-enhancing effect, beyond the time when digitalis treatment is begun (Gold et al. 1939), explains why ventricular rate may be satisfactorily controlled at rest, when vagal tone is high, but not during exercise, when vagal tone is withdrawn and sympathetic tone is increased (Beasley et al. 1985). Similarly, if serious illness is present, particularly infection or hypoxia, digitalis may fail to reduce the ventricular rate even at rest, in spite of serum levels considered to be "therapeutic" (Goldman et al. 1975), presumably because vagal tone is absent.

Many patients with atrial fibrillation have impaired ventricular function, and some may have congestive cardiac failure; such patients may benefit from the positive inotropic properties of digoxin. By alleviating heart failure, digoxin may reduce sympathetic drive, resulting in a reduction of ventricular rate additional to that brought about by its actions on the conducting system. The effect of digitalis on contractility is independent of autonomic tone, and is dissociated from that on the conduction system (Kim et al. 1975). It has been suggested that hydrophilic preparations of digitalis such as ouabain may be preferable for control of ventricular rate in atrial fibrillation, while a lipophilic preparation such as digitoxin may be more appropriate in heart failure, particularly if there is a tendency to atrioventricular block (Joubert and van der Meer 1979).

Pharmacokinetics

There are reports of patients (and rats) with unusually high digitalis requirements, particularly in hyperthyroidism (Luchi and Gruber 1968; Huffman et al. 1977), but generally the pharmacokinetics of digoxin are well understood. When the effect of an infusion of 1.5 mg digoxin over 6 hours was compared with 3 boluses of 0.5 mg 8 hours apart, the results were in agreement with a three compartment model (Smit et al. 1990). The maximal fall of ventricular rate in patients with rapid atrial fibrillation occurred after 8–9 hours with the infusion, and after 19–20 hours with the boluses. Jelliffe and Brooker (1974) have described a nomogram for digoxin therapy, based on known pharmacokinetics. The size of a loading dose is determined by body weight, and the maintenance dose by body weight, sex and renal function. The nomogram is generally an excellent guide to dose requirements, but is unreliable when quinidine and some other drugs are prescribed together with digoxin; quinidine and verapamil produce significant increases in serum digoxin levels (Reid and Meek 1979; Mungall et al. 1980; Klein and Kaplinsky 1982).

Serum digoxin assays may be helpful in confirming digitalis toxicity, and in identifying patients who are non-compliant, or who have malabsorption (Selzer 1985); care is needed to standardise the conditions of blood sampling (Jogestrand et al. 1989). However, in most patients with atrial fibrillation, simple but careful clinical monitoring is an adequate basis for the selection of a suitable dose of digoxin (Savill et al. 1985; Anderson et al. 1988).

During maintenance therapy with digoxin the ventricular rate does not follow the fluctuations in plasma levels, and following withdrawal of digoxin a significant chronotropic effect persists for more than 6 days, by which time the plasma concentration is negligible (Zener et al. 1973). Plasma digoxin concentrations show a poor correlation with resting ventricular rates between subjects (Chamberlain et al. 1970; Penchas and Zajicek 1978), but the correlation is better within subjects (Redfors 1972).

The concentrations of digoxin in serum and in atrial myocardium are correlated in sinus rhythm, but they are unrelated in atrial fibrillation, in which atrial myocardial concentrations are higher than in sinus rhythm due to increased atrial binding (Jogestrand 1980).

Toxicity

Kastor and Yurchak (1967) suggest that clinical indications of the correct dose of digitalis are a reduction of ventricular rate to 70–90 at rest, little rise with effort, and an absent pulse deficit. It has been customary to withhold the agent if the ventricular rate falls below 50–60. However, Williams et al. (1978) showed that a slow pulse is not a reliable sign of digitalis toxicity.

The gastrointestinal, neuropsychiatric and visual disturbances due to digitalis toxicity have been well documented by Chung (1969). The effect of a toxic dose of digitalis on the heart is to increase the intrinsic rate of the atrioventricular node, leading to nodal escape rhythm or non-paroxysmal nodal tachycardia; both of these are characterised by a regular ventricular rhythm (Chung and Dean 1970; Kastor 1973). Conversion of atrial fibrillation to flutter or paroxysmal atrial tachycardia with block may occur as a result of digitalis toxicity, which is particularly likely in the latter case (Taguchi and Ryan 1969).

Clinical Studies of Digitalis

It is generally accepted that digitalis is beneficial in reducing the ventricular rate in atrial fibrillation, and indeed, many believe that this is the principal role for the drug. Because of its obvious efficacy in this regard, there have been few placebo-controlled trials of digitalis in atrial fibrillation; one of these is described in Chapter 7.

Aberg et al. (1972) conducted exercise tests in patients with atrial fibrillation at two different doses of digitalis. The ventricular rate was lower at a given load with the higher dose. The coefficient of variation of R-R intervals was not affected.

In a dose–response study, Redfors (1971a,b) demonstrated that ventricular rate in atrial fibrillation increased linearly with work load, and was reduced at rest and at each level of exercise in proportion to the dose of digoxin over a range of 0.125–1.00 mg/day. As the dose was raised, the subjects' sense of well-being improved, and the work load achieved on a bicycle ergometer increased. The average optimum dose was 0.44 mg/day. However, the study was only single-blind, the doses were not randomised, and a training effect could have been responsible for the results.

Atrial fibrillation is a common problem after cardiac surgery, particularly of the mitral valve; preoperative treatment with digitalis has been used to lower the ventricular rate when atrial fibrillation commences. Average ventricular rates at the onset of atrial fibrillation were 163 without digitalis, 138 when patients were incompletely digitalised and 121 when they were fully digitalised (Selzer and Walter 1966). On the other hand, in paroxysmal atrial fibrillation, pretreatment with digoxin did not affect the ventricular rate at the onset of a paroxysm, and paroxysms were more likely to be prolonged in patients taking digoxin (Rawles et al. 1990).

Beta-adrenergic Blockers

Digitalis has an anti-adrenergic action, but this is weak with non-toxic dosages, and the drug may fail to prevent ventricular rate from rising during conditions of high sympathetic drive such as exercise. However, the ventricular rate in atrial fibrillation needs to be higher than in sinus rhythm to achieve a given cardiac output because of the absence of an appropriately timed atrial contraction. Thus, the use of beta-adrenergic blockers in addition to, or instead of, digoxin, while limiting ventricular rate in exercise, may lead to a decline of exercise tolerance. Further, if underlying ventricular function is impaired there is the risk of deterioration by removing sympathetic support. Some studies have used ventricular rate during or after exercise as the main outcome measure, setting arbitrary target rates without regard to the effect of heart rate on exercise tolerance or cardiac output.

A beta-adrenergic blocker with no intrinsic sympathetic activity, such as propranolol, is most effective in reducing exercise tachycardia, but its use may be associated with bradycardia during sleep (Leclercq et al. 1981); night-time bradycardia may be prevented with a beta-adrenergic blocker with high intrinsic sympathetic activity, or a partial beta-agonist.

David et al. (1979) demonstrated a ventricular rate that was higher than in sinus rhythm in patients with atrial fibrillation exercised for 3 minutes while taking digitalis. The addition of timolol normalised the heart rate response to exercise, and this was claimed to be beneficial. In another study of atrial fibrillation, reduced heart rate variability was obtained by the addition of pindolol to digoxin, and this, too, was said to be an improvement (James et al. 1989).

Gibson and Coltart (1972) demonstrated the haemodynamic effects of practolol in 6 patients with atrial fibrillation. Ventricular rate was reduced and stroke volume was increased, but not enough to counteract the fall in rate, so there was a non-significant reduction of cardiac output; a reduction of contractility was shown.

In 6 patients with atrial fibrillation taking digitalis, administration of propranolol resulted in a lower pulse rate at rest and on exercise, but deterioration of exercise tolerance occurred in 3 (Brown and Goble 1969). Similar results were obtained with metoprolol (Khalsa et al. 1978), which led to reduced cardiac output after exercise compared with control subjects or those given verapamil (Maier et al. 1983).

DiBianco et al. (1984) reported a randomised double blind trial of nadolol in 20 patients with atrial fibrillation receiving digoxin. While nadolol reduced average and exercise ventricular rates, the duration of exercise was also significantly decreased, though this was not the case in another study of nadolol of similar design (Zoble et al. 1987). Atwood et al. (1987) demonstrated a reduction of oxygen uptake and a reduced duration of exercise with celiprolol.

Lewis et al. (1989) compared the effects of atenolol, verapamil and xamoterol in digitalised patients with chronic atrial fibrillation. Post-exercise heart rates were reduced by all three treatments compared with placebo, and minimum heart rates of 45 or less were found with all treatments except xamoterol, a partial beta-agonist. Exercise tolerance was impaired by atenolol, and patients taking xamoterol, while not showing such extremes of heart rate, had poorer exercise tolerance than when they were taking verapamil.

Xamoterol, while limiting the range of ventricular rates found on Holter recording, was not associated with any adverse events or worsenening symptoms in 6 digitalised patients with poor left ventricular function investigated by Furniss et al. (1989), and Molajo et al. (1984) reported improved exercise tolerance with the use of xamoterol in addition to digitalis in atrial fibrillation. Ang et al. (1990) reported symptomatic improvement with xamoterol compared with digoxin, patients taking xamoterol having less dizziness and a less extreme range of ventricular rates.

In chronic cardiac failure, xamoterol improved exercise tolerance and reduced breathlessness and tiredness compared with both placebo and digoxin (The German and Austrian Xamoterol Study Group 1988), but was associated with significantly higher mortality than placebo in patients with heart failure of NYHA classes 3 or 4 (The Xamoterol in Severe Heart Failure Study Group 1990).

In summary, beta-adrenergic blockade, while reducing the ventricular rate in atrial fibrillation, especially during exercise, is not generally associated with any symptomatic improvement, but may reduce exercise capacity. Xamoterol, exceptionally among beta-blockers, has been found to be beneficial in some studies. However, caution is needed in patients with atrial fibrillation and poor cardiac function since the use of xamoterol is associated with increased mortality in higher degrees of heart failure.

Calcium Antagonists

Verapamil

The immediate effect of intravenous verapamil on atrial fibrillation is to slow the ventricular rate and regularise the ventricular rhythm; in a few subjects sinus rhythm may be restored (Schamroth 1971; Khalsa et al. 1979; Aronow et al. 1979; Waxman et al. 1981; Hwang et al. 1984; Krikler 1986). A similar antiarrhythmic efficacy throughout 24 hours is achieved by conventional and slow-release preparations of verapamil (Molgaard et al. 1987), the time-course of the dynamic effect of the drug lagging behind the plasma concentration (Neuss 1982). Verapamil is safe combined with digitalis (Sloman et al. 1975; Panidis et al. 1983), and may be effective in thyrotoxicosis if digoxin fails and beta-adrenergic blockers are contraindicated because of heart failure (Dahlstrom and Ladefoged 1987).

Verapamil, a slow calcium channel blocking drug, decreases the amplitude of action potentials in the atrioventricular node and increases the effective refractory period of nodal tissue (Klein and Kaplinsky 1982). Regularisation of ventricular rhythm in atrial fibrillation is ascribed to atrioventricular nodal block with junctional escape rhythm (Neuss 1982; Neuss et al. 1984). In many subjects the amplitude of the f wave is diminished (Schamroth et al. 1972), and both intravenous and oral calcium antagonists sustain electrically induced atrial fibrillation (Shenasa et al. 1988), so the regularisation of ventricular rhythm could result from failure of penetration of the node by weaker and more frequent atrial wavefronts (Theisen et al. 1985). Exercise, isoprenaline and atropine all cause a gradual loss of the ventricular regularity induced by verapamil (Khalsa and Olsson 1979). If adrenergic tone is normal, verapamil has only a modest effect on atrioventricular junctional automaticity, but if adrenergic tone is reduced by beta-blockade, calcium channel blockade then causes profound depression of the junctional escape rate (Lupi et al. 1979); the combined use of a beta-adrenergic blocker and a calcium antagonist may result in asystole.

Lang et al. (1983a) reported a double-blind crossover study of verapamil in digitalised patients with atrial fibrillation. In patients on verapamil, maximal exercise capacity improved, and ventricular rate was reduced at rest and during exercise. And in a randomised, placebo-controlled crossover comparison of digoxin and verapamil, Klein and Kaplinsky (1982) demonstrated that the effect of verapamil on ventricular rate is accompanied by a significant improvement in maximal exercise capacity, which is not found with digoxin. The report of a placebo-controlled trial comparing digoxin, oral verapamil and a combination of the two concluded that verapamil alone is superior to digoxin alone (Lang et al. 1983b). However, in the latter study, drug treatment appears not to have been blinded, and it is not stated whether the differences in exercise tolerance achieved statistical significance. Lewis et al. (1987) did not find that increased exercise tolerance was associated with the lower post-exercise heart rates on verapamil, but their patients reported more constipation. In a placebo controlled trial of verapamil, digoxin and a combination of the two, Pomfret et al. (1988) showed that verapamil alone or in combination with digoxin resulted in lower ventricular rates during exercise than did digoxin alone, and claimed that this was not associated with any deterioration of left ventricular function.

Diltiazem

Like verapamil, diltiazem is safe and effective in reducing ventricular rate in atrial fibrillation, both at rest and during exercise, and either alone or in combination with digoxin (Roth et al. 1986; Steinberg et al. 1987; Rohl and Schulz 1987;

Maragno et al. 1988; Salerno et al. 1989). However, reduction of exercise-induced tachycardia *per se* may not be of benefit in most patients with atrial fibrillation (Lewis and McDevitt 1988a). As discussed in Chapter 6, reduction of ventricular rate may be associated with increased stroke volume, but this is offset by a rate-related reduction of cardiac output (Lewis et al. 1988a). Side effects are a problem with both verapamil and diltiazem, but treatment of atrial fibrillation with diltiazem rather than digoxin is associated with fewer ventricular premature beats (Lewis and McDevitt 1988b).

Diltiazem, while reducing heart rate at all stages of a treadmill exercise test, did not change maximal exercise capacity (Atwood et al. 1988b). In a randomised double-blind trial, digoxin, diltiazem and a combination of the two were compared (Lewis et al. 1988b); the mean post-exercise heart rate was 15% lower with combination therapy than with digoxin alone, but there was no evidence that this reduction of ventricular rate was associated with any improvement in exercise tolerance, as assessed by a 6-minute walking test. But maximal exercise capacity was significantly improved by diltiazem in a double-blind crossover study in digitalised patients reported by Vitale et al. (1989), and by both diltiazem and verapamil in a study reported by Lundstrom and Ryden (1990). The latter authors concluded that patients with chronic atrial fibrillation have a modest improvement in exercise tolerance with calcium channel blockade; the effects on ventricular rate and physical performance are of similar magnitude for verapamil and diltiazem. These calcium antagonists are satisfactory alternatives to digoxin in the treatment of atrial fibrillation.

Cardioversion

In 1962, Lown et al. described a new device for terminating cardiac arrhythmias, and its successful use in ventricular tachycardia, atrial flutter and atrial fibrillation; the term "cardioversion" applied to atrial fibrillation was used the following year (Lown et al. 1963). The equipment consists of an inductor and a capacitor carrying a high-voltage electrical charge, discharge of which is triggered by the R wave of the surface electrocardiogram. By timing the shock to avoid the vulnerable period of the myocardium, the risk of precipitating ventricular fibrillation is minimised. Synchronised direct current cardioversion rapidly became the preferred means of restoring sinus rhythm from atrial fibrillation, it being simpler, safer and more effective than quinidine (Lown et al. 1963); intravenous diazepam provides a convenient sedative (Winters et al. 1968). A low-energy (5 joules) discharge is used initially, the energy being increased until reversion occurs or until other arrhythmias are seen (DeSilva and Lown 1982). The size and location of the electrode paddles is not critical, provided the current pathway traverses the long axis of the heart (Kerber et al. 1981). Anticoagulation is recommended if atrial fibrillation has been present for more than a week (DeSilva and Lown 1982).

Restoration of sinus rhythm is achieved in about 80% of cases (Halmos 1966b; Szekely et al. 1966; Lown 1967; Scott and Pantridge 1968), but many patients soon relapse into atrial fibrillation, especially if the atrial fibrillation is of long standing, the left atrium is enlarged, or there is congestive cardiac failure (Wikland et al. 1967; Futral and McGuire 1967; Fisher et al. 1968; Upton and Honey 1971).

Selection of Patients for Cardioversion

Because of the high risk of early relapse, it is generally not worth while attempting cardioversion in patients who have had atrial fibrillation for more than 1 year, who have mitral valve disease with a very large left atrium, or who have left ventricular failure (DeSilva and Lown 1982). Lone atrial fibrillation – atrial fibrillation in the absence of demonstrable heart disease or thyrotoxicosis – is a relative contraindication to cardioversion because of the risk of relapse, though this has to be balanced against the severity of symptoms during the arrhythmia, and the tolerance of drug therapy to prevent relapse.

Pregnancy is not a contraindication to cardioversion, and published cases are reviewed by Cullhed (1983).

Where the ventricular rate in atrial fibrillation is already slow without treatment there may be asystole or extreme sinus bradycardia after cardioversion, so atrial fibrillation is preferable and cardioversion is contraindicated.

Complications of Cardioversion

Complications occur in a substantial proportion of cardioversions, and include ventricular fibrillation, myocardial damage, pulmonary oedema, hypotension and systemic embolism (Turner and Towers 1965; Navab and La Due 1965; Eliot et al. 1966); death occurred in 4 out of 220 patients reported by Resnekov and McDonald (1967). Complications are more likely if the contraindications are set aside and cardioversion is attempted in longstanding or lone atrial fibrillation, or in atrial fibrillation occurring in association with ischaemic heart disease or cardiomyopathy.

Ventricular fibrillation may occur if the defibrillator is improperly adjusted, when the shock may be synchronised with a tall T wave or an artefact rather than the R wave. But there are occasional reports of delayed ventricular fibrillation after successful cardioversion (Furman et al. 1966; Sheils 1966), and equally puzzling is a report of delayed conversion to sinus rhythm up to 105 seconds after a countershock (Duvernoy and Anbe 1976). Szekely et al. (1970b) made the point that immediate or delayed post-shock rhythm disorders may be drug related, and emphasised that great caution should be exercised in the use of antidysrhythmic drugs in conjunction with electrical cardioversion; cardioversion is particularly dangerous in the presence of digitalis toxicity (Kleiger and Lown 1966).

Following cardioversion, transient ST-segment elevation or conduction disorders may be observed, and there may be elevation of non-specific enzymes such as creatine kinase, the degree of elevation being related to the energy of the shock (Chun et al. 1981; Ovsyshcher et al. 1984). The concentration of the myocardial isoenzyme of creatine kinase may exceed the level considered diagnostic for acute myocardial infarction (Jakobsson et al. 1990), and there is obvious concern that these enzyme rises might result from myocardial damage. However, the isoenzyme is believed to originate from skeletal muscle, and scintigraphy, while showing uptake of radioisotope by chest wall, does not show uptake by myocardium (Metcalfe et al. 1988).

Pulmonary oedema sometimes follows cardioversion (Lindsay 1967), and in one case where 5 shocks totalling 1280 joules had been delivered, direct myocardial injury seems the most likely mechanism (Gomaa et al. 1972). In other cases it is postulated that the preferential return of right more than left atrial function leads to right ventricular output exceeding that from the left, with consequent pulmonary congestion and oedema (Budow et al. 1971).

Systemic embolism after cardioversion is well recognised, and acute myocardial infarction possibly due to coronary artery embolism has been reported (Gonzalez et al. 1981).

Physiological Consequences of Cardioversion

The elective conversion of atrial fibrillation to sinus rhythm provides a much-exploited opportunity to assess retrospectively the effect of atrial fibrillation on atrial and ventricular function, overall cardiac performance and exercise capacity. In some patients atrial transport function is immediately restored, but in many it gradually improves over several weeks, and in a few there is no recovery at all. Atrial function has been studied by apex-cardiography (Ikram et al. 1968), phonocardiography (O'Rourke 1970), kinetocardiography (Mahlich et al. 1973), echocardiography (Orlando et al. 1979), using systolic time intervals (Van Herick et al. 1979), by cardiac catheterisation (Grover et al. 1971), and by cardiac catheterisation combined with pacing to vary the P-R interval (Thompson et al. 1972). The P wave of the electrocardiogram recorded after cardioversion may reflect persisting left atrial enlargement (Hendrix and Davis 1973).

In patients with coronary artery disease the "a" wave in the right atrial pressure trace returned immediately after cardioversion in 75% of cases, but the left atrial "a" wave returned in only 55%; those patients without an immediate return of the left atrial "a" wave reverted to atrial fibrillation within 3 weeks (Khaja and Parker 1972).

An echocardiographic study undertaken 2 hours after cardioversion showed a decrease in left atrial dimension, an increase in left ventricular end-diastolic dimension, and an increase in the cardiac index. These changes were more pronounced in patients with mitral valve disease than in those with ischaemic heart disease (Ieri et al. 1982). Pulsed Doppler examination of atrial mechanical function showed that the peak velocity of the atrial filling wave of the left ventricle did not return to normal until 3 weeks after cardioversion. The atrial contribution to left ventricular filling was normal at this time, but the left atrial dimension continued to decrease over a 3-month study period (Manning et al. 1989). Shapiro et al. (1988), using Doppler ultrasound to assess ventricular filling, showed that atrial function soon after cardioversion was normal if the duration of atrial fibrillation was less than a week, but was reduced or absent if the arrhythmia had been sustained for longer. In such patients atrial function improved over a 48 day follow-up period. In 14 patients studied by O'Neill et al. (1990), the atrial filling fraction determined by Doppler echocardiography was 15%, 13% and 22% at 5 minutes, 30 minutes and 24 hours after cardioversion. An embolic event occurred in one patient who had immediate return of atrial mechanical activity.

The ultimate expression of cardiac function is cardiac output. Halmos and Patterson (1965) measured cardiac output by dye dilution before and 1 week after cardioversion; cardiac output rose by 53%. Following cardioversion, an increase of cardiac output has been demonstrated after exercise but not at rest (Resnekov 1967), and in patients with mitral and aortic valve disease but not in those with lone atrial fibrillation (Killip and Baer 1966); pyruvate and lactate levels improved in those with rheumatic but not hypertensive or arteriosclerotic heart disease (Kaplan et al. 1968). Five to ten days after cardioversion by quinidine, cardiac output was increased by 22%, and in the 3 hours after electrical cardioversion there was a gradual rise of cardiac output of 12% (Rodman et al. 1966). Scott and Patterson (1969) found no significant change in cardiac output 3 minutes or 3 hours after cardioversion, but cardiac output was up by 29% 3 days later.

Lipkin et al. (1988) demonstrated a delayed improvement in exercise capacity after cardioversion of atrial fibrillation to sinus rhythm. Maximal oxygen consumption and anaerobic threshold were not significantly changed on the first day, but had increased by 1 month in all patients.

After a mean interval of 39 days following cardioversion in 11 patients, Atwood et al. (1989b) demonstrated a significantly increased oxygen uptake at maximal exercise, with increased ventilation efficiency once oxygen uptake was more than 60% of maximal. Heart rates at submaximal and maximal exercise were about 50 beats per minute less than they had been in atrial fibrillation.

Long-Term Results of Cardioversion

One of the first studies of the long-term results of cardioversion, while showing an 81% conversion rate, reported that only 22% of patients were in sinus rhythm 1 year later (Aberg and Cullhed 1968); other workers have reported similarly disappointing results (Radford and Evans 1968; McCarthy et al. 1969; Takkunen et al. 1970; Resnekov and McDonald 1971).

The most significant predictor of relapse is atrial fibrillation lasting for more than 1 year before cardioversion is attempted (Jensen et al. 1965; Szekely et al. 1970a; Hansen et al. 1979; Dittrich et al. 1989). Also, an enlarged left atrium with a left atrial dimension of more than 45 mm makes relapse more likely (Hoglund and Rosenhamer 1985; Dethy et al. 1988). After conversion, failure of the atrial contribution to ventricular filling to increase by more than 10% between 4 and 24 hours is highly predictive of relapse (Dethy et al. 1988). Good prognostic factors are atrial fibrillation of short duration, normal atrial size, spontaneous or drug-induced conversion, and absence of mitral valve disease (Halmos 1966a; Brodsky et al. 1989).

Chemical Cardioversion and Prevention of Atrial Fibrillation

Quinidine

Quinidine is the archetypal Class 1 antiarrhythmic agent. It causes reduction of the amplitude and slope of phase 0 of the action potential by blocking the fast inward sodium channel. Automaticity is decreased, conduction velocity slowed, and the refractory period prolonged (Bellet 1971 p 980). In the atria the conditions conducive to fibrillation are reversed, and the threshold for fibrillation is increased (Gold et al. 1985). But in the ventricles prolongation of the action potential and the Q-T interval may result in ventricular fibrillation or torsades de pointes, these complications being known euphemistically as "quinidine syncope" (Selzer and Wray 1964; Jenzer and Hagemeijer 1976). Systemic embolism, a feared complication of electrical cardioversion, may occur after cardioversion by pharmacological means (Yapa and Green 1990).

The use of quinidine to effect cardioversion resulted from observations that when extract of cinchona bark is given for malaria, atrial fibrillation frequently reverts to sinus rhythm. Following the introduction of electrical cardioversion for atrial fibrillation, quinidine continued in use for the subsequent maintenance of sinus rhythm; there are many studies which attest to its value in preventing recurrence of atrial fibrillation (Byrne-Quinn and Wing 1970; Hartel et al. 1970; Hillestad et al. 1971, 1972; Resnekov et al. 1971; Sodermark et al. 1975; Boissel et al. 1981). Initiation of quinidine therapy before cardioversion does not increase the proportion of patients who are successfully restored to sinus rhythm, nor the incidence of post-conversion arrhythmias. But the occasional sudden deaths of patients taking quinidine have been reported from time to time.

Levi and Proto (1970, 1972) used the combination of propranolol and quinidine for the prevention of atrial fibrillation, claiming that propranolol provides a safeguard against quinidine syncope due to ventricular fibrillation. Although this contention appears to be supported by Stern (1971), a report by Hillestad and Storstein (1969) suggests that the combination treatment is dangerous, at least in chemical cardioversion.

Coplen et al. (1990) conducted a meta-analysis of six randomised controlled trials of quinidine therapy for the maintenance of sinus rhythm after cardioversion. The proportion of patients in sinus rhythm 12 months after cardioversion was 50% with quinidine and 25% with placebo – a highly significant difference in favour of quinidine ($p<0.001$ at all times during follow-up). However, the total mortality rate in patients taking quinidine was 2.9% compared with 0.8% with placebo, an odds ratio of about 3 ($p<0.05$). Thus, quinidine treatment is more effective than no antiarrhythmic therapy in suppressing recurrences of atrial fibrillation but appears to be associated with increased total mortality. This conclusion necessitates great caution in the use of quinidine, but the same injunction may also apply to other, newer Class 1 antiarrhythmic agents (Cardiac Arrhythmia Suppression Trial Investigators 1989). For example, in a comparative study of quinidine and flecainide, adverse events, although more frequent with quinidine, were considered more severe with flecainide (Borgeat et al. 1986).

Lundstrom and Ryden (1988) reported the long-term results of their management policy for chronic atrial fibrillation. The first attempt at cardioversion is made without drugs. If conversion is unsuccessful or atrial fibrillation recurs then further attempts are made with quinidine or disopyramide prophylaxis. Twenty-three per cent of patients were in sinus rhythm without prophylactics 1 year after the first cardioversion, and 54% were in sinus rhythm after up to 12 conversions; the proportion of patients in sinus rhythm at 2 years was 41%.

Calcium Antagonists

In the search for a safer drug than quinidine for the prevention of atrial fibrillation after cardioversion, Rasmussen et al. (1981) conducted a controlled trial comparing quinidine with verapamil. Quinidine was found to have a greater ability than verapamil to induce conversion and maintain sinus rhythm, but two patients on quinidine died outside hospital.

Bepridil is a calcium antagonist which also has a depressant effect on the fast sodium channel. In a comparative study with amiodarone in established atrial fibrillation, ventricular rates at rest and during exercise were equally well controlled by both agents, but 9 of 14 patients converted to sinus rhythm while on bepridil, compared with 4 out of 10 on amiodarone. However, 8 of 14 patients on bepridil had ventricular arrhythmias, and 2 patients had torsades de pointes degenerating to fatal ventricular fibrillation (Perelman et al. 1987).

Disopyramide

Hartel et al. (1974) described the use of disopyramide for the maintenance of sinus rhythm after electrical cardioversion of atrial fibrillation. During a 3-month follow-up period a significantly higher proportion of patients taking the drug remained in sinus rhythm. The main side effects were due to the drug's anticholinergic properties, to which it may owe its effectiveness in preventing relapse. In another trial of similar design the drug was very well tolerated, and at the end of 1 year 54% of patients taking disopyramide were in sinus rhythm compared with 30% on placebo ($p<0.01$) (Karlson et al. 1988). On the strength of these results and the absence of reported pro- arrhythmic effects, disopyramide appears to be suitable for prevention of relapse after cardioversion.

Propafenone

Trials of prophylaxis against paroxysmal atrial fibrillation are extraordinarily difficult because of the irregularity and infrequency of attacks (Hammill et al. 1988; Kerr et al. 1988). Connolly and Hoffert (1989) recruited 11 patients who were randomised to alternate between propafenone and placebo every month for 4 months. The percentage of days with an attack of atrial fibrillation was significantly reduced by propafenone compared with placebo.

Connolly et al. (1987) conducted a placebo-controlled trial of propafenone for treatment of atrial tachyarrhythmias after cardiac surgery. Propafenone proved useful for both control of ventricular rate and conversion to sinus rhythm.

Digoxin

Atrial fibrillation is common after acute myocardial infarction, though frequently it terminates spontaneously and does not require antiarrhythmic therapy (Sanzobrino and Lemberg 1979). Response to treatment at this time may be difficult to interpret, and it may be unwise to extrapolate the findings to atrial fibrillation occurring in other circumstances. Nevertheless, acute myocardial infarction has provided the clinical setting for several pharmacological studies of conversion to sinus rhythm or control of ventricular rate. Cowan et al. (1986) described a comparison of intravenous amiodarone and intravenous digoxin in the treatment of atrial fibrillation complicating suspected acute myocardial infarction. After 24 hours a similar proportion of patients in each group had reverted to sinus rhythm, though the mean time to conversion was shorter in those treated with amiodarone.

In a placebo-controlled trial of conversion of atrial fibrillation complicating acute myocardial infarction, digoxin did not affect the likelihood of reversion to sinus rhythm (Falk et al. 1987). The mean time to conversion was 5.1 hours with digoxin and 3.3 hours with placebo. In both these studies the longer duration of atrial fibrillation with digoxin may be a manifestation of its vagomimetic action, which would tend to perpetuate atrial fibrillation. Most patients with acute atrial fibrillation who revert to sinus rhythm after administration of digoxin do not show any important degree of ventricular slowing before conversion (Weiner et al. 1983).

Beta-adrenergic Blockers

Another common setting for the onset of atrial fibrillation is cardiac surgery. In a trial of prophylactic digitalisation for coronary artery bypass surgery, supraventricular tachyarrhythmias were significantly fewer in patients given digitalis (Johnson et al. 1976), but in another trial of similar design Tyras et al. (1979) came to the opposite conclusion. In a study comparing placebo, digoxin and propranolol there was no difference in the incidence of atrial fibrillation between placebo and digoxin, but there was a lower incidence with propranolol (Rubin et al. 1987). Stephenson et al. (1980) concluded after a randomised study that propranolol is effective in the prevention of cardiac arrhythmias following coronary artery bypass grafting.

The efficacy of sotalol in treating acute atrial fibrillation and flutter after open heart surgery was compared with that of a combination of digoxin and disopyramide by Campbell et al. (1985). Sotalol was as effective as the combination and acted significantly faster; several patients given sotalol showed an undue response to beta blockade.

A variable proportion of patients with acute atrial fibrillation revert to sinus rhythm after beta-adrenergic blocking agents given in uncontrolled studies (Wolfson et al. 1967; Martinoli et al. 1975; Rehnqvist 1981; Stroobandt and ·Kesteloot 1981).

Amiodarone

Amiodarone prolongs the action potential and the duration of the refractory period. There are clinical reports of its sucessful use in atrial fibrillation for reduction of ventricular rate (Blevins et al. 1987), for prophylaxis against paroxysmal atrial fibrillation (Horowitz et al. 1985), or for relapse after cardioversion (Blomstrom et al. 1984; Brodsky et al. 1987).

Success has been claimed for amiodarone in supraventricular tachyarrhythmias resistant to other antiarrhythmic drugs, but side effects are common and troublesome (Gold et al. 1986; Emmertsen et al. 1987).

Flecainide and Other Class 1 Agents

Clinical reports are available of reversion to sinus rhythm after procainamide (Halpern et al. 1980; Fenster et al. 1983), pirmenol (Toivonen et al. 1986), encainide (Pool 1986) and flecainide (Goy et al. 1985, 1988; Nathan et al. 1987; Berns et al. 1987) Flecainide appears very effective in restoring sinus rhythm, but pro-arrhythmic side effects occur.

In a comparison of flecainide and quinidine for conversion of atrial fibrillation to sinus rhythm the overall conversion rate was 63%; gastrointestinal symptoms were more frequent in patients taking quinidine, but more serious conduction disturbances were seen in patients taking flecainide (Borgeat et al. 1986). Flecainide is more effective than verapamil in converting atrial fibrillation to sinus rhythm (Suttorp et al. 1989).

Atrial fibrillation occurring after cardiopulmonary bypass reverted to sinus rhythm sooner after flecainide than after a combination of digoxin and disopyramide; monitoring blood flecainide was considered essential in the presence of poor left ventricular or hepatic function (Gavaghan et al. 1988).

In paroxysmal atrial fibrillation the combinations of digoxin and flecainide or digoxin and quinidine were more effective than digoxin alone; flecainide caused fewer side effects than quinidine (Steinbeck et al. 1988). Combined with amiodarone, flecainide allows the dose of amiodarone to be reduced (Chouty and Coumel 1988).

Anticoagulant Therapy

Rheumatic Valve Disease

Epidemiological studies, discussed in the previous chapter, have amply confirmed the clinical impression that atrial fibrillation is associated with thromboembolism, the association being strongest for patients with mitral stenosis. However, the strongly held conviction that anticoagulant therapy reduces the embolic risk, particularly in mitral stenosis, has not been supported by evidence from randomised, controlled clinical trials, because none has been carried out.

Abernathy and Willis (1973) reviewed eight studies of the effect of anticoagulant treatment on the frequency of thromboembolism in patients with rheumatic heart disease. The selection of participants was not random, study groups were not matched, a variable proportion of patients had had previous embolic events, and the studies did not deal specifically with atrial fibrillation. However, in each study the frequency of embolic events was several-fold higher in those not on anticoagulant treatment, in spite of the tendency for patients with adverse histories to be included in the groups given anticoagulants (Nordlander 1982).

Roy et al. (1986) reported their experience with anticoagulant therapy in 254 patients with atrial fibrillation due to a variety of underlying conditions, including mitral valve disease. Of 833 patient-years of follow-up, 284 were on anticoagulants and 549 were not. Thirty out of a total of 32 systemic emboli – of which 21 were cerebral – occurred while patients were not on anticoagulants. The embolism rates were 5.5 versus 0.7 per 100 patient-years, a ratio of nearly 8 : 1; the rate of haemorrhagic complications in patients on anticoagulants was 2.1 per 100 patient-years.

Studies such as this (Szekely 1964), though poorly controlled, have contributed such a weight of evidence in favour of anticoagulant therapy that it would now be considered unethical to conduct a randomised, placebo-controlled trial of such treatment in atrial fibrillation with mitral stenosis. It is less certain that anticoagulation is beneficial in patients with pure mitral regurgitation, or mitral stenosis and sinus rhythm, but it seems unlikely that trials of anticoagulant therapy will ever be performed in rheumatic heart disease, except perhaps in less developed countries and with antiplatelet agents (Nunain et al. 1990).

Non-rheumatic Atrial Fibrillation

While the evidence of benefit from anticoagulation in atrial fibrillation with rheumatic heart disease is convincing, and has long been apparent to clinicians, it is only relatively recently that the embolic risk of non-rheumatic atrial fibrillation has been identified by epidemiologists. The absolute risk of stroke in non-rheumatic atrial fibrillation is so low that it is unlikely to be appreciated by any but the most discerning clinician with an extensive practice. Large clinical trials are therefore required to demonstrate whether anticoagulant therapy can reduce the embolic risk in this group of patients. One subgroup with an increased risk of thromboembolic events consists of those who present with stroke (Easton and Sherman 1980). Clinical evidence suggests that recurrent strokes may be prevented by anticoagulant therapy (Carter 1965; Hart et al. 1983), which may be commenced as warfarin as soon as cerebral haemorrhage has been excluded (Fisher 1982). However, a comparison of patients with stroke and non-rheumatic atrial fibrillation in Oxfordshire, who were not on anticoagulants, with a similar group in Maastricht, who were on anticoagulants, showed no difference in survival or recurrence (Lodder et al. 1988). These data suggest that even in this high-risk group any benefit from anticoagulation is modest. Sandercock et al. (1986) argued for a trial of secondary prevention in all patients with stroke, not just those with atrial fibrillation.

The complications of anticoagulant therapy during 818 atrial fibrillation patient-years of follow-up have been reported by Lundstrom and Ryden (1989). About three major peripheral haemorrhages per 100 patient-years were recorded in patients with atrial fibrillation without previous thromboembolic events; the incidence was similar in patients with and without mitral valve disease. The incidence of haemorrhage in patients with previous embolic events was 8 per 100 patient-years, reflecting instability of anticoagulant control in these patients. The incidence of breakthrough cerebrovascular events was about 3 per 100 patient-years.

The benefits of anticoagulant therapy have to set against the risks, and it is likely that the risk–benefit ratio will be finely balanced for full, effective anticoagulation. However, while the benefits would be less certain, by using antiplatelet drugs instead of oral anticoagulants the risks of treatment could most probably be reduced.

Preliminary reports are available from two placebo-controlled trials of aspirin versus warfarin for the prevention of thromboembolism in non-rheumatic atrial fibrillation. In the AFASAK study from Copenhagen, thromboembolic events were less frequent with warfarin than with either aspirin or placebo; the efficacy of aspirin in preventing thromboembolic events did not differ significantly from placebo (Petersen et al. 1989).

The Stroke Prevention in Atrial Fibrillation Study Group Investigators (1990) documented event rates of 8.3% per year on placebo versus 1.6% per year on treatment with either warfarin or aspirin, a risk reduction of 81%. Compared with placebo, aspirin brought about a risk reduction of 49%, though not in patients over 75 years old. The trial continues, dropping the placebo arm but comparing warfarin and aspirin.

On the basis of the available evidence, anticoagulant treatment with warfarin is recommended for patients with non-rheumatic chronic atrial fibrillation (Petersen 1990; Wipf and Lipsky 1990). The role of aspirin in the prevention of thromboembolism remains uncertain.

Cardioversion

The percentage of patients with systemic embolism after cardioversion is low, ranging from 1% to 7% (Bjerkelund and Orning 1969). Nevertheless embolism, especially if it results in stroke, is particularly tragic since it is iatrogenic, and appears to be preventable. However, as in rheumatic atrial fibrillation, the effectiveness of anticoagulant therapy in preventing embolism has not been confirmed by a properly designed clinical trial. The best available evidence that anticoagulation protects against thromboembolism after cardioversion comes from Bjerkelund and Orning (1969). In 186 patients who were receiving long-term anticoagulation there were two embolic events following cardioversion (1.1%), but in 162 patients not on anticoagulant there were 11 (6.8%). The difference was most marked in high risk patients with a history of previous embolism. None of 55 patients on anticoagulants had new embolic episodes after cardioversion, while 3 out of 11 not on anticoagulants did. Fewer events occurred in those on anticoagulant therapy in spite of there being a higher proportion of patients with mitral stenosis and previous embolism in that group.

Embolism may occur up to a week after cardioversion, so the current recommendation, based on clinical experience rather than clinical trial, is for therapy to commence 2–3 weeks before cardioversion and be continued for at least two weeks afterwards (Stein et al. 1990).

The Wolff–Parkinson–White Syndrome

Atrial fibrillation occurring in the Wolff–Parkinson–White syndrome may be associated with a very rapid ventricular rate and circulatory collapse; control of ventricular rate is then a matter of extreme urgency. The ventricular complexes on the electrocardiogram are broad and bizarre, and atrial fibrillation with pre-excitation may not be recognised, or may be misdiagnosed as ventricular tachycardia (Garratt et al. 1989). Digoxin is contraindicated as it may reduce the refractory period of the accessory pathway and increase the ventricular rate (Sellers et al. 1977; Guaragna et al. 1982) and its intravenous injection has been followed by ventricular fibrillation (Dreifus et al. 1971; Papa et al. 1978). The use of verapamil may also result in acceleration of the ventricular response (Gulam-husein et al. 1982; Harper et al. 1982; Rowland 1983), with further circulatory collapse (Garratt et al. 1989), or ventricular fibrillation (Jacob et al. 1985); diltiazem appears to have a similar action on the accessory pathway, as pre-excited R-R intervals are shortened (Shenasa et al. 1987). Beta-blockers are ineffective on the accessory pathway (Morady et al. 1987).

Class 1 agents that have been shown to prolong the refractory period of the accessory pathway or reduce the ventricular rate in atrial fibrillation are ajmaline (Sclarovsky et al. 1981), disopyramide (Spurrell et al. 1975; Kerr et al. 1982), encainide (Markel et al. 1986; Chimienti et al. 1987), flecainide (Kappenberger et al. 1985; Neuss et al. 1983), procainamide (Mandel et al. 1975; Guaragna et al. 1982; Wellens et al. 1982) and propafenone (Breithardt et al. 1984), but not lignocaine (Barrett et al. 1980), although successful use of this drug has been reported (Josephson et al. 1976).

The Class 3 agent, amiodarone, impairs conduction across the accessory pathway and reduces the ventricular rate in atrial fibrillation, as well as making atrial fibrillation less likely to occur (Wellens et al. 1976; Feld et al. 1988). However, its action on the accessory pathway is inconsistent, and enhanced accessory pathway conduction has been reported (Kappenberger et al. 1984; Schutzenberger et al. 1987).

Understandably, there have been no trials of emergency treatment of the life-threatening tachycardia resulting from atrial fibrillation in the Wolff–Parkinson–White syndrome; flecainide is particularly effective and may be the preferred treatment. But the safest and most reliable way of treating atrial fibrillation with pre-excitation is by direct current cardioversion, particularly where the differen-tial diagnosis is polymorphic ventricular tachycardia, since no single drug is effective in both conditions (Garratt et al. 1989).

Much effort has been expended in the identification of patients with the Wolff–Parkinson–White syndrome who are at risk of sudden death from ventricular fibrillation. A low risk is associated with absence of symptoms (Milstein et al. 1986), intermittent pre-excitation (Klein and Gulamhusein 1983), and blocking of the accessory pathway with ajmaline (Wellens et al. 1980), procainamide (Bar et al. 1982) or exercise (Morady et al. 1983; Critelli et al. 1984). High risk is associated with an accessory pathway refractory period of less than 210 millise-conds, or a high ventricular rate during atrial fibrillation induced in an electrop-hysiological study (Bar et al. 1982). However, the conducting properties of the accessory pathway are not constant and may alter with changing autonomic tone, so that the assessment of risk by any of these methods is not completely reliable.

In patients troubled with re-entry tachycardia or atrial fibrillation, selection of an appropriate prophylactic drug may be assisted by electrophysiological studies (Wellens 1975).

For the high-risk patient, or the patient for whom prophylaxis is unsuccessful or associated with side effects, surgical treatment is indicated (Sadick et al. 1978).

Surgical ablation of the accessory pathways is undertaken after preliminary electrophysiological study and epicardial mapping at the time of operation, since pathways are often multiple (Guiraudon et al. 1990). Surgical treatment of the high-risk or symptomatic patient offers a good chance of cure (Sharma et al. 1985; Menasche et al. 1990).

New and Experimental Techniques in Atrial Fibrillation Pacing

When the ventricular rate in atrial fibrillation is slow and regular, the diagnosis of atrioventricular block may be made, provided treatment with digitalis or beta-adrenergic blockers can be excluded. The indications for pacing are as in sinus rhythm; the decision to recommend pacing may be helped by the presence of bi- or tri-fascicular block (Reid et al. 1973).

When intermittent heart block is suspected in atrial fibrillation, Holter monitoring may be used to assist in diagnosis. The problem then arises of distinguishing between ventricular pauses that represent the extremity of the range of instantaneous ventricular rates that characterises atrial fibrillation, and pauses due to heart block. Ventricular pauses of 2 seconds or more were found in about 80% of subjects with chronic atrial fibrillation by Uebis et al. (1985). Pauses this long were more common by night than by day and were considered normal. Ventricular pauses of more than 4 seconds showed no day– night variation in incidence, were associated with symptoms of dizziness, and were considered an indication for pacemaker implantation. Ector et al. (1983) proposed that ventricular asystole lasting 3 seconds was an indication for pacing, while Rebello and Brownlee (1987) recommended pacing for pauses of 2 seconds if associated with a convincing history of near- syncope.

Coumel et al. (1983) described the use of atrial pacing at a rate of 90 per minute to prevent vagally induced atrial fibrillation. In 6 cases this variety of paroxysmal atrial fibrillation had been resistant to quinidine, and aggravated by digitalis, beta- blockers and verapamil.

Atrial fibrillation may be preferable to atrial flutter or atrial tachycardia in some patients in whom the ventricular rate in these arrythmias is difficult to control. Atrial fibrillation may be induced by rapid atrial stimulation (Pittman et al. 1974) or a synchronised shock (Guiney and Lown 1972), the ventricular rate then responding to digoxin. Moreira et al. (1989) reported the use of of an implanted atrial pacemaker with a rate of 375 stimuli per minute in a patient with intractable atrial tachycardia. This device induced atrial fibrillation when atrial tachycardia was detected, and the ventricular rate was then 70–80 on digoxin.

Lau et al. (1990b) described a novel technique for control of ventricular rate in atrial fibrillation. A single pacing stimulus was delivered to the right ventricle following every spontaneous beat after an average delay of 232 ms. In 15 patients, this intercalated pacing resulted in a fall of the mean ventricular rate from 137 to 75, with enhancement of stroke volume, pulse pressure and regularity. Witt-kampf and De Jongste (1986) had previously described a reduction in the irregularity of ventricular rate by demand pacing of the right ventricle, at the cost of a slight increase in mean heart rate. The underlying electrophysiological mechanisms have been discussed in Chapter 3.

Two studies published in 1986 compared atrial and ventricular pacing for sinus node disease (Markewitz et al. 1986; Rosenqvist et al. 1986). Both showed that chronic atrial fibrillation is much more likely with ventricular than with atrial pacing, and that atrioventricular block of some degree develops often enough that atrial pacing alone may be inadvisable, though this is a matter of debate (Frank et al. 1990). Protection against atrioventricular block is provided by more complex and expensive dual-chamber pacemakers, but these introduce new problems. There is insufficient difference between the amplitude of atrial potentials during sinus rhythm and atrial fibrillation for them to be differentiated reliably by a dual-chamber pacemaker (Kerr and Mason 1985), so the ventricular rate may track the atrial rate should atrial fibrillation develop (McCabe 1986). There is also the potential for the development of a reciprocating tachycardia, the pacemaker acting as the antegrade limb and retrograde conduction taking place through the atrioventricular node (Kerr et al. 1983).

Internal Cardioversion

Nathan et al. (1984) reported that low energy endocardial cardioversion of atrial fibrillation was not universally successful at the highest energies tolerable, and that some patients suffered considerable discomfort. But Levy et al. (1988, 1990) have reported some success with high-energy endocardial cardioversion. Patients selected for this procedure were disabled by a rapid ventricular response resistant to treatment with digoxin, beta-blocking agents, verapamil or amiodarone, alone or in combination. Only if external cardioversion failed was internal cardioversion then attempted. Two electrode catheters were used, one in the apex of the right ventricle, and the other in the right atrial cavity; up to three shocks of 200–400 joules were delivered. Sinus rhythm was restored in 20 of 24 patients, and 13 were in sinus rhythm at the time of discharge from hospital. The authors tentatively suggest that internal cardioversion may have superior efficacy to external cardioversion; it may have a role in some intractable cases which would otherwise require His bundle ablation and pacemaker implantation.

His Bundle Ablation

Production of complete atrioventricular block and control of the ventricular rate by an implanted pacemaker is a management option for a few patients with atrial fibrillation who have refractory tachycardia; the pacemaker should preferably adjust ventricular rate automatically in response to physical activity (Sowton and Holt 1986; Daley et al. 1987). Such treatment may result in reversal of cardiomyopathy brought on by persistent tachycardia (Lemery et al. 1987).

His bundle ablation is performed using direct current from a defibrillator, or alternating current at radio frequency. The current is applied between an electrode over the left scapula and a catheter electrode positioned across the tricuspid valve close to the bundle of His. The results of 405 attempts at atrioventricular junction ablation have been documented by the Percutaneous Cardiac Mapping and Ablation Registry (Evans et al. 1990). The procedure was considered completely successful in 64% of patients who at a mean follow-up period of 19 months had third degree atrioventricular block, were asymptomatic, and were taking no antiarrhythmic drugs; in 17% the attempt was unsuccessful and patients still had symptoms or had died.

Surgical Treatment for Atrial Fibrillation

For multiple re-entry to occur in atrial fibrillation, the atria have to exceed a certain "critical mass". Of course it is not the mass that is critical, but the atrial area that will accommodate a sufficient length of re-entry pathways for fibrillation not to be extinguished. The surgical approach to atrial fibrillation is to reduce the length of these pathways, but at the same time to preserve as much atrial mass as possible. The mass of atrial myocardium determines the strength of the haemodynamically important atrial kick.

Surgical section of the bundle of His merely enables the ventricular rate to be controlled; the patient requires a pacemaker, and the thromboembolic risk is still present.

Surgical isolation of the left atrium leads to inability to sustain atrial fibrillation, preservation of right atrial contractile function, but loss of left atrial kick; there is still the risk of systemic embolism from the left atrium (Cox et al. 1982).

In the corridor operation the sinoatrial and atrioventricular nodes and a strip of atrial septum joining them are insulated from the rest of the fibrillating atria (Guiraudon et al. 1990). By this means sinus rhythm is maintained, but the booster function of both atria is lost and the atria may still harbour thrombus.

Of the surgical treatments for atrial fibrillation, the maze operation offers the most promise (Cox 1990). Multiple incisions in the atria heal by scar tissue which interrupts re-entry circuits, and atrial fibrillation is unsustainable. However, depolarisation and contraction of atrial myocardium still occurs, so atrial kick is preserved and the embolic risk is eliminated.

References

Abelmann WH, Fareeduddin K (1969) Increased tolerance of orthostatic stress in patients with heart disease. Am J Cardiol 23:354–63

Aberg H, Cullhed I (1968) Direct current conversion of atrial fibrillation – long-term results. Acta Med Scand 184:433–40

Aberg H, Nordgren L (1970) The effect of digitalis on the atrial activity in atrial fibrillation. Acta Soc Med Ups 75:13–18

Aberg H, Strom G, Werner I (1972) The effect of digitalis on the heart rate during exercise in patients with atrial fibrillation. Acta Med Scand 191:441–5

Abernathy WS, Willis PW (1973) Thromboembolic complications of rheumatic heart disease. Cardiovasc Clin 5:131–75

Anderson CM, Chambers S, Clamp M et al. (1988) Can audit improve patient care? Effects of studying use of digoxin in general practice. Br Med J 297:113–14

Ang EL, Chan WL, Cleland JGF et al. (1990) Placebo-controlled trial of xamoterol versus digoxin in chronic atrial fibrillation. Br Heart J 64:256–60

Aronow WS, Landa D, Plasencia G, Wong R, Karlsberg RP, Ferlinz J (1979) Verapamil in atrial fibrillation and atrial flutter. Clin Pharmacol Ther 26:578–83

Atwood JE, Sullivan M, Forbes S et al. (1987) Effect of beta- adrenergic blockade on exercise performance in patients with chronic atrial fibrillation. J Am Coll Cardiol 10:314–20

Atwood JE, Myers J, Sullivan M et al. (1988a) Maximal exercise testing and gas exchange in patients with chronic atrial fibrillation. J Am Coll Cardiol 11:508–13

Atwood JE, Myers JN, Sullivan MJ, Forbes SM, Pewen WF, Froelicher VF (1988b) Diltiazem and exercise performance in patients with chronic atrial fibrillation. Chest 93:20–5

Atwood JE, Myers J, Sandhu S et al. (1989a) Optimal sampling interval to estimate heart rate at rest and during exercise in atrial fibrillation. Am J Cardiol 63:45–8

Atwood JE, Myers J, Sullivan M et al. (1989b) The effect of cardioversion on maximal exercise capacity in patients with chronic atrial fibrillation. Am Heart J 118:913–18

Bacharach SL, Green MV, Bonow RO, Findley SL, Ostrow HG, Johnston GS (1981) Measurement of ventricular function by ECG gating during atrial fibrillation. J Nucl Med 22:226–31

Bar FW, Brugada P, Wellens HJJ (1982) Atrial fibrillation and pre-excitation. In: Kulburtus HE, Olsson SB, Schlepper M (eds) Atrial fibrillation. AB Hassle, Molndal, pp 179–87

Barrett PA, Laks MM, Mandel WJ, Yamaguchi I (1980) The electrophysiologic effects of intravenous lidocaine in the Wolff- -Parkinson–White syndrome. Am Heart J 100:23–33

Beasley R, Smith DA, McHaffie DJ (1985) Exercise heart rate at different serum digoxin concentrations in patients with atrial fibrillation. Br Med J 290:9–11

Bellet S (1971) Clinical disorders of the heart beat, 3rd edn. Lea & Febiger, Philadelphia

Berns E, Rinkenberger RL, Jeang MK, Dougherty AH, Jenkins M, Naccarelli GV (1987) Efficacy and safety of flecainide acetate for atrial tachycardia or fibrillation. Am J Cardiol 59:1337–41

Bjerkelund CJ, Orning OM (1969) The efficacy of anticoagulant therapy in preventing embolism related to DC electrical cardioversion of atrial fibrillation. Am J Cardiol 23:208–16

Blevins RD, Kerin NZ, Benaderet D et al. (1987) Amiodarone in the management of refractory atrial fibrillation. Arch Intern Med 147:1401–4

Blomstrom P, Edvardsson N, Olsson SB (1984) Amiodarone in atrial fibrillation. Acta Med Scand 216:517–24

Boissel JP, Wolf E, Gillet J et al. (1981) Controlled trial of a long-acting quinidine for maintenance of sinus rhythm after conversion of sustained atrial fibrillation. Eur Heart J 2:49– 55

Bonner AJ, Stewart J, Tavel ME (1976) "Presystolic" augmentation of diastolic heart sounds in atrial fibrillation. Am J Cardiol 37:427–31

Borgeat A, Goy JJ, Maendly R, Kaufmann U, Grbic M, Sigwart U (1986) Flecainide versus quinidine for conversion of atrial fibrillation to sinus rhythm. Am J Cardiol 58:496–8

Breithardt G, Borggrefe M, Wiebringhaus E, Seipel L (1984) Effect of propafenone in the Wolff–Parkinson–White syndrome: electrophysiologic findings and long-term follow-up. Am J Cardiol 54:29D–39D

Brodsky MA, Allen BJ, Walker CJ, Casey TP, Luckett CR, Henry WL (1987) Amiodarone for maintenance of sinus rhythm after conversion of atrial fibrillation in the setting of a dilated left atrium. Am J Cardiol 60:572–5

Brodsky MA, Allen BJ, Capparelli EV, Luckett CR, Morton R, Henry WL (1989) Factors determining maintenance of sinus rhythm after chronic atrial fibrillation with left atrial dilatation. Am J Cardiol 63:1065–8

Brown RW, Goble AJ (1969) Effect of propranolol on exercise tolerance of patients with atrial fibrillation. Br Med J ii:279–80

Budow J, Natarajan P, Kroop IG (1971) Pulmonary edema following direct current cardioversion for atrial arrhythmias. JAMA 218:1803–5

Byrne-Quinn E, Wing AJ (1970) Maintenance of sinus rhythm after DC reversion of atrial fibrillation. A double-blind controlled trial of long-acting quinidine bisulphate. Br Heart J 32:370–6

Campbell TJ, Gavaghan TP, Morgan JJ (1985) Intravenous sotalol for the treatment of atrial fibrillation and flutter after cardiopulmonary bypass. Comparison with disopyramide and digoxin in a randomised trial. Br Heart J 54:86–90

Cardiac Arrhythmia Suppression Trial (CAST) Investigators (1989) Preliminary report: effect of encainide and flecainide on mortality in a randomized trial of arrhythmia suppression after myocardial infarction. N Engl J Med 321:406–12

Carter AB (1965) Prognosis of cerebral embolism. Lancet ii:514–9

Chamberlain DA, White RJ, Howard MR, Smith TW (1970) Plasma digoxin concentrations in patients with atrial fibrillation. Br Med J iii:429–32

Chamberlain DA, Coltart DJ, Ead HA (1971) Conversion of atrial flutter or tachycardia to atrial fibrillation by rapid atrial stimulation. Br Heart J 33:613

Chevalier H (1966) A plea for atrial fibrillation. Am Heart J 72:423–5

Chimienti M, Moizi M, Salerno JA et al. (1987) Electrophysiologic and clinical effects of intravenous and oral encainide in patients with WPW syndrome and paroxysmal AF. Eur Heart J 8:282–90

Chouty F, Coumel P (1988) Oral flecainide for prophylaxis of paroxysmal atrial fibrillation. Am J Cardiol 62:35D–37D

Chun PK, Davia JE, Donohue DJ (1981) ST-segment elevation with elective DC cardioversion. Circulation 63:220–4

Chung EK (1969) Digitalis intoxication. Exerpta Medica Foundation, Amsterdam

Chung EK, Dean HM (1970) A recognition of digitalis intoxication in the presence of atrial fibrillation. Cardiology 55:22–7

Connolly SJ, Hoffert DL (1989) Usefulness of propafenone for recurrent paroxysmal atrial fibrillation. Am J Cardiol 63:817–19

Connolly SJ, Mulji AS, Hoffert DL, Davis C, Shragge BW (1987) Randomized placebo-controlled trial of propafenone for treatment of atrial tachyarrhythmias after cardiac surgery. J Am Coll Cardiol 10:1145–8

Coplen SE, Antman EM, Berlin JA, Hewitt P, Chalmers TC (1990) Efficacy and safety of quinidine therapy for maintenance of sinus rhythm after cardioversion. A meta-analysis of randomized control trials. Circulation 82:1106–16

Corbelli R, Masterson M, Wilkoff BL (1990) Chronotropic response to exercise in patients with atrial fibrillation. PACE 13:179–87

Coumel P, Friocourt P, Mugica J, Attuel P, Leclercq JF (1983) Long-term prevention of vagal atrial arrhythmias by atrial pacing at 90/minute: experience with 6 cases. PACE 6:552–60

Cowan JC, Gardiner P, Reid DS, Newell DJ, Campbell RW (1986) A comparison of amiodarone and digoxin in the treatment of atrial fibrillation complicating suspected acute myocardial infarction. J Cardiovasc Pharmacol 8:252–6

Cox JL, Williams JM, Smith PK, Sabiston DC (1982) The surgical management of atrial fibrillation. In: Kulbertus HE, Olsson SB, Schlepper M (eds) Atrial fibrillation. AB Hassle, Molndal, pp 321–35

Cox J (1990) Paper presented at the British Cardiac Society meeting, Torbay, UK

Critelli G, Gallagher JJ, Perticone F, Coltorti F, Monda V, Condorelli M (1984) Evaluation of noninvasive tests for identifying patients with preexcitation syndrome at risk of rapid ventricular response. Am Heart J 108:905–9

Cullhed I (1983) Cardioversion during pregnancy. A case report. Acta Med Scand 214:169–72

Cushny AR (1925) The action and uses in medicine of digitalis and its allies. Longman, Green, London

Dahlstrom CG, Ladefoged SD (1987) Verapamil in atrial fibrillation in hyperthyroidism. Br Med J 294:1384

Daley PJ, Chapman PD, Troup PJ (1987) Catheter ablation of the atrioventricular junction and activity responsive pacing. Effect on refractory atrial fibrillation with hypertrophic cardiomyopathy. Chest 91:461–2

David D, Di Segni E, Klein HO, Kaplinsky E (1979) Inefficacy of digitalis in the control of heart rate in patients with chronic atrial fibrillation: beneficial effects of an added beta adrenergic blocking agent. Am J Cardiol 44:1378–82

Davies SW, Gardener JE, Bowker TJ, Timmis AD, Balcon R (1990) A new method of haemodynamic assessment of mitral stenosis in atrial fibrillation: construction of a nomogram. Br Heart J 64:395–9

DeSilva RA, Lown B (1982) Cardioversion for atrial fibrillation – indications and complications. In: Kulburtus HE, Olsson SB, Schlepper M (eds) Atrial fibrillation. Hassle, Molndal, pp 231–239

Dethy M, Chassat C, Roy D, Mercier LA (1988) Doppler echocardiographic predictors of recurrence of atrial fibrillation after cardioversion. Am J Cardiol 62:723–6

DiBianco R, Morganroth J, Freitag JA et al. (1984) Effects of nadolol on the spontaneous and exercise-provoked heart rate of patients with chronic atrial fibrillation receiving stable dosages of digoxin. Am Heart J 108:1121–7

Dittrich HC, Erickson JS, Schneiderman T, Blacky AR, Savides T, Nicod PH (1989) Echocardiographic and clinical predictors for outcome of elective cardioversion of atrial fibrillation. Am J Cardiol 63:193–7

Dreifus LS, Haiat R, Watanabe Y, Arriaga J, Reitman N (1971) Ventricular fibrillation. A possible mechanism of sudden death in patients with Wolff–Parkinson–White syndrome. Circulation 43:520–7

Dreifus LS, Wellens HJ, Watanabe Y, Kimbiris D, Truex R (1976) Sinus bradycardia and atrial fibrillation associated with the Wolff–Parkinson–White syndrome. Am J Cardiol 38:149–56

Drinkovic N (1982) Subcostal M-mode echocardiography of the right atrial wall in the diagnosis of cardiac arrhythmias. Am J Cardiol 50:1104–8

Duvernoy WF, Anbe DT (1976) Delayed conversion to sinus rhythm after direct-current countershock. Heart Lung 5:465–70

Easton JD, Sherman DG (1980) Management of cerebral embolism of cardiac origin. Stroke 11:433–42

Ector H, Rolies L, De Geest H (1983) Dynamic electrocardiography and ventricular pauses of 3 seconds and more: etiology and therapeutic implications. PACE 6:548–51

Eliot RS, Lillehei CW, Ferlic R, Sellers RD (1966) Unexpected events attending elective D.C. cardioversion for atrial fibrillation. Geriatrics 21:171–6

Emmertsen K, Bjerregaard P, Andreasen F (1987) Amiodarone for refractory supraventricular tachycardias. Acta Med Scand 221:435- -9

Engel TR, Gonzalez AD (1978) Effects of digitalis on atrial vulnerability. Am J Cardiol 42:570–6

Evans GT, Huang WH, Scheinman (1990) His bundle ablation for atrial tachyarrhythmias. In: Touboul P, Waldo AL (eds) Atrial arrhythmias. Current concepts and management. Mosby Year Book, St Louis, pp 439–57

Falk RH, Knowlton AA, Bernard SA, Gotlieb NE, Battinelli NJ (1987) Digoxin for converting recent-onset atrial fibrillation to sinus rhythm. A randomized, double-blinded trial. Ann Intern Med 106:503–6

Farah S, Loomis TA (1950) The action of cardiac glycosides on experimental auricular flutter. Circulation 2:742–8

Feld GK, Nademanee K, Stevenson W, Weiss J, Klitzner T, Singh BN (1988) Clinical and electrophysiologic effects of amiodarone in patients with atrial fibrillation complicating the Wolff–Parkinson–White syndrome. Am Heart J 115:102–7

Fenster PE, Comess KA, Marsh R, Katzenberg C, Hager WD (1983) Conversion of atrial fibrillation to sinus rhythm by acute intravenous procainamide infusion. Am Heart J 106:501–4

Fisher CM (1982) Embolism in atrial fibrillation. In: Kulburtus HE, Olsson SB, Schlepper M (eds) Atrial fibrillation. Hassle, Molndal, pp 192–207

Fisher RD, Mason DT, Morrow AG (1968) Restoration of sinus rhythm after mitral valve replacement. Correlations with left atrial pressure and size. Circulation 37(Suppl 2):173–7

Frank R, Petitot JC, Touil F, Lascault G, Fontaine G (1990) Pacemaker indications and selection in sick sinus syndrome. In: Touboul P, Waldo AL (eds) Atrial arrhythmias. Current concepts and management. Mosby Year Book, St Louis, pp 392–9

Friedman HZ, Weber-Bornstein N, Deboe SF, Mancini GBJ (1987) Cardiac care unit admission criteria for suspected acute myocardial infarction in new-onset atrial fibrillation. Am J Cardiol 59:866–9

Fujiki A, Tani M, Mizumaki K, Yoshida S, Sasayama S (1990) Quantification of human concealed atrioventricular nodal conduction: Relation to ventricular response during atrial fibrillation. Am Heart J 120:598–603

Furman S, Rubin I, Frieden J (1966) Delayed ventricular fibrillation following direct-current conversion of atrial fibrillation to regular sinus rhythm. N Y State J Med 66:271–2

Furniss SS, Beat KJ, Reid DS (1989) Effectiveness of addition of xamoterol to digoxin in patients with atrial fibrillation and impaired left ventricular function – a placebo controlled study. Clin Sci 77 (Supplement 21):12P

Futral AA, McGuire LB (1967) Reversion of chronic atrial fibrillation. JAMA 199:885–8

Garratt C, Antoniou A, Ward D, Camm AJ (1989) Misuse of verapamil in pre-excited atrial fibrillation. Lancet ii:367–9

Gavaghan TP, Koegh AM, Kelly RP, Campbell TJ, Thorburn C, Morgan JJ (1988) Flecainide compared with a combination of digoxin and disopyramide for acute atrial arrhythmias after cardiopulmonary bypass. Br Heart J 60:497–501

Gibson DG, Coltart DJ (1972) Haemodynamic effects of practolol. Br Heart J 34:95–9

Gold H, Kwit NT, Otto H, Fox T (1939) On vagal and extravagal factors in cardiac slowing by digitalis in patients with auricular fibrillation. J Clin Invest 18:429–37

Gold RL, Bren GB, Katz RJ, Varghese PJ, Ross AM (1985) Independent and interactive effects of digoxin and quinidine on the atrial fibrillation threshold in dogs. J Am Coll Cardiol 6:119–23

Gold RL, Haffajee CI, Charos G, Sloan K, Baker S, Alpert JS (1986) Amiodarone for refractory atrial fibrillation. Am J Cardiol 57:124–7

Goldman S, Probst P, Selzer A, Cohn K (1975) Inefficacy of "therapeutic" serum levels of digoxin in controlling the ventricular rate in atrial fibrillation. Am J Cardiol 35:651–5

Gomaa M, Dayem MK, Eissa A, Ateyya M (1972) Pulmonary edema after direct current countershock. Chest 62:623–5

Gonzalez M, Hernandez E, Aranda JM, Linares E, Cortes F, Cintron G (1981) Acute myocardial infarction due to intracoronary occlusion after elective cardioversion for atrial fibrillation in a patient with angiographic nearly normal coronary arteries. Am Heart J 102:932–4

Goodman DJ, Rossen RM, Ingham R, Rider AK, Harrison DC (1975a) Sinus node function in the denervated heart: Effects of digitalis. Br Heart J 37:612–18

Goodman DJ, Rossen RM, Cannom DS, Rider AK, Harrison DC (1975b) Effect of digoxin on atrioventricular conduction. Studies in patients with and without cardiac autonomic innervation. Circ 51:251–61

Goy JJ, Grbic M, Hurni M (1985) Conversion of supraventricular arrhythmia to sinus rhythm using flecainide. Eur Heart J 6:518–24

Goy JJ, Kaufmann U, Kappenberger L, Sigwart U (1988) Restoration of sinus rhythm with flecainide in patients with atrial fibrillation. Am J Cardiol 62:38D–40D

Greenspan K, Lord TJ (1973) Digitalis and vagal stimulation during atrial fibrillation: effects on atrioventricular conduction and ventricular arrhythmias. Cardiovasc Res 7:241–6

Grover DN, Mathur VS, Shrivastava S, Roy SB (1971) Electromechanical correlation of left atrial function after cardioversion. Br Heart J 33:226–32

Guaragna RF, Capucci A, Sangiorgio P, Bracchetti D (1982) [Drug treatment of a trial fibrillation in patients with Wolff- Parkinson–White syndrome.] G Ital Cardiol 12:284–91

Guiney TE, Lown B (1972) Electrical conversion of atrial flutter to atrial fibrillation. Flutter mechanism in man. Br Heart J 34:1215–24

Guiraudon GM, Klein GJ, Sharma AD, Yee R (1990) Surgical alternatives for supraventricular tachycardias. In: Touboul P, Waldo AL (eds) Atrial arrhythmias. Current concepts and management. Mosby Year Book, St Louis, pp 488–97

Gulamhusein S, Ko P, Carruthers SG, Klein GJ (1982) Acceleration of the ventricular response during atrial fibrillation in the Wolff–Parkinson–White syndrome after verapamil. Circulation 65:348–54

Halmos PB (1966a) The prognosis of atrial fibrillation after mitral valvotomy. Am Heart J 72:30–4

Halmos PB (1966b) Direct current conversion of atrial fibrillation. Br Heart J 28:302–8

Halmos PB, Patterson GC (1965) Effect of atrial fibrillation on cardiac output. Br Heart J 27:719–23

Halpern SW, Ellrodt G, Singh BN, Mandel WJ (1980) Efficacy of intravenous procainamide infusion in converting atrial fibrillation to sinus rhythm. Relation to left atrial size. Br Heart J 44:589–95

Hammill SC, Wood DL, Gersh BJ, Osborn MJ, Holmes DR (1988) Propafenone for paroxysmal atrial fibrillation. Am J Cardiol 61:473–4

Hansen JF, Andersen ED, Qlesen KH et al. (1979) DC-conversion of atrial fibrillation after mitral valve operation. An analysis of the long-term results. Scand J Thorac Cardiovasc Surg 13:267–70

Harper RW, Whitford E, Middlebrook K, Federman J, Anderson S, Pitt A (1982) Effects of verapamil on the electrophysiologic properties of the accessory pathway in patients with the Wolff- Parkinson–White syndrome. Am J Cardiol 50:1323–30

Hart RG, Coull BM, Hart D (1983) Early recurrent embolism associated with nonvalvular atrial fibrillation: a retrospective study. Stroke 14:688–93

Hartel G, Louhija A, Konttinen A, Halonen PI (1970) Value of quinidine in maintenance of sinus rhythm after electric conversion of atrial fibrillation. Br Heart J 32:57–60

Hartel G, Louhija A, Konttinen A (1974) Disopyramide in the prevention of recurrence of atrial fibrillation after electroversion. Clin Pharmacol Therap 15:551–55

Hendrix GH, Davis RE (1973) Transient electrocardiographic changes of left atrial hypertrophy following successful cardioversion for atrial fibrillation. South Med J 66:729–30

Hertzeanu H, Aron L (1985) Holter monitoring for dizziness and syncope in old age. Acta Cardiol 40:291–9

Hillestad L, Storstein O (1969) Conversion of chronic atrial fibrillation to sinus rhythm with combined propranolol and quinidine treatment. Am Heart J 77:137–9

Hillestad L, Bjerkelund C, Dale J, Maltau J, Storstein O (1971) Quinidine in maintenance of sinus rhythm after electroconversion of chronic atrial fibrillation. A controlled clinical study. Br Heart J 33:518–21

Hillestad L, Dale J, Storstein O (1972) Quinidine before direct current countershock. A controlled study. Br Heart J 34:139–42

Hoffman BF, Singer DH (1964) Effects of digitalis on electrical activity of cardiac fibers. Prog Cardiovasc Dis 7:226–60

Hoglund C, Rosenhamer G (1985) Echocardiographic left atrial dimension as a predictor of maintaining sinus rhythm after conversion of atrial fibrillation. Acta Med Scand 217:411–5

Hornsten TR, Bruce RA (1968) Effects of atrial fibrillation on exercise performance in patients with cardiac disease. Circulation 37:543–8

Horowitz LN, Spielman SR, Greenspan AM et al. (1985) Use of amiodarone in the treatment of persistent and paroxysmal atrial fibrillation resistant to quinidine therapy. J Am Coll Cardiol 6:1402–7

Huffman DH, Klaassen CD, Hartman CR (1977) Digoxin in hyperthyroidism. Clin Pharmacol Ther 22:533–8

Hwang MH, Danoviz J, Pacold I, Rad N, Loeb HS, Gunnar RM (1984) Double-blind crossover randomized trial of intravenously administered verapamil. Its use for atrial fibrillation and flutter following open heart surgery. Arch Intern Med 144:491–4

Ieri A, Zipoli A, Bartoli P, Marmugi P, Morelli G (1982) Improvement of the cardiac function after electrical cardioversion of atrial fibrillation. Echocardiographic study in patients with and without mitral stenosis. G Ital Cardiol 12:91–5

Ikram H, Nixon PG, Arcan T (1968) Left atrial function after electrical conversion to sinus rhythm. Br Heart J 30:80–3

Jacob AS, Nielsen DH, Gianelly RE (1985) Fatal ventricular fibrillation following verapamil in Wolff–Parkinson–White syndrome with atrial fibrillation. Ann Emerg Med 14:159–60

Jakobsson J, Odmansson I, Nordlander R (1990) Enzyme release after elective cardioversion. Eur Heart J 11:749–52

James MA, Channer KS, Papouchado M, Rees JR (1989) Improved control of atrial fibrillation with combined pindolol and digoxin therapy. Eur Heart J 10:83–90

Jelliffe RW, Brooker G (1974) A nomogram for digoxin therapy. Am J Med 57:63–8

Jensen JB, Humphries JO, Kouwenhoven WB, Jude JR (1965) Electroshock for atrial flutter and atrial fibrillation. Follow- up studies on 50 patients. JAMA 194:1181–4

Jenzer HR, Hagemeijer F (1976) Quinidine syncope: torsade de pointes with low quinidine plasma concentrations. Eur J Cardiol 4:447–51

Jogestrand T (1980) Digoxin concentration in right atrial myocardium, skeletal muscle and serum in man: Influence of atrial rhythm. Eur J Clin Pharmacol 17:243–50

Jogestrand T, Edner M, Haverling M (1989) Clinical value of serum digoxin assays in outpatients: improvement by the standardization of blood sampling. Am Heart J 117:1076–83

Johnson LW, Dickstein RA, Fruehan CT et al. (1976) Prophylactic digitalisation for coronary artery bypass surgery. Circulation 53:819–22

Josephson ME, Kastor JA, Kitchen JG (1976) Lidocaine in Wolff- Parkinson–White syndrome with atrial fibrillation. Ann Intern Med 84:44–5

Joubert P, van der Meer L (1979) A specific cardiac glycoside for cardiac failure and another for atrial fibrillation? S Afr Med J 56:1040–2

Kaplan MA, Gray RE, Iseri LT (1968) Metabolic and hemodynamic responses to exercise during atrial fibrillation and sinus rhythm. Am J Cardiol 22:543–9

Kappenberger LJ, Fromer MA, Steinbrunn W, Shenasa M (1984) Efficacy of amiodarone in the Wolff–Parkinson–White syndrome with rapid ventricular response via accessory pathway during atrial fibrillation. Am J Cardiol 54:330–5

Kappenberger LJ, Fromer MA, Shenasa M, Gloor HO (1985) Evaluation of flecainide acetate in rapid atrial fibrillation complicating Wolff–Parkinson–White syndrome. Clinical cardiology 8:321–6

Karlson BW, Torstensson I, Abjorn C, Jansson SO, Peterson LE. Disopyramide in the maintenance of sinus rhythm after electroconversion of atrial fibrillation. A placebo-controlled one-year follow-up study. Eur Heart J 1988;9:284–90

Kastor JA (1973) Digitalis intoxication in patients with atrial fibrillation. Circulation 47:888–96

Kastor JA, Yurchak PM (1967) Recognition of digitalis intoxication in the presence of atrial fibrillation. Ann Intern Med 67:1045–54

Kerber RE, Jensen SR, Grayzel J, Kennedy J, Hoyt R (1981) Elective cardioversion: influence of paddle-electrode location and size on success rates and energy requirements. N Engl J Med 305:658–62

Kerr CR, Mason MA (1985) Amplitude of atrial electrical activity during sinus rhythm and during atrial flutter-fibrillation. PACE 8:348–55

Kerr CR, Prystowsky EN, Smith WM, Cook L, Gallagher JJ (1982) Electrophysiologic effects of disopyramide phosphate in patients with Wolff–Parkinson–White syndrome. Circulation 65:869–78

Kerr CR, Cooper JA, Wallace T, Tyers GF (1983) Two mechanisms of arrhythmia induction by a DDD pacemaker: a case report. PACE 6:820–4

Kerr CR, Klein GJ, Axelson JE, Cooper JC (1988) Propafenone for prevention of recurrent atrial fibrillation. Am J Cardiol 61:914–6

Khaja F, Parker JO (1972) Hemodynamic effects of cardioversion in chronic atrial fibrillation. Special reference to coronary artery disease. Arch Intern Med 129:433–40

Khalsa A, Edvardsson N, Olsson SB (1978) Effects of metoprolol on heart rate in patients with digitalis treated chronic atrial fibrillation. Clin Cardiol 1:91–5

Khalsa A, Olsson SB (1979) Verapamil-induced ventricular regularity in atrial fibrillation. Effects of exercise, isoproterenol, atropine and conversion to sinus rhythm. Acta Med Scand 205:509–15

Khalsa A, Olsson B, Henriksson BA (1979) Effect of oral verapamil on ventricular irregularity in long-standing atrial fibrillation. Acta Med Scand 205:39–47

Killip T, Baer RA (1966) Hemodynamic effects after reversion from atrial fibrillation to sinus rhythm by precordial shock. J Clin Invest 45:658–71

Kim YI, Noble RJ, Zipes DP (1975) Dissociation of the inotropic effect of digitalis from its effect on atrioventricular conduction. Am J Cardiol 36:459–67

Kleiger R, Lown B (1966) Cardioversion and digitalis. II. Clinical studies. Circulation 33:878–87

Klein GJ, Gulamhusein SS (1983) Intermittent preexcitation in the Wolff–Parkinson–White syndrome. Am J Cardiol 52:292–6

Klein HO, Kaplinsky E (1982) Verapamil and digoxin: their respective effects on atrial fibrillation and their interaction. Am J Cardiol 50:894–902

Kraemer MD, Sullivan M, Atwood JE, Forbes S, Myers J, Froelicher V (1989) Reproducibility of treadmill exercise data in patients with atrial fibrillation. Cardiology 76:234–42

Krikler DM (1986) Verapamil in arrhythmia. Br J Clin Pharmacol 21 Suppl 2:183S–9S

Lang R, Klein HO, Di Segni E et al. (1983a) Verapamil improves exercise capacity in chronic atrial fibrillation: double-blind crossover study. Am Heart J 105:820–5

Lang R, Klein HO, Weiss E et al. (1983b) Superiority of oral verapamil therapy to digoxin in treatment of chronic atrial fibrillation. Chest 83:491–9

Lau CP, Leung WH, Wong CK, Tai YT, Cheng CH (1990b) A new pacing method for rapid regularization and rate control in atrial fibrillation. Am J Cardiol 65:1198–203

Leclercq JF, Rosengarten MD, Kural S, Attuel P, Coumel P (1981) Effects of intrinsic sympathetic activity of beta-blockers on SA and AV nodes in man. Eur J Cardiol 12:367–75

Lemery R, Brugada P, Cheriex E, Wellens HJJ (1987) Reversibility of tachycardia-induced left ventricular dysfunction after closed- chest catheter ablation of the atrioventricular junction for intractable atrial fibrillation. Am J Cardiol 60:1406–8

Levi GF, Proto C (1970) Combined treatment of atrial fibrillation with propranolol and quinidine. Cardiology 55:249–54

Levi GF, Proto C (1972) Combined treatment of atrial fibrillation with quinidine and beta-blockers. Br Heart J 34:911–14

Levy S, Lacombe P, Cointe R, Bru P (1988) High energy transcatheter cardioversion of chronic atrial fibrillation. J Am Coll Cardiol 12:514–18

Levy S, Dolla E, Ebagosti A, Bru P (1990) Internal cardioversion for chronic atrial fibrillation. In: Touboul P, Waldo AL (eds) Atrial arrhythmias. Current concepts and management. Mosby Year Book, St Louis, pp 411–8

Lewis RV, McDevitt DG (1988a) Factors affecting the clinical response to treatment with digoxin and two calcium antagonists in patients with atrial fibrillation. Br J Clin Pharmacol 25:603–6

Lewis RV, McDevitt DG (1988b) The relative effects of digoxin and diltiazem upon ventricular ectopic activity in patients with chronic atrial fibrillation. Br J Clin Pharmacol 26:327–9

Lewis R, Lakhani M, Moreland TA, McDevitt DG (1987) A comparison of verapamil and digoxin in the treatment of atrial fibrillation. Eur Heart J 8:148–53

Lewis RV, Irvine N, McDevitt DG (1988a) Relationships between heart rate, exercise tolerance and cardiac output in atrial fibrillation: the effects of treatment with digoxin, verapamil and diltiazem. Eur Heart J 9:777–81

Lewis RV, Laing E, Moreland TA, Service E, McDevitt DG (1988b) A comparison of digoxin, diltiazem and their combination in the treatment of atrial fibrillation. Eur Heart J 9:279–83

Lewis RV, McMurray J, McDevitt DG (1989) Effects of atenolol, verapamil, and xamoterol on heart rate and exercise tolerance in digitalised patients with chronic atrial fibrillation. J Cardiovasc Pharmacol 13:1–6

Lewis T, Drury AN, Wedd AM, Iliescu CC (1921) Observations upon the action of certain drugs upon fibrillation of the auricles. Heart 9:207–67

Lindsay J (1967) Pulmonary edema following cardioversion. Am Heart J 74:434–5

Lipkin DP, Frenneaux M, Stewart R, Joshi J, Lowe T, McKenna WJ (1988) Delayed improvement in exercise capacity after cardioversion of atrial fibrillation to sinus rhythm. Br Heart J 59:572–7

Lodder J, Dennis MS, Van Raak L, Jones LN, Warlow CP (1988) Cooperative study on the value of long term anticoagulation in patients with stroke and non-rheumatic atrial fibrillation. Br Med J 296:1435–8

Lown B (1967) Electrical reversion of cardiac arrhythmias. Br Heart J 29:469–89

Lown B, Amarasingham R, Neuman J (1962) New method for terminating cardiac arrhythmias; use of synchronised capacitor discharge. JAMA 182:548–55

Lown B, Perlroth MG, Kaidbey S, Abe T, Harken DE (1963) "Cardioversion" of atrial fibrillation. A report on the treatment of 65 episodes in 50 patients. N Engl J Med 269:325–31

Luchi RJ, Gruber JW (1968) Unusually large digitalis requirements. A study of altered digoxin metabolism. Am J Med 45:322–8

Lundstrom T, Ryden L (1988) Chronic atrial fibrillation. Long- term results of direct current conversion. Acta Med Scand 223:53–9

Lundstrom T, Ryden L (1989) Haemorrhagic and thromboembolic complications in patients with atrial fibrillation on anticoagulant prophylaxis. J Intern Med 225:137–42

Lundstrom T, Ryden L (1990) Ventricular rate control and exercise performance in chronic atrial fibrillation: effects of diltiazem and verapamil. J Am Coll Cardiol 16:86–90

Lupi GA, Urthaler F, James TN (1979) Effects of verapamil on automaticity and conduction with particular reference to tachyphylaxis. Eur J Cardiol 9:345–68

Mackenzie J (1911) Digitalis. Heart 2:273–389

Mahlich J, Schweizer W, Burkart F (1973) Atrial function after cardioversion for atrial fibrillation. Br Heart J 35:24–7

Maier WD, Neuss H, Bilgin Y, Gigler G, Thormann J, Schlepper M (1983) [Modification of hemodynamics in tachycardiac atrial fibrillation by metoprolol and verapamil] Z Kardiol 72:465–70

Mandel WJ, Laks MM, Obayashi K, Hayakawa H, Daley W (1975) The Wolff–Parkinson–White syndrome: pharmacologic effects of procaine amide. Am Heart J 90:744–54

Manning WJ, Leeman DE, Gotch PJ, Come PC (1989) Pulsed Doppler evaluation of atrial mechanical function after electrical cardioversion of atrial fibrillation. J Am Coll Cardiol 13:617–23

Maragno I, Santostasi G, Gaion R et al. (1988) Low- and medium-dose diltiazem in chronic atrial fibrillation: Comparison with digoxin and correlation with drug plasma levels. Am Heart J 116:385–92

Markel ML, Prystowsky EN, Heger JJ, Miles WM, Fineberg N, Zipes DP (1986) Encainide for treatment of supraventricular tachycardias associated with the Wolff–Parkinson–White syndrome. Am J Cardiol 58:41C–8C

Markewitz A, Schad N, Hemmer W, Bernheim C, Ciavolella M, Weinhold C (1986) What is the most appropriate stimulation mode in patients with sinus node dysfunction? PACE 9:1115–20

Martinoli E, Medugno G, Crepaldi L, Scardi S, Camerini F (1975) [Oxprenolol in the treatment and prevention of cardiac arrhythmias.] Minerva Med 66:3566–73

McCabe JB (1986) Pacemaker-mediated tachycardia: tracking of atrial fibrillation during DDD pacing. Ann Emerg Med 15:83–5

McCarthy C, Varghese PJ, Barritt DW (1969) Prognosis of atrial arrhythmias treated by electrical counter shock therapy. A three- year follow-up. Br Heart J 31:496–500

Meijler FL (1985) An "account" of digitalis and atrial fibrillation. J Am Coll Cardiol 5 (Suppl A):60A–8A

Menasche P, Leclercq JF, Cauchemez B et al. (1990) Surgery for the Wolff–Parkinson–White syndrome in 73 consecutive patients: what have we learnt from intraoperative mapping? Eur J Cardiothorac Surg 3:387–90

Mendez C, Mendez R (1957) The action of cardiac glycosides on the excitability and conduction velocity of the mammalian atrium. J Pharmacol Exp Ther 121:402–13

Mendez C, Aceves J, Mendez R (1961a) Inhibition of adrenergic cardiac acceleration by cardiac glycosides. J Pharmac Exp Ther 131:191–8

Mendez C, Aceves J, Mendez R (1961b) The anti-adrenergic action of digitalis on the refractory period of the A-V transmission system. J Pharmacol Exp Ther 131:199–204

Metcalfe MJ, Smith F, Jennings K, Paterson N (1988) Does cardioversion of atrial fibrillation result in myocardial damage? Br Med J 296:1364

Miller PH, Carleton RA (1968) Hemodynamic assessment during atrial fibrillation. Value of ventricular pacing. Am J Cardiol 22:568–71

Milstein S, Sharma AD, Klein GJ (1986) Electrophysiologic profile of asymptomatic Wolff–Parkinson–White pattern. Am J Cardiol 57:1097–100

Moe GK, Abildskov JA (1964) Observations on the ventricular dysrhythmia associated with atrial fibrillation in the dog heart. Circ Res 14:447–60

Moe GK, Abildskov JA, Mendez C (1964) An experimental study of concealed conduction. Am Heart J 67:338–56

Molajo AO, Coupe MO, Bennett DH (1984) Effect of Corwin (ICI 118587) on resting and exercise heart rate and exercise tolerance in digitalised patients with chronic atrial fibrillation. Br Heart J 52:392–5

Molgaard H, Bjerregaard P, Jorgensen HS, Klitgaard NA (1987) 24- hour antiarrhythmic effect of conventional and slow-release verapamil in chronic atrial fibrillation. Eur J Clin Pharmacol 33:447–53

Morady F, Sledge C, Shen E, Sung RJ, Gonzales R, Scheinman MM (1983) Electrophysiologic testing in the management of patients with the Wolff–Parkinson–White syndrome and atrial fibrillation. Am J Cardiol 51:1623–8

Morady F, Dicarlo LA, Baerman JM, De Buitleir M (1987) Effect of propranolol on ventricular rate during atrial fibrillation in the Wolff–Parkinson–White syndrome. PACE 10:492–6

Moreira DA, Shepard RB, Waldo AL (1989) Chronic rapid atrial pacing to maintain atrial fibrillation: use to permit control of ventricular rate in order to treat tachycardia induced cardiomyopathy. PACE 12:761–75

Mungall DR, Robichaux RP, Perry W et al. (1980) Effects of quinidine on serum digoxin concentration: a prospective study. Ann Intern Med 93:689–93

Nadeau RA, James TN (1963) Antagonistic effects on the sinus node of acetylstrophanthidin and adrenergic stimulation. Circ Res 13:388–91

Nathan AW, Bexton RS, Spurrell RA, Camm AJ (1984) Internal transvenous low energy cardioversion for the treatment of cardiac arrhythmias. Br Heart J 52:377–84

Nathan AW, Camm AJ, Bexton RS, Hellestrand KJ (1987) Intravenous flecainide acetate for the clinical management of paroxysmal tachycardias. Clin Cardiol 10:317–22

Navab A, La Due JS (1965) Postconversion systemic arterial embolism. Am J Cardiol 16:452–3

Neporent LM, Da Silva JA (1967) Heart sounds in atrial flutter- fibrillation. Am J Cardiol 19:301–4

Neuss H (1982) Control of ventricular rate in atrial fibrillation: role of verapamil. In: Kulbertus HE, Olsson SB, Schlepper M (eds) Atrial fibrillation. Hassle, Molndal, pp 294– 306

Neuss H, Buss J, Schlepper M et al. (1983) Effects of flecainide on electrophysiological properties of accessory pathways in the Wolff–Parkinson–White syndrome. Eur Heart J 4:347–53

Neuss H, Golling FR, Schlepper M, Thormann J, Weissmuller P, Kindler M (1984) [Regularisation of ventricular intervals in atrial fibrillation – electrophysiologic findings on the underlying mechanism.] Z Kardiol 73:106–12

Nordlander R (1982) Anticoagulation in atrial fibrillation. In: Kulbertus HE, Olsson SB, Schlepper M (eds) Atrial fibrillation. Hassle, Molndal, pp 251–9

Nunain SO, Debbas NMG, Camm AJ (1990) Determinants of the course and prognosis of atrial fibrillation. In: Touboul P, Waldo AL (eds) Atrial arrhythmias. Current concepts and management. Mosby Year Book, St Louis, pp 350–8

O'Neill PG, Puleo PR, Bolli R, Rokey R (1990) Return of atrial mechanical function following electrical conversion of atrial dysrhythmias. Am Heart J 120:353–9

Orlando JR, van Herick R, Aronow WS, Olson HG (1979) Hemodynamics and echocardiograms before and after cardioversion of atrial fibrillation to normal sinus rhythm. Chest 76:521–6

O'Rourke RA (1970) Appearance of atrial sound after reversion of atrial fibrillation. Br Heart J 32:597–9

Ovsyshcher IA, Ilia R, Wanderman KL (1984) Conduction disturbance and ST elevation following cardioversion. Isr J Med Sci 20:736–8

Panidis IP, Morganroth J, Baessler C (1983) Effectiveness and safety of oral verapamil to control exercise-induced tachycardia in patients with atrial fibrillation receiving digitalis. Am J Cardiol 52:1197–201

Papa LA, Saia JA, Chung EK (1978) Ventricular fibrillation in Wolff–Parkinson–White syndrome, type A. Heart Lung 7:1015–19

Penchas S, Zajicek G (1978) Plasma digoxin levels and the interbeat interval signal in atrial fibrillation. Z Kardiol 67:104–8

Perelman MS, McKenna WJ, Rowland E, Krikler DM (1987) A comparison of bepridil with amiodarone in the treatment of established atrial fibrillation. Br Heart J 58:339–44

Petersen P (1990) Thromboembolic complications of atrial fibrillation and their prevention: a review. Am J Cardiol 65:24C–28C

Petersen P, Boysen G, Godtfredsen J, Andersen ED, Andersen B (1989) Placebo-controlled, randomised trial of warfarin and aspirin for prevention of thromboembolic complications in chronic atrial fibrillation. Lancet ii:175–9

Pittman DE, Gay TC, Patel II, Joyner CR (1974) Termination of atrial flutter and atrial tachycardia with rapid atrial stimulation. Angiology 26:784–802

Pomfret SM, Beasley CRW, Challenor V, Holgate ST. (1988) Relative efficacy of oral verapamil and digoxin alone and in combination for the treatment of patients with chronic atrial fibrillation. Clin Sci 74:351–7

Pool PE (1986) Treatment of supraventricular arrhythmias with encainide. Am J Cardiol 58:55C–7C

Prinzmetal M, Corday E, Brill IC, Oblath RW, Kruger HE (1952) The auricular arrhythmias. Thomas, Springfield, Illinois

Radford MD, Evans DW (1968) Long-term results of DC reversion of atrial fibrillation. Br Heart J 30:91–6

Rasmussen K, Wang H, Fausa D (1981) Comparative efficiency of quinidine and verapamil in the maintenance of sinus rhythm after DC conversion of atrial fibrillation. A controlled clinical trial. Acta Med Scand Suppl 645:23–8

Rawles JM, Metcalfe MJ, Jennings K (1990) Time of occurrence, duration and ventricular rate of paroxysmal atrial fibrillation: the effect of digoxin. Br Heart J 63:225–7

Rebello R, Brownlee WC (1987) Intermittent ventricular standstill during chronic atrial fibrillation in patients with dizziness or syncope. PACE 10:1271–6

Redfors A (1971a) The effect of different digoxin doses on subjective symptoms and physical working capacity in patients with atrial fibrillation. Acta Med Scand 190:307–20

Redfors A (1971b) Digoxin dosage and ventricular rate at rest and exercise in patients with atrial fibrillation. Acta Med Scand 190:321–33

Redfors A (1972) Plasma digoxin concentration – its relation to digoxin dosage and clinical effects in patients with atrial fibrillation. Br Heart J 34:383–91

Rehnqvist N (1981) Clinical experience with intravenous metoprolol in supraventricular tachyarrhythmias. A multicentre study. Ann Clin Res 13 (Suppl 30):68–72

Reid DS, Jachuck SJ, Henderson CB (1973) Cardiac pacing in incomplete atrioventricular block with atrial fibrillation. Br Heart J 35:1154–60

Reid PR, Meek AG (1979) Digoxin–quinidine interaction. Johns Hopkins Med J 145:227–9

Resnekov L (1967) Haemodynamic studies before and after electrical conversion of atrial fibrillation and flutter to sinus rhythm. Br Heart J 29:700–8

Resnekov L, McDonald L (1967) Complications in 220 patients with cardiac dysrhythmias treated by phased direct current shock, and indications for electroconversion. Br Heart J 29:926–36

Resnekov L, McDonald L (1971) Electroversion of lone atrial fibrillation and flutter including haemodynamic studies at rest and on exercise. Br Heart J 33:339–50

Resnekov L, Gibson D, Waich S, Muir J, McDonald L (1971) Sustained-release quinidine (Kinidin Durules) in maintaining sinus rhythm after electroversion of atrial dysrhythmias. Br Heart J 33:220–5

Ricci DR, Orlick AE, Reitz BA, Mason JW, Stinson EB, Harrison DC (1978) Depressant effect of digoxin on atrioventricular conduction in man. Circulation 57:898–903

Robinson K, Rowland E, Krikler DM (1988) Wolff–Parkinson–White syndrome: atrial fibrillation as the presenting arrhythmia. Br Heart J 59:578–80

Rodman T, Pastor BH, Figueroa W (1966) Effect on cardiac output of conversion from atrial fibrillation to normal sinus mechanism. Am J Med 41:249–58

Rohl D, Schulz D (1987) Die Wirkung von Diltiazem auf die Herzfrequenz in Ruhe und unter Belastung bei Patienten mit Vorhofflimmern. Herz Kreislauf 19:365–9

Rosenqvist M, Brandt J, Schuller H (1986) Atrial versus ventricular pacing in sinus node disease: a treatment comparison study. Am Heart J 111:292–7

Roth A, Harrison E, Mitani G, Cohen J, Rahimtoola SH, Elkayam U (1986) Efficacy and safety of medium- and high-dose diltiazem alone and in combination with digoxin for control of heart rate at rest and during exercise in patients with chronic atrial fibr Circulation 73:316–24

Rowland TW (1983) Augmented ventricular rate following verapamil treatment for atrial fibrillation with Wolff–Parkinson–White syndrome. Pediatrics 72:245–6

Roy D, Marchand E, Gagne P, Chabot M, Cartier R (1986) Usefulness of anticoagulant therapy in the prevention of embolic complications of atrial fibrillation. Am Heart J 112:1039–43

Rubin DA, Nieminski KE, Reed GE, Herman MV (1987) Predictors, prevention, and long-term prognosis of atrial fibrillation after coronary artery bypass graft operations. J Thorac Cardiovasc Surg 94:331–5

Sadick N, Baird DK, Uther JB (1978) Surgical division of an accessory atrioventricular connection in the posterior septal region following ventricular fibrillation in the Wolff– Parkinson–White syndrome. Aust N Z J Med 8:652–5

Salerno DM, Dias VC, Kleiger RE et al. (1989) Efficacy and safety of intravenous diltiazem for treatment of atrial fibrillation and atrial flutter. The Diltiazem–Atrial Fibrillation/Flutter Study Group. Am J Cardiol 63:1046–51

Sandercock P, Bamford J, Warlow C, Peto R, Starkey I (1986) Is a controlled trial of long-term oral anticoagulants in patients with stroke and non-rheumatic atrial fibrillation worthwhile? Lancet i:788–92

Sanzobrino B, Lemberg L (1979) Acute problems in the coronary care unit that do not require intervention. II. Atrial fibrillation with optimal heart rates. Heart Lung 8:575–7

Savill J, Mitchell M, Wood D, Krikler DM (1985) Rapid plasma digoxin assay in outpatients – a useful routine technique? Br Heart J 54:248–50

Schamroth L (1971) Immediate effects of intravenous verapamil on atrial fibrillation. Cardiovasc Res 5:419–24

Schamroth L, Krikler DM, Garrett C (1972) Immediate effects of intravenous verapamil in cardiac arrhythmias. Br Med J i:660–2

Schutzenberger W, Leisch F, Gmeiner R (1987) Enhanced accessory pathway conduction following intravenous amiodarone in atrial fibrillation. A case report. Int J Cardiol 16:93–5

Sclarovsky S, Kracoff OH, Strasberg B, Lewin RF, Agmon J (1981) Paroxysmal atrial flutter and fibrillation associated with preexcitation syndrome: treatment with ajmaline. Am J Cardiol 48:929–33

Scott ME, Pantridge JF (1968) The value of direct current conversion of atrial fibrillation. Am Heart J 75:579–81

Scott ME, Patterson GC (1969) Cardiac output after direct current conversion of atrial fibrillation. Br Heart J 31:87–90

Sellers TD, Bashore TM, Gallagher JJ (1977) Digitalis in the pre- excitation syndrome. Analysis during atrial fibrillation. Circulation 56:260–7

Selzer A (1985) Role of serum digoxin assay in patient management. J Am Coll Cardiol 5(Suppl A):106A–10A

Selzer A, Walter RM (1966) Adequacy of preoperative digitalis therapy in controlling ventricular rate in postoperative atrial fibrillation. Circulation 34:119–22

Selzer A, Wray HW (1964) Quinidine syncope. Paroxysmal ventricular fibrillation occurring during treatment of chronic atrial arrhythmias. Circulation 30:17–26

Shapiro EP, Effron MB, Lima S, Ouyang P, Siu CO, Bush D (1988) Transient atrial dysfunction after conversion of chronic atrial fibrillation to sinus rhythm. Am J Cardiol 62:1202–7

Sharma AD, Klein GJ, Guiraudon GM, Milstein S (1985) Atrial fibrillation in patients with Wolff–Parkinson–White syndrome: incidence after surgical ablation of the accessory pathway. Circulation 72:161–9

Sheils WS (1966) Ventricular fibrillation following cardioversion. W V Med J 62:438–40

Shenasa M, Fromer M, Faugere G et al. (1987) Efficacy and safety of intravenous and oral diltiazem for Wolff–Parkinson–White syndrome. Am J Cardiol 59:301–6

Shenasa M, Kus T, Fromer M, LeBlanc RA, Dubuc M, Nadeau R (1988) Effect of intravenous and oral calcium antagonists (diltiazem and verapamil) on sustenance of atrial fibrillation. Am J Cardiol 62:403–7

Shlofmitz RA, Hirsch BE, Meyer BR (1986) New-onset atrial fibrillation: is there need for emergent hospitalization? J Gen Intern Med 1:139–42

Sloman G, Spokes J, Ramshaw J, Vohra J (1975) Haemodynamic effect of intravenous verapamil in controlled atrial fibrillation. Aust NZ J Med 5:420–3

Smit AJ, Scaf AH, van Essen LH, Lie KI, Wesseling H (1990) Digoxin infusion versus bolus injection in rapid atrial fibrillation: relation between serum level and response. Eur J Clin Pharmacol 38:335–41

Sodermark T, Jonsson B, Olsson A et al. (1975) Effect of quinidine on maintaining sinus rhythm after conversion of atrial fibrillation or flutter. A multicentre study from Stockholm. Br Heart J 37:486–92

Sowton E, Holt P (1986) Strategies for the management of tachyarrhythmias. PACE 9:1304–8

Spurrell RA, Thorburn CW, Camm J, Sowton E, Deuchar DC (1975) Effects of disopyramide on electrophysiological properties of specialized conduction system in man and on accessory atrioventricular pathway in Wolff–Parkinson–White syndrome. Br Heart J 37:861–7

Stein B, Halperin JL, Fuster V (1990) Should patients with atrial fibrillation be anticoagulated prior to and chronically following cardioversion? Cardiovasc Clin 21:231–47

Steinbeck G, Doliwa R, Bach P (1988) [Therapy of paroxysmal atrial fibrillation. Cardiac glycosides alone or combined with anti-arrhythmia agents?.] Dtsch Med Wochenschr 113:1867–71

Steinberg JS, Katz RJ, Bren GB, Buff LA, Varghese PJ (1987) Efficacy of oral diltiazem to control ventricular response in chronic atrial fibrillation at rest and during exercise. J Am Coll Cardiol 9:405-11

Stephenson LW, MacVaugh H III, Tomasello DN, Josephson ME (1980) Propranolol for prevention of postoperative cardiac arrhythmias: a randomized study. Ann Thorac Surg 29:113-16

Stern S (1971) Treatment and prevention of cardiac arrhythmias with propranolol and quinidine. Br Heart J 33:522-5

Stroke Prevention in Atrial Fibrillation Study Group Investigators (1990) Preliminary report of the Stroke Prevention in Atrial Fibrillation Study. N Engl J Med 322:863-8

Stroobandt R, Kesteloot H (1981) Intravenous metoprolol for the treatment of acute supraventricular tachyarrhythmias. Acta Cardiol 36:155-65

Suttorp MJ, Kingma JH, Lie-A-Huen L, Mast EG (1989) Intravenous flecainide versus verapamil for acute conversion of paroxysmal atrial fibrillation or flutter to sinus rhythm. Am J Cardiol 63:693-6

Sykes D, Dewar R, Mohanaruban K et al. (1990) Measuring blood pressure in the elderly: does atrial fibrillation increase observer variability? Br Med J 300:162-3

Szekely P (1964) Systemic embolism and anticoagulant prophylaxis in rheumatic heart disease. Br Med J i:1209-12

Szekely P, Batson GA, Stark DC (1966) Direct current shock therapy of cardiac arrhythmias. Br Heart J 28:366-73

Szekely P, Sideris DA, Batson GA (1970a) Maintenance of sinus rhythm after atrial defibrillation. Br Heart J 32:741-6

Szekely P, Wynne NA, Pearson DT, Batson GA, Sideris DA (1970b) Direct current shock and antidysrhythmic drugs. Br Heart J 32:209-18

Taguchi JT, Ryan JM (1969) Spontaneous conversion of established atrial fibrillation. Clinical significance of a change to atrial flutter or to paroxysmal atrial tachycardia with AV block. Arch Intern Med 124:468-76

Takkunen J, Oilinki O, Salokannel J, Vuopala U (1970) Mortality of patients with atrial fibrillation before and after introduction of DC counter shock therapy. Acta Med Scand 188:127-31

The German and Austrian Xamoterol Study Group (1988) Double-blind placebo-controlled comparison of digoxin and xamoterol in chronic heart failure. Lancet i:489-93

The Xamoterol in Severe Heart Failure Study Group (1990) Xamoterol in severe heart failure. Lancet 336:1-6

Theisen K, Haufe M, Peters J, Theisen F, Jahrmarker H (1985) Effect of the calcium antagonist diltiazem on atrioventricular conduction in chronic atrial fibrillation. Am J Cardiol 55:98-102

Thibault GE (1985) Making the coronary care unit cost-effective. Am J Cardiol 56:35C-9C

Thompson ME, Metzger CC, Shaver JA, Heidenreich FP, Martin CE, Leonard JJ (1972) Assessment of left atrial transport function immediately after cardioversion. Am J Cardiol 29:481-9

Toivonen LK, Nieminen MS, Manninen V, Frick MH (1986) Conversion of paroxysmal atrial fibrillation to sinus rhythm by intravenous pirmenol. A placebo controlled study. Br Heart J 55:176-80 Turner JR, Towers JR (1965) Complications of cardioversion. Lancet ii:612-4

Tyras DH, Stothert JC, Kaiser GC, Barner HB, Codd JE, Willman VL (1979) Supraventricular tachyarrhythmias after myocardial revascularisation: A randomized trial of prophylactic digitalization. J Thorac Cardiovasc Surg 77:310-4

Uebis R, Merx W, Fritzsche V (1985) [Asystolic pauses in atrial fibrillation. Incidence, dependence on the underlying disease and significance for pacemaker therapy.] Dtsch Med Wochenschr 110:1157-60

Upton AR, Honey M (1971) Electroconversion of atrial fibrillation after mitral valvotomy. Br Heart J 33:732-8

Van Herick R, Orlando J, Aronow WS (1979) Systolic time intervals before and after cardioversion of atrial fibrillation. Chest 75:359-61

Vera Z, Mason DT, Awan NA et al. (1977) Improvement of symptoms in patients with sick sinus syndrome by spontaneous development of stable atrial fibrillation. Br Heart J 39:160-7

Vitale P, Auricchio A, De Stefano R et al. (1989) Effectiveness of diltiazem in controlling ventricular response and improving exercise capacity in chronic atrial fibrillation. Double-blind, cross-over study. Cardiologia 34:73-81

Watt JH, Donner AP, McKinney CM, Klein GJ (1984) Atrial fibrillation: minimal sampling interval to estimate average rate. J Electrocardiol 17:153-6

Waxman HL, Myerburg RJ, Appel R, Sung RJ (1981) Verapamil for control of ventricular rate in paroxysmal supraventricular tachycardia and atrial fibrillation or flutter: a double-blind randomized cross-over study. Ann Intern Med 94:1-6

Weiner P, Bassan MM, Jarchovsky J, Iusim S, Plavnick L (1983) Clinical course of acute atrial fibrillation treated with rapid digitalization. Am Heart J 105:223-7

Wellens HJJ (1975) Contribution of cardiac pacing to our understanding of the Wolff–Parkinson–White syndrome. Br Heart J 37:231–41

Wellens HJ, Lie KI, Bar FW et al. (1976) Effect of amiodarone in the Wolff–Parkinson–White syndrome. Am J Cardiol 38:189–94

Wellens HJ, Bar FW, Gorgels AP, Vanagt EJ (1980) Use of ajmaline in patients with the Wolff–Parkinson–White syndrome to disclose short refractory period of the accessory pathway. Am J Cardiol 45:130–3

Wellens HJ, Braat S, Brugada P, Gorgels AP, Bar FW (1982) Use of procainamide in patients with the Wolff–Parkinson–White syndrome to disclose a short refractory period of the accessory pathway. Am J Cardiol 50:1087–9

WHO/ISC Task Force (1978) Definitions of terms related to cardiac rhythm. Am Heart J 95:796–806

Wiener L, Dwyer EM (1968) Electrical induction of atrial fibrillation. An approach to intractable atrial tachycardia. Am J Cardiol 21:731–4

Wikland B, Edhag O, Eliasch H (1967) Atrial fibrillation and flutter treated with synchronized DC shock. A study on immediate and long-term results. Acta Med Scand 182:665–71

Williams P, Aronson J, Sleight P (1978) Is a slow pulse-rate a reliable sign of digitalis toxicity? Lancet ii:1340–2

Winters WL Jr, McDonough MT, Hafer J, Dietz R (1968) Diazepam. A useful hypnotic drug for direct-current cardioversion. JAMA 204:926–8

Wipf JE, Lipsky BA (1990) Atrial fibrillation. Thromboembolic risk and indications for anticoagulation. Arch Intern Med 150:1598–603

Wittkampf FHM, De Jongste MJL (1986) Rate stabilisation by right ventricular pacing in patients with atrial fibrillation. PACE 9:1147–53

Wolfson S, Herman MV, Sullivan JM, Gorlin R (1967) Conversion of atrial fibrillation and flutter by propranolol. Br Heart J 29:305–9

Yapa RS, Green GJ (1990) Embolic stroke following cardioversion of atrial fibrillation to sinus rhythm with oral amiodarone therapy [letter]. Postgrad Med J 66:410

Yee R, Klein GJ (1984) Syncope in the Wolff–Parkinson–White syndrome: incidence and electrophysiologic correlates. PACE 7:381–8

Yin FC, Liu ZR (1989) Estimating arterial resistance and compliance during transient conditions in humans. Am J Physiol 257:H190–7

Zener JC, Anggard EE, Harrison DC, Kalman SM (1973) Persistence of digoxin effect in atrial fibrillation. JAMA 224:239–41

Zoble RG, Brewington J, Qlukotun AY, Gore R (1987) Comparative effects of nadolol-digoxin combination therapy and digoxin monotherapy for chronic atrial fibrillation. Am J Cardiol 60:39D–45D

Appendix

A Simple Computer Model of the Electrophysiological System of the Heart, Considering the Sinoatrial and Atrioventricular Nodes as Sine Wave Oscillators

The program is written in Microsoft GW-BASIC and is intended to run on a personal computer with an IBM Enhanced Graphics Adaptor (EGA). As it stands the program runs about one third as fast as a normal heart if a time-step of 50 milliseconds is used. When compiled with QuickBASIC 4.0, a rate roughly equivalent to that of the normal heart is achieved with a time-step of 10 milliseconds. Appropriate values for the other parameters in the model are discussed in Chapter 3.

```
10 REM "Cora – the very model of a heart"
20 KEY OFF: SCREEN 9: CLS : CLEAR
30 TWOPI = 6.283185307: PI = 3.141592654: QUADSAN = 1: QUADAVN =
1: YON = 315
40 INPUT "SA node cycle length or mean A-A, msec (<CR> to exit) ";
SANCL: PRINT
50 IF SANCL = 0 THEN SYSTEM
60 INPUT "Sinus rhythm or atrial fibrillation: <CR> or AF"; AF$: PRINT
70 IF AF$ = "AF" THEN RANDOMIZE: PRINT
80 INPUT "AV node cycle length, msec "; AVNCL: PRINT
90 INPUT "AV node coupling "; AVNRESET: AVRESET = AVNRESET *
65: PRINT
100 INPUT "Time step, msec "; TSTEP: PRINT : CLS
110 FOR I = 0 TO 640 STEP 4: PSET (I, 325), 2: NEXT I
120 FOR I = 0 TO 640 STEP 20: CIRCLE (I, 325), 1, 2: NEXT I
130 FOR I = 0 TO 640 STEP 100: CIRCLE (I, 325), 2, 2: NEXT I
140 CIRCLE (159, 85), 60, 1: CIRCLE (479, 85), 60, 1
150 IF AF$ = "AF" THEN GOTO 170
160 LOCATE 6, 19: PRINT "SAN"
170 LOCATE 6, 59: PRINT "AVN"
```

```
180 IF INKEY$ <> "" THEN GOTO 10
190 T = T + TSTEP
200 IF AF$ <> "AF" THEN GOTO 230
210 IF RND < TSTEP / SANCL THEN P = 1 ELSE P = 0
220 GOTO 280
230 SAN = SAN + TSTEP / SANCL * TWOPI
240 CIRCLE (SX, 85 + (SY - 85) * .75), 4, 0
250 SX = 159 + 65 * COS(SAN): SY = 85 - 65 * SIN(SAN)
260 IF SY < 85 THEN P = 1 - QUADSAN ELSE P = 0
270 CIRCLE (SX, 85 + (SY - 85) * .75), 4, 15
280 AVN = AVN + TSTEP / AVNCL * TWOPI
290 CIRCLE (AX, 85 + (AY - 85) * .75), 4, 0
300 IF P = 1 THEN CIRCLE (159, 85), 60, 4: CIRCLE (159, 85), 65, 4: CIRCLE
(159, 85), 65, 0: CIRCLE (159, 85), 60, 1
310 IF P = 1 THEN SHIFT = AVRESET ELSE SHIFT = 0
320 IF AY < 86 THEN SHIFT = 0
330 IF SHIFT = 0 THEN GOTO 370
340 PSET (250, 85), 4: DRAW "R=" + VARPTR$(SHIFT): PSET (250, 65), 4:
DRAW "R=" + VARPTR$(SHIFT): PSET (250, 105), 4: DRAW "R=" +
VARPTR$(SHIFT)
350 FOR I = 1 TO 100: NEXT I
360 PSET (250, 85), 0: DRAW "R=" + VARPTR$(SHIFT): PSET (250, 65), 0:
DRAW "R=" + VARPTR$(SHIFT): PSET (250, 105), 0: DRAW "R=" +
VARPTR$(SHIFT)
370 IF SY < 85 THEN QUADSAN = 1 ELSE QUADSAN = 0
380 IF SHIFT = 0 THEN GOTO 460
390 ANG = ABS(ATN((AY - 85) / (AX + SHIFT - 479)))
400 IF AY > 85 THEN GOTO 440
410 IF AX + SHIFT < 479 THEN GOTO 430
420 AVN = ANG: GOTO 460
430 AVN = PI - ANG: GOTO 460
440 IF AX + SHIFT < 479 THEN AVN = PI + ANG: GOTO 460
450 AVN = TWOPI - ANG
460 AX = 479 + 65 * COS(AVN): AY = 85 - 65 * SIN(AVN)
470 IF AY < 85 THEN QRS = 1 - QUADAVN ELSE QRS = 0
480 IF QRS = 1 THEN CIRCLE (479, 85), 60, 4: CIRCLE (479, 85), 65, 4:
CIRCLE (479, 85), 65, 0: CIRCLE (479, 85), 60, 1
490 IF AY < 85 THEN QUADAVN = 1 ELSE QUADAVN = 0
500 CIRCLE (AX, 85 + (AY - 85) * .75), 4, 15
510 PSET (EXON, YON), 15
520 XON = XON + TSTEP / 10
530 IF XON > 639 THEN XON = XON - 640: EXON = 0: GOTO 550
540 GOTO 560
550 LINE (0, 320)-(649, 290), 0, BF: PSET (EXON, YON), 15
560 IF AF$ <> "AF" THEN GOTO 610
570 YON = 315 + 6 * (.5 - RND)
580 IF QRS = 1 THEN LINE -(XON, YON): LINE -(XON + 2, YON - 20):
LINE -(XON + 4, YON): EXON = XON + 4: GOTO 670
590 IF XON > EXON THEN LINE -(XON, YON): EXON = XON
600 GOTO 180
```

610 IF P = 1 THEN LINE -(XON, 315): LINE -(XON + 5, 311): LINE -(XON + 10, 315): EXON = XON + 10: PP = T
620 IF QRS = 1 THEN GOTO 650
630 IF XON > EXON THEN LINE -(XON, YON): EXON = XON
640 GOTO 180
650 IF XON < EXON THEN LINE -(XON, YON), 0: LINE -(XON + 2, YON - 20), 15: LINE -(XON + 4, YON): EXON = XON + 4: GOTO 670
660 LINE -(XON, YON): LINE -(XON + 2, YON - 20): LINE -(XON + 4, YON): EXON = XON + 4
670 PR = T - PP: RR = T - PREVRR: PREVRR = T: LOCATE 20, 1: PRINT RR= "; RR; "   ": IF AF$ <> "AF" THEN LOCATE 19, 1: PRINT "PR= "; PR; "   "
680 GOTO 180

Subject Index